THIRD EDITION

Essentials of Intentional Interviewing

Counseling in a Multicultural World

Allen E. Ivey
*Courtesy Professor,
University of South Florida,
Tampa*

Mary Bradford Ivey
*Vice President,
Microtraining Associates*

Carlos P. Zalaquett
*University of South Florida,
Tampa*

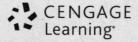

CENGAGE
Learning®

Australia • Brazil • Mexico • Singapore • United Kingdom • United States

Essentials of Intentional Interviewing;
Counseling in a Multicultural World,
Third Edition
Allen E. Ivey, Mary Bradford Ivey,
Carlos P. Zalaquett

Product Director: Jon-David Hague

Product Manager: Julie Martinez

Content Developer: Amelia Blevins

Product Assistant: Nicole Richards

Marketing Manager: Shanna Shelton

Content Project Manager: Rita Jaramillo

Art Director: Vernon Boes

Manufacturing Planner: Judy Inouye

Production Service: MPS Limited

Photo Researcher: Sofia Priya Dharshini

Text Researcher: Sowmya Mathialagan

Copy Editor: Margaret C. Tropp

Text and Cover Designer: Lisa Delgado

Cover Image: © Alhovik/Shutterstock.com

Compositor: MPS Limited

For product information and technology assistance, contact us at
Cengage Learning Customer & Sales Support, 1-800-354-9706.

For permission to use material from this text or product,
submit all requests online at **www.cengage.com/permissions.**
Further permissions questions can be e-mailed to
permissionrequest@cengage.com.

Library of Congress Control Number: 2014950259

Student Edition:
ISBN-13: 978-1-305-08733-0

Loose-leaf Edition:
ISBN-13: 978-1-305-39955-6

Cengage Learning
20 Channel Center Street
Boston, MA 02210
USA

Cengage Learning is a leading provider of customized learning solutions with office locations around the globe, including Singapore, the United Kingdom, Australia, Mexico, Brazil, and Japan. Locate your local office at **www.cengage.com/global.**

Cengage Learning products are represented in Canada by Nelson Education, Ltd.

To learn more about Cengage Learning Solutions, visit **www.cengage.com.**

Purchase any of our products at your local college store or at our preferred online store **www.cengagebrain.com.**

Printed in the United States of America
Print Number: 02   Print Year: 2015

Love is listening.
Paul Tillich

To James E. Lanier, Ph.D.,
Retired Professor, University of Illinois, Springfield.

A model of love and listening, James Lanier has a significant influence in this book. He taught boxing to African-American youth and through his relationship and listening abilities, they started coming to him for help with their troubles. Very shortly, he became aware that the word problem was a problem for the youth. "I don't have a problem," they said. He found they responded much better to words such as concerns, issues, and, perhaps most of all challenges. You will find his influence on us specifically in our use of language, but also he was important in our movement toward positive psychology, wellness, and Therapeutic Lifestyle Changes (TLC).

Dedicated to my family
Jenifer Zalaquett
Andrea Zalaquett and Christine Zalaquett

Carlos P. Zalaquett

ABOUT THE AUTHORS

Allen E. Ivey is Distinguished University Professor (Emeritus), University of Massachusetts, Amherst, and Courtesy Professor, University of South Florida, Tampa. Allen is a Diplomate in Counseling Psychology and a Fellow of the American Counseling Association. He has received many awards, but is most proud of being named a Multicultural Elder at the National Multicultural Conference and Summit. Allen is author or coauthor of more than 45 books and 200 articles and chapters, translated into 26 languages. He has lectured & key-noted conferences around the world. He is the originator of the microskills approach, which is fundamental to this book.

Allen Ivey

Mary Bradford Ivey is Courtesy Professor of Counseling, University of South Florida, Tampa. A former school counselor, she has served as visiting professor at the University of Massachusetts, Amherst; University of Hawai'i; and Flinders University, South Australia. Mary is the author or coauthor of 14 books, translated into multiple languages. She is a Nationally Certified Counselor (NCC) and a licensed mental health counselor (LMHC), and she has held a certificate in school counseling. She is also known for her work in promoting and explaining development counseling in the United States and internationally. Her elementary counseling program was named one of the ten best in the nation at the Christa McAuliffe Conference. She is one of the first 15 honored Fellows of the American Counseling Association.

Mary Bradford Ivey

Carlos P. Zalaquett is a Professor and Coordinator of Clinical Mental Health Counseling in the Department of Psychological and Social Foundations at the University of South Florida. He is also the Director of the USF Successful Latina/o Student Recognition Awards Program, Executive Secretary for the United States and Canada of the Society of Interamerican Psychology, and President of the Florida Mental Health Counseling Association and the Florida Behavioral Health Alliance. Carlos is the author or coauthor of more than 50 scholarly publications and four books, including the Spanish version of *Basic Attending Skills*. He has received many awards, including the USF Latinos Association's Faculty of the Year and Tampa Hispanic Heritage's Man of Education Award. He is an internationally recognized expert on mental health, psychotherapy, diversity, and education and has conducted workshops and lectures in seven countries.

Carlos P. Zalaquett

CONTENTS

VIDEO TABLE OF CONTENTS

The following videos will be available to view in the MindTap for *Essentials of Intentional Interviewing*, 3e. The list includes time tested videos as well as 15 new videos developed to enhance your learning. You can access MindTap at CengageBrain.com by searching by title or ISBN.

CHAPTER	Video
1	Getting to Know the Authors
	Getting to Know Our Coauthor
	Neuroscience and Counseling: Interviewing and counseling changes the brain
2	A Graduate Student in Counseling: An Afrocentric Approach–video: A graduate student in counseling. An Afrocentric Approach.
	Toward Multicultural Competence: A Gay Client Coming Out.
	Toward Multicultural Competence: Counseling a Child Being Teased
	Multicultural Video. Basic Listening Skills and Diversity Issues
	Neuroscience and Counseling: Using positive psychology and TLCs for wellness
	Informed Consent
	Toward Multicultural Competence. Self-Disclosure and Race/Ethnicity
	Educating Agents of Change. A Review of the Educational Goals of the Psychology Department in the Central American University.
	Psychology of Liberation: The Legacy of Ignacio Martín Baró, SJ., Ph.D.
3	A Negative and Positive Example of Attending
	Sharpening Your Observation Skills
	Discerning Levels of Empathy I: A Demonstration with Mary Bradford Ivey and Allen E. Ivey
	Discerning Levels of Empathy II: A Demonstration with Allen E. Ivey and Mary Bradford Ivey
	Neuroscience and Counseling: Listen to your clients with empathic awareness
4	The Power of Questions: An exercise to help you think critically about questions.
5	Toward Multicultural Competence
	Encouraging, Paraphrasing, and Summarizing: Can you identify these important skills?
	Neuroscience and Counseling: Client stories develop in a social context

PREFACE

Essentials of Intentional Interviewing: Counseling in a Multicultural World, Third Edition, is based on the most researched framework on helping skills—microcounseling. These helping skills serve as the foundation of counseling, psychology, social work, and education. Students enjoy the focus on specific skills and identified competencies, the emphasis on diversity and positive psychology, and the many practice exercises. Many students develop an individual Portfolio of Competencies that can be shared as they move to practicum and internship settings. This book also provides the basics of interviewing in many other fields, ranging from nursing to business to communication studies.

This book was conceived because we wanted to write a competency-based brief text for courses in basic interviewing skills. This book contains the most important information from our larger book, *Intentional Interviewing and Counseling,* and can produce similar results. A key aspect of our vision is ensuring that students learn concepts that they can immediately take into direct practice in role-plays—and later in actual practice.

Through the step-by-step procedures of microcounseling, students will learn to predict the impact of diverse skills on client thinking, feeling, and behavior and will be able to intentionally flex and use another skill when the prediction does not hold. This is a results-oriented approach, thoroughly tested and researched in thousands of settings throughout the world and in more than 450 data-based research articles.

At the same time, this book moves beyond being a skills text. As students become competent in skills, they are constantly encouraged to generalize the skills beyond the textbook into daily life and interviewing practice. Equally important, students see very specifically in the text how the microskills are played out in actual counseling and clinical practice through transcripts of interviews showing decisional counseling and crisis counseling, while transcripts and details about other popular theories will be found on MindTap, the accompanying website.

Microcounseling introduced diversity and multicultural issues to our field as important issues in 1974, and we have continually expanded that awareness over the years. We believe that multicultural factors enrich our understanding of individual uniqueness and that all interviewing involves multicultural issues. All of us are cultural beings with unique individual differences that contribute to our broader humanity.

Neuroscience and neurobiology take a more integrated place in this new edition and provide scientific reasons for the microskills approach. Recent studies add further support to the extensively researched microskills approach. Neuroscience research validates the centrality of empathy and relationship. Skills such as attending behavior, reflection of feeling, supportive confrontation, and others are the building blocks of a successful working alliance.

The media offer findings from neuroscience on a daily basis, and we include this foundational information because your clients likely will ask questions and neuroscience findings strengthen our interviewing and counseling practice. Counseling, the talk cure, does change the mind and brain!

Thus, the central emphasis of this text is on students' developing interviewing competence, knowing that their actions affect the whole human being in front of them. Helping is a true mind/body intervention. Attending behavior and listening at one level are easy to read about, but it is essential to understand the profound effect and necessity for dedicated practice that leads to competency in skills.

▶ Who Is This Book for?

This book is designed to meet the needs of both beginners and more advanced students. It will clarify interviewing and counseling skills for community college students, undergraduates, and graduate students in courses oriented toward the helping professions such as human services, all types of mental health work, behavioral health, counseling, medicine, nursing, psychology, coaching, and social work.

In addition, the microskills approach has proven effective in training for many other groups, including communication students, nutrition counselors, AIDS counselors and refugees in Africa, and trainees in management and leadership. The teaching of microskills extends even to agricultural extension and library work. All of us can profit from gaining more skill in communication.

Some of you may be interested in examining and considering the larger book, *Intentional Interviewing and Counseling.* There you will find a full chapter on observational skills, details of motivational interviewing, more information on research background, and detailed material outlining the relationship of neuroscience to counseling and interviewing.

▶ What Fundamental Outcomes Can Students Expect?

Although this edition includes many innovations, discussed below, the central core of *Essentials of Intentional Interviewing* remains the same. This book will enable students to:

▶ Draw out client stories and understand client thoughts, feelings, and behaviors. In addition, look for underlying meaning issues.

▶ Intentionally anticipate how clients will respond to their use of interviewing skills. Because clients will often surprise us with their responses, we need to be able to flex intentionally and meet client needs by spontaneously moving to another skill, strategy, or even theory.

▶ Complete a full interview using only listening skills by the time they are halfway through this book. Students will be able to analyze their behavior and determine how clients are responding generally to their work, but also which skills appear to work best with each client.

▶ Understand and apply three foundations of effective helping: multicultural competence, ethics, and a strength-based positive psychology. All three sections have been rewritten or updated.

▶ Develop beginning competence in three approaches to the interview: decisional counseling, person-centered counseling, and crisis counseling. In addition, the web resources include detailed information on cognitive behavioral therapy, brief solution-oriented counseling, and coaching.

▶ Define their natural style of helping and their own integration of helping skills. Students will be able to analyze their own interviewing behavior and its effectiveness with clients.

▶ *Empathic relationship—story and strengths—goals—restory—action.* The five-stage structure of the interview remains, but with a new set of labels that clarify the meaning and intent of each stage. As students master the stages of the interview, they are better prepared to become competent in the many strategies and theories of our field.

▶ *Crisis counseling and assessing suicide potential.* All of us will encounter individual and family crises. Likely many of us will be called at some time to assist in a major crisis, ranging from floods and hurricanes to school shootings.

▶ Develop a Portfolio of Competencies, bringing together learning and specific competencies. The MindTap web resources bring the cognitive concepts to action with many practical exercises and videos.

▶ New Competency Features in This Third Edition of *Essentials of Intentional Interviewing*

This thoroughly revised and updated edition contains many important new features. Key among them are videos and interactive exercises, which supplement and support learning.

▶ *Empathy and empathic understanding.* We know that empathic relationships contribute 30% to our success with clients. Starting with Chapter 3, you will find new explanations showing how to integrate empathy with the teaching and learning of listening and influencing microskills. Each chapter's demonstration interviews now illustrate how we can identify and measure empathic understanding throughout the session. Neuroscience has clarified empathy through research on mirror neurons. Our most central concept is no longer just theory—it exists, can be measured, and is more important than ever.

▶ *Positive psychology, resilience, and therapeutic lifestyle changes (TLC).* Our traditional emphasis on client strengths has become more central. Special attention has been given to the importance of focusing on the positive and building on strengths to build resilience.

▶ *New organization of all chapters.* Students prefer books that have a recognizable layout of presentation. *Defining* the counseling skill begins each chapter, followed by *Discerning* the basic techniques and strategies. *Observing* the skill in action via a transcript is next. Then comes *Practicing*, designed to make the concepts immediately real. Elaboration of the skill comes in the *Refining* section, where additional details of skill mastery are presented. *Summarizing*, of course, outlines the main points of the chapter.

▶ *Major updating to chapters on paraphrasing and reflection of feelings, as well as almost totally rewritten material on the influencing skills.* Paraphrasing and summarizing are basic to understanding cognitive executive brain functioning. The material on reflection of feeling explores new findings from neuroscience on emotional regulation. The value of influencing skills such as directives, logical consequences, and psychoeducation is enhanced by increased emphasis on resilience and positive psychology, while therapeutic lifestyle changes (TLC) counseling provides a new system for integrating wellness into the interview.

▶ *Further integration of neuroscience as a rationale for existing interviewing and counseling practice.* Neuroscience research informs us that most of what we are already doing is right and enables us to do it even better. Because of constant media attention to revolutionary new brain discoveries, clients now come to us with some awareness that they can change neural networks and improve cognitive/emotional understanding. We can use information on the brain at appropriate times to explain the value that can be obtained from the session. Early clinical evidence suggests that clients are more likely to take action after the session as they become aware of what positive steps can do to change not only their thoughts and feelings but also their brains.

MindTap for *Essentials of Intentional Interviewing* MindTap for *Essentials of Intentional Interviewing* engages and empowers students to produce their best work—consistently. By seamlessly integrating course material with videos, activities, apps, and much more, MindTap creates a unique learning path that fosters increased comprehension and efficiency.

For students:

▶ MindTap delivers real-world relevance with activities and assignments that help students build critical thinking and analytic skills that will transfer to other courses and their professional lives.

▶ MindTap helps students stay organized and efficient with a single destination that reflects what's important to the instructor, along with the tools students need to master the content.

▶ MindTap empowers and motivates students with information that shows where they stand at all times—both individually and compared to the highest performers in class.

Additionally, for instructors, MindTap allows you to:

▶ Control what content students see and when they see it with a learning path that can be used as is or matched to your syllabus exactly.

▶ Create a unique learning path of relevant readings, multimedia, and activities that move students up the learning taxonomy from basic knowledge and comprehension to analysis, application, and critical thinking.

▶ Integrate your own content into the MindTap Reader using your own documents or pulling from sources such as RSS feeds, YouTube videos, websites, Google Docs, and more.

▶ Use powerful analytics and reports that provide a snapshot of class progress, time in course, engagement, and completion.

In addition to the benefits of the platform, MindTap for *Essentials of Intentional Interviewing* includes:

▶ New video demonstration examples
▶ Interactive learning activities
▶ Case studies
▶ The downloadable interview evaluation forms
▶ Web resources for further reading
▶ The Portfolio of Competencies, and much more

 Throughout the text you will see the icon to the left next to each MindTap activity, indicating that you can go to CengageBrain.com to access these valuable web resources.

▶ The Most Up-to-Date Ancillary System Available

A robust set of student and faculty resources is available for *Essentials of Intentional Interviewing: Counseling in a Multicultural World,* Third Edition. Visit CengageBrain.com to learn more or to obtain access to the ancillary materials.

Resources for instructors include an Instructor's Guide with a sample syllabus, PowerPoints® for flexible teaching, and an online Test Bank with many possibilities for testing and evaluation.

Finally, both student and instructor ancillaries are available in a course cartridge that can be loaded into a course management system such as Blackboard. If you are doing online teaching, this package has just about everything you'll need.

Phone and online help for using these materials is available. Contact your local Cengage Learning representative for ordering options and for access to important resources. For technical support, visit Cengage Learning online at www.cengage.com/support. In addition, the authors are available and eager to be of assistance with any problems or issues that may arise: Allen Ivey (allenivey@gmail.com), Mary Bradford Ivey (mary.b.ivey@gmail.com), and Carlos P. Zalaquett (carlosz@usf.edu).

Skype to your classes? Allen, Mary, and Carlos enjoy participating online with students. As time is available for both professors and the authors, feel free to contact us at the above email addresses.

▶ Acknowledgments

This book has been dedicated to James Lanier, Professor at University of Illinois, Spring-field. He has been an important part of our constant movement toward a healthier, more positive orientation to human change processes. Thomas Daniels, Memorial University, Cornerbrook, has been central to the development of the microskills approach for many years. His summary of research on more than 450 data-based studies is available by request.

Weijun Zhang's writing and commentaries are central to this book. We also thank Courtland Lee, Robert Manthei, Mark Pope, Kathryn Quirk, Azara Santiago-Rivera, Sandra Rigazio-DiGilio, and Derald Wing Sue for their contributions.

Discussions with Viktor Frankl helped clarify the presentation of reflection of meaning. William Matthews was especially helpful in formulating the five-stage model of the interview. Machiko Fukuhara, Professor Emeritus, Tokiwa University, and president of the Japanese Association of Microcounseling, has been our friend, colleague, and coauthor for many years. Her understanding and guidance have contributed in many direct ways to the clarity of our concepts and to our understanding of multicultural issues. We give special thanks and recognition to this wise partner.

David Rathman, chief executive officer of Aboriginal Affairs, South Australia, has constantly supported and challenged this book, and his influence shows in many ways. Matthew Rigney, also of Aboriginal Affairs, was instrumental in introducing us to new ways of thinking. These two people first showed us that traditional, individualistic ways of thinking are incomplete, and therefore they were critical in the development of the focusing skill with its emphasis on the cultural/environmental context. Lia and Zig Kapelis of Flinders University and Adelaide University are thanked for their support and participation while Allen and Mary served as visiting professors in South Australia.

The skills and concepts of this book rely on the work of many different individuals over the past 40 years, most notably Eugene Oetting, Dean Miller, Cheryl Normington, Richard Haase, Max Uhlemann, and Weston Morrill at Colorado State University, who were there at the inception of the microtraining framework.

Many of our students at the University of South Florida in Tampa, the University of Massachusetts in Amherst, the University of Hawai'i in Manoa, and Flinders University in South Australia also contributed in important ways through their reactions, questions, and suggestions.

We would like to pay special tribute to Anne Draus of Scratchgravel Publishing Services, our longtime publication coordinator. Anne passed on to another place while we were completing this book. She was a warm and wonderful person, a pleasure to work with, and a topnotch project manager. We would also like to give special thanks to Lynn Lustberg and Peggy Tropp, who came aboard quickly and made the final publication of this book possible.

Finally, it is always a pleasure to work with the group at Cengage Learning, particularly our wonderful editors, Julie Martinez and Amelia Blevins. We appreciate the entire team that worked on this edition: Brenda Ginty, Rita Jaramillo, Caryl Gorska, Deanna Ettinger, Brittani Hall, Julie Geagan-Chevez, Sean Cronin, Kyra Kane and Charoma Blyden. We are grateful to the following reviewers for their valuable suggestions and comments: Dena Abbott, Texas Woman's University; Laurence Botnick, Community College of Denver; Jeff Harris, Texas Woman's University; Acacia Parks, Hiram College; Lori A. Russell-Chapin, Bradley University; and Kimberly Turnblom, Boise State University.

We would be happy to hear from readers with suggestions and ideas. We appreciate the time that you as a reader are willing to spend with us. You will be listened to and your thoughts will enrich future editions.

Allen E. Ivey (allenivey@gmail.com)
Mary Bradford Ivey (mary.b.ivey@gmail.com)
Carlos P. Zalaquett (carlosz@usf.edu)

SECTION I
Overview and Competencies of Culturally Sensitive Listening

The first thing I'd say . . . is . . . "Listen." It's the second thing I'd say too, and the third, and the fourth. . . . And if you do people will talk. They'll always talk. Why? Because no one has ever listened to them before in all their lives. Perhaps they've not ever even listened to themselves.

—*Studs Terkel*

What major competencies can you expect to achieve by the time you finish studying *Essentials of Intentional Interviewing*? By the time you are halfway through the text, you will be able to complete a full interview using only the microskills of empathic listening. Consider this your first and major goal.

Other awareness, knowledge, skills, and action competencies that you can anticipate include the ability to:

1. Identify specific components of the interview and master them step-by-step.
2. Understand and practice multicultural competence as a necessary base for interviewing practice.
3. Practice interviewing by listening empathically and ethically.
4. Employ positive psychology to build resilience and counteract the overly negative tone of some interviewing and counseling theories and practices.
5. Discover how the listening and influencing skills of this book can be applied immediately in virtually all methods of interviewing, counseling, and psychotherapy—e.g., decisional, person-centered, narrative, cognitive behavioral, psychodynamic, brief counseling, and many others.
6. Learn and practice the basics of crisis counseling with added information on trauma and suicide.
7. Discover some brain basics as they relate to interviewing and counseling practice.
8. Perhaps most important, *practice* the microskills to full mastery with the ability to anticipate how the client will respond. Reading is only a beginning—*practice with feedback* is what will make this book come alive long after you have completed the course.

Chapter 1, Introduction: Foundations of Interviewing and Counseling, offers an overview and a road map of what this book can do for you. The route toward competency starts when you audio or video record a live interview with a volunteer client.

Chapter 2, Multicultural Competence, Ethics, Positive Psychology, and Resilience, presents three crucial aspects of all interviewing and counseling. A strength-based "can do" approach enables clients to resolve issues rather than focusing on what they "can't do."

Chapter 3, Attending, Empathy, and Observation Skills: Fundamentals of All Interviewing and Counseling Approaches, will enable you to become competent in the most essential behaviors of empathic listening—the foundation skills for all interviewing, counseling, and psychotherapy.

Chapter 1
Introduction
Foundations of Interviewing and Counseling

We humans are social beings. We come into the world as the result of others' actions. We survive here in dependence on others. Whether we like it or not, there is hardly a moment of our lives when we do not benefit from others' activities. For this reason it is hardly surprising that most of our happiness arises in the context of our relationships with others.

—The Dalai Lama

Chapter Goals

Awareness, knowledge, skills, and actions developed through the concepts of this chapter and this book will enable you to:

▶ Most important, reflect on your goals for helping and your natural style of being with others. You will be asked to record and document a brief session as a baseline of the expertise you bring to the helping relationship.

▶ Define and discuss similarities and differences among interviewing, counseling, and psychotherapy.

▶ Focus on the client story to identify strengths and positive resources, and use these for building resiliency.

▶ Gain knowledge of the microskills approach to the interview, a step-by-step approach that provides a flexible base on which to build your personal style and theory of counseling.

▶ Become aware that interviewing and counseling can change the brain in positive ways and that neuroscience and neurobiology are now recognized as the cutting edge of our field, leading us toward greater understanding and competence.

▶ Consider significant factors related to the place you conduct your interviews, whether in an office or in the community. Special attention is given to online counseling and the impact of the Internet on clients.

▶ You as Helper, Your Goals, Your Competencies

Here is a real first interview as summarized by Kathryn Quirk, a student like you. Listening and relationship come first. Listen to Sienna, and think what you might respond if you were the counselor. Compare your thoughts with those of Kathryn.

> Sienna, 16 years old, is 8 months pregnant with her first child. She says, "I wonder when I'll be able to see Freddy [baby's father] again. I mean, I want him involved; he wants to be with me, and the baby. But my mom wants me home. His mom said she's looking for a three-bedroom apartment so we could possibly live there, but I know my mom will never go for it. She wants me to stay with her until I graduate high school and, well, to be honest, so that this never happens again [she points to her belly]."
>
> I listen carefully to her story and then respond, "I'm glad to hear that Freddy wants to be involved in the care of his son and maintain a relationship with you. What are your goals? How do you feel about talking through this with your mom?"
>
> "I don't know. We don't really talk much anymore," she says as she slumps down in her chair and picks away at her purple nail polish.
>
> She then describes her life before Freddy, focused mainly on the crowd she hung around, a group of girls whom she says were wild, mean, and tough. Her mood is melancholy, and she seems anxious and discouraged. I say, "Well, it seems that there's a lot to talk about. How do you feel about continuing our conversation before sitting down with your mom?"
>
> Surprisingly, she says, "No. Let's talk next week with her. The baby is coming and, well, it'll be harder then." As we close the session, I ask her, "How has this session been helpful for you?" Sienna responds, "Well, I guess you're going to help me talk about some difficult issues with my mom, and I didn't think I could do that."

This was the first step in a series of five interviews. As the story evolved, we invited Freddy for a session. He turned out to be employed and was anxious to meet his responsibilities, although finances remained a considerable challenge. A meeting followed this with both mothers in which a workable action plan for all families was generated. I helped Sienna find a school with a special program for pregnant teens.

This case exemplifies the reality of helping that you and Kathryn face. We often face complex issues with no clear positive ending. However, if we can develop an empathic relationship and listen to the story carefully, clearer goals develop and solutions usually follow.

> **EXERCISE 1.1 Love Is Listening**
> *Listening, love, caring,* and *an empathic relationship* are central. Listening is the core of the helping process. What are your reactions and thoughts about the centrality of listening? What relevance does Paul Tillich's affirmation, "Love is listening," have in the interview with Sienna?
>   Please take a few moments to reflect on a time when you needed help and someone listened. How did you feel inside as they stayed with you as you told your story? Perhaps you can stop for just a moment and imagine the situation as it was. Can you possibly see or even hear the listener now? What was he or she doing, and how did it help?

▶ Online Resources for *Essentials of Intentional Interviewing*

We have created optional interactive exercises and resources to help you further understand major concepts and sharpen your interviewing competencies. Additional case studies, video clips of sessions, and practice tests are designed to help you master the skills of interviewing and counseling. Simply go to www.cengagebrain.com and use the search field to find the

MindTap for *Essentials of Intentional Interviewing,* Third Edition to access these valuable resources.

We will place icons like this one nearby to reference these digital resources, along with suggested websites that you can search for in your web browser. We designed them to make your learning more meaningful and to help ensure competence in interviewing and counseling practice.

If you did not receive an Access Card packaged with your book, you can purchase access to these web resources at CengageBrain.com. Check with your professor first to find out if the online resources are required.

Assess your current level of knowledge and competence as you begin the chapter:

1. **Self-Assessment Quiz:** The chapter quiz will help you determine your current level of knowledge. You can take it before and after reading the chapter.

2. **Portfolio of Competencies:** Before you read the chapter, please fill out the Self-Evaluation Checklist to assess your existing knowledge and competence on the ideas and concepts presented in this chapter. Then, at the end of the chapter, complete the checklist again to summarize your competencies after study and practice.

Video: Getting to Know the Authors
Interactive Self-Assessment: Personal Strengths and Your Natural Style
Interactive Self-Assessment: Self-Awareness and Emotional Understanding
Case Study: What Do You Say Next? Working With a Difficult Case

---

▶ Interviewing, Counseling, Psychotherapy, and Related Fields

The terms *counseling* and *interviewing* are used interchangeably in this text. The overlap is considerable (see Figure 1.1), and at times interviewing will touch briefly on counseling and psychotherapy. Both counselors and psychotherapists typically draw on the interview in the early phases of their work. You cannot become a successful counselor or therapist unless you have solid interviewing skills.

Interviewing is the more basic process used for gathering data, providing information, and helping clients resolve issues. Interviewers may be found in many settings, including employment offices, schools, hospitals, businesses, law offices, and others. Professional coaches use interviewing skills in career coaching, spiritual coaching, executive coaching, and other specialties such as retirement coaching for elders. All counselors and therapists are continually involved in interviewing.

Counseling is a more intensive and personal process than interviewing. Although interviewing and information gathering are essential, counseling is more about listening to and understanding a client's life challenges and, with the client, developing strategies for change and growth. Counseling is most often associated with the professional fields of human relations, counseling and clinical psychology, pastoral counseling, and social work; it is also part of the role of medical personnel and psychiatrists.

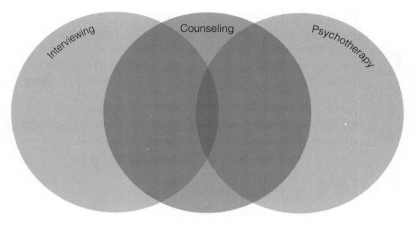

FIGURE 1.1 The interrelationship of interviewing, counseling, and psychotherapy.

Psychotherapy focuses on more deep-seated personality or behavioral difficulties, and may take longer. Psychotherapists interview clients to obtain basic information and often use counseling as well. The skills and concepts of intentional interviewing are necessary for the successful conduct of longer-term psychotherapy.

Figure 1.1 illustrates the considerable overlap among the three major forms of helping. For example, a personnel manager may interview a candidate for a job, but in the next hour may counsel an employee who is deciding whether to accept a promotion that requires a move to another city. A school guidance counselor may interview a student to check on course selection, but may counsel the next student about college choice or a conflict with a friend. A psychologist may interview one person to obtain research data, and in the next hour counsel another concerned with an impending divorce. In the course of a single contact, a social worker may interview a client to obtain financial data, and then move on to counsel the same client about personal relationships.

Interviewing is such a broad area that literally all fields of business, administration, government, medicine, mental health, coaching, and many others use the skills of this book. More than a million counselors, psychologists, social workers, and human service workers in the United States perform primarily interviewing and counseling tasks, although many will engage part- or full-time in psychotherapy. All are concerned with building client resilience in one way or another.

Internet counseling is becoming more frequent, but its place in the helping fields has not yet been resolved. How can ethical and professional standards be met? If you are reading this book as part of an online course, then you understand the potential. With laptop video cameras, face-to-face sessions are possible via Skype, Facetime, and other services. Observing nonverbal communication thus becomes possible, although only to a limited extent. Also, let's not forget the issue of confidentiality and security. There will be other concerns as this form of helping develops.

What does all this mean to you, the student? The basic skills you will find here are used again and again regardless of where you go or how you are employed. The work settings and titles will vary, but it is you who will have the primary responsibility for crisis counseling, work with children and youth in schools and community agencies, family counseling and therapy. In addition, you may encounter much of the challenging work with more severely distressed individuals, those who have been diagnosed with depression or anxiety or have suffered traumatic experience.

Interactive Exercise: Interviewing, Counseling, and Psychotherapy

► The Microskills Approach

Microskills are communication skill units that help you to interact more effectively with a client, whether you are an interviewer, counselor, coach, or psychotherapist. These same skills are what you will use in all advanced theories and strategies. In addition, microskills are basic to fields that communicate with clients, including business, sales, medicine, physical therapy, public relations, and many others. Developing competence in interviewing skills will enable you to plan and structure a session to make a difference in the life of your client.

The microskills model was developed through years of study and analysis of hundreds of interviews. It is designed to show you how to use specific skills to master the art of communicating with a client. Your natural talent and style will be greatly enhanced by practicing and becoming competent in these skills.

Why the term *microskills*? "Micro" means breaking down complex, larger components into more precise, small, and manageable components. Breaking down the skills of effective counseling into a clear step-by-step process makes them easier to learn and teach.

Knowing how is not enough; doing it well with expertise is very different. The microskills of the interview are critical dimensions of effective interviewing and counseling, just as the specific skills of tennis are to Roger Federer and Maria Sharapova. Rounding a corner for a NASCAR driver, or mastering harmony and style for a new pop singer or band, takes persistence and extensive practice to become truly expert. Practice and feedback are key to mastering all skills, including the microskills!

As you become competent with each microskill, you are prepared for learning the next level. This approach is similar to that employed by professionals of all kinds—from Olympic athletes and musicians to clothing designers, journalists, and craftspeople. World-famous golf pros and tennis stars, as well as accomplished musicians and dancers, begin with basics and continue to higher levels, eventually integrating them into a holistic style. Indeed there are natural athletes and interviewers, but they all benefit from analyzing what they are doing in order to become more effective.

THE MICROSKILLS HIERARCHY

The specific skill steps of the **microskills hierarchy** are outlined in Figure 1.2. These steps provide the specifics that will enable you to work with a multitude of clients and *anticipate which skill will likely be most effective with what anticipated result*. In addition, competence in the microskills hierarchy will give you a clear sense of how to organize and structure a session. This five-stage structure will be useful with multiple theories of interviewing, counseling, and psychotherapy, as well as coaching.

Multicultural competence, ethics, and a positive psychology/resilience approach form the foundation (Chapter 2). We live in a multicultural world where every client that you encounter will have a different life experience and every life will be different from yours. Without a basic understanding of and sensitivity to a client's uniqueness, the

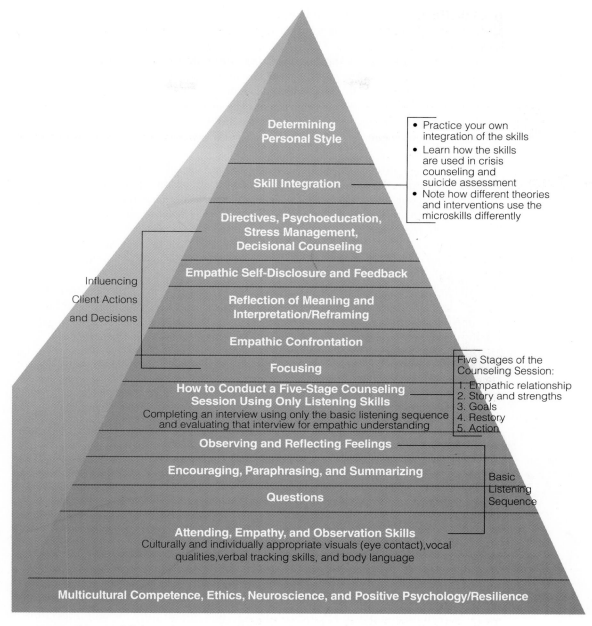

FIGURE 1.2 The microskills hierarchy: A pyramid for building cultural intentionality.

Source: Based on The Microskills Hierarchy: A pyramid for building cultural intentionality, Allen E. Ivey

interviewer will fail to establish a relationship and truly grasp a client's issues. Throughout, this book will examine the multicultural issues and opportunities we all experience.

Ethical standards are discussed in Chapter 2, as well as how you can use positive psychology principles to promote client resilience. Special attention is paid to therapeutic lifestyle changes, a new and important addition to every interviewer's and counselor's knowledge base.

Attending, observation, and empathy skills are key to the listening process and cultural intentionality (Chapter 3). Attending behavior is critical for developing a working relationship and drawing out client stories. Attending and observation skills provide a foundation for empathic understanding. Together, these three areas represent the essence of the relationship and the working alliance—all essential for a multiculturally sensitive session.

Questioning, paraphrasing, summarizing, and reflection of feeling are discussed in Chapters 4, 5, and 6. These skills together are termed the *basic listening sequence (BLS)*. They elaborate attending and observation and enable you to draw out client stories and concerns fully. Careful use of questions fills out the story and provides a way to integrate missing links. Paraphrasing, reflection of feeling, and summarization are most important in clarifying client cognitive/emotional stories, as well as demonstrating to clients that they have been heard.

The five-stage interview and skill integration are the subject of Chapter 7. With a solid background in the basic listening sequence, you will be able to conduct a complete interview using only listening skills. The five stages of the interview process are: *empathic relationship—story and strengths—goals—restory—action*. These five stages provide an encompassing framework that can serve as a model and checklist in many settings, with varying clients, and with different theoretical approaches. Note our use of the term *action* instead of the traditional words "termination" or "ending." *Action* based on interview learning promotes successful change.

An empathic relationship has been found to be responsible for 30% or more of the change observed in effective counseling sessions. This is where your unique personhood and listening skills set the stage for the rest of the interview. Listening skills are basic to getting the story out, but we also need to pay attention to client strengths and supports.

Active strategies for change form the second half of the book and focus on microskills that are used primarily in stages 4 and 5 of the interview. Here you will find the *focusing* skill (Chapter 8) used to expand and elaborate client concerns. *Empathic confrontation* (Chapter 9) help clients discover and resolve inconsistencies and conflict. The influencing skills of *reflection of meaning, interpretation/reframing, empathic self-disclosure*, and *feedback* (Chapters 10 and 11) help clients restory and think about their issues in new ways, which in turn lead to new thoughts, feelings, and behaviors.

Chapter 12 on *directives, psychoeducation*, and *natural and logical consequences* will enable you to put an active "positive spin" on client change. Decisional counseling will be introduced as the most basic helping approach as, in one way or another, all clients are making decisions.

Skill integration (Chapter 13) is your place to blend the microskills, while also addressing case management and treatment planning for future sessions. Key here is the suggestion that you make a video of a full session and analyze what happened and its impact on the client. A full career counseling session transcript with Allen and Mary Ivey, available in the web resources, will provide a model for your own self-examination.

Crisis counseling (Chapter 14) provides a second opportunity for skill integration. Inevitably, you will work with clients in crisis, and this application of the five stages, coupled with some basics of crisis work, provides an introduction for further study. In addition, you will find more theories presented in the web resources, including client-centered counseling, brief solution-based counseling, cognitive behavioral therapy, and others.

Determining personal style and theory is at the apex of the microskills hierarchy. Chapter 15 asks you to review your work with this book and the competencies you have developed. At this point, you will have another opportunity to write your own narrative, your own personal story about interviewing and counseling. This chapter is designed to help you "put it together" in your own way.

Box 1.1 summarizes relevant research findings regarding the microskills.

BOX 1.1 Microskills Research and Related Evidence That You Can Use

More than 450 microskills research studies have been conducted (Daniels, 2014; Daniels & Ivey, 2006). The model has been tested nationally and internationally in more than 1,000 clinical and teaching programs. This was the first skills book based on research, and the microskills have now been translated into 25 languages.

Microcounseling was the first systematic video-based counseling model to identify specific observable interviewing skills. It was also the first skills training program that emphasized multicultural issues. Some of the most significant research findings include the following:

▶ *You can anticipate how clients will respond to you when you use microskills in the session.* Several critical reviews have found microtraining an effective framework for teaching skills to a wide variety of people, from beginning interviewers to experienced professionals, who need to relate to patients and clients more effectively. Teaching your clients many of the microskills will facilitate their personal growth and ability to communicate with their families or coworkers.

▶ *Practice is essential.* Practice the skills to mastery if the skills are to be maintained and used after training. *Use it or lose it!* Complete practice exercises and generalize what you learn to real life. Whenever

possible, audio or video record your practice sessions.

▶ *Multicultural differences are real.* People from different cultural groups (e.g., ethnicity/race, gender) have different patterns of skill usage. **Cultural intentionality** is essential—learn about people different from you, and use skills in a culturally appropriate manner.

▶ *Different counseling theories have varying patterns of skill usage, as well as different client issues requiring differential use of skills.* Developing competence in microskills and the five-stage interview model will enable you to rapidly understand and perform widely varying theories of counseling and therapy. For example, you will find the listening skills central in all, while influencing skills will differ widely.

▶ *Flexibility.* Cultural intentionality prepares you for the unexpected and enables you to flex another microskill.

▶ *Use feedback as your compass.* Client and colleague feedback, whether spontaneous or requested, will help guide your interventions throughout your career. Seek feedback when implementing a microskill. Kenneth Blanchard has said, "Feedback is the breakfast of champions." Feedback provides the building blocks of positive relationships, effective interventions, and successful change.

Interactive Exercise: The Microskills Hierarchy
Group Practice Exercise: The Microskills Hierarchy

▶ Cultural Intentionality

Microskills are based on seeking cultural expertise and multicultural competence. The goal is to provide the interviewer and counselor with a high flexible set of observable skills that can make a difference for all involved.

Many, even most, of our clients come to us feeling that they are not functioning effectively and are focused on what's wrong with them. They are *stressed*. Clients may feel *stuck*, *overwhelmed*, and *unable to act*. Frequently, they will be unable to make a career or life decision. Often, they will have a *negative self-concept*, or they may be full of *anger toward themselves and/or others*. This focus on the negative is what we want to combat as we emphasize developing client intentionality, resilience, and self-actualization. See Box 1.2 for a review of counseling and therapy historical emphasis on clients' problems and deficits.

At the same time, you, the interviewer, may find yourself stuck and stressed in the session, wondering what to do next. Meeting that challenge requires that you be resilient and able to think through and actualize new and useful responses for your clients.

What does cultural intentionality mean for you? First, it means that clients exist in a multicultural situation and context. Interviewing and counseling do not exist in a vacuum. Becoming cultural competent has become basic to interviewing and counseling practice

National and International Perspectives on Counseling Skills

Problems, Concerns, Issues, and Challenges—How Shall We Talk About the Story?

James Lanier, University of Illinois, Springfield

Counseling and therapy historically have tended to focus on client problems. The word *problem* implies difficulty and the necessity of eliminating or solving the problem. Problem may imply deficit. Traditional diagnoses such as those found in the *Diagnostic and Statistical Manual of Mental Disorders, 5th Edition* (DSM-5) (American Psychiatric Association, 2013) carry the idea of problem a bit further, using the word *disorder* in such terms as *panic disorder, conduct disorder, obsessive-compulsive disorder*, and many other highly specific *disorders*. The way we use these words often defines how clients see themselves. Professionals who label clients make their definitions come to pass whether they were accurate or not.

I'm not fond of problem-oriented language, particularly that word *disorder*. I often work with African American youth. If I asked them, "What's your problem?" they likely would reply, "I don't have a problem, but I do have a concern." The word *concern* suggests something we all have all the time. The word also suggests that we

can deal with it—often from a more positive standpoint. Defining *concerns* as *problems* or *disorders* leads to placing the blame and responsibility for resolution almost solely on the individual.

Finding a more positive way to discuss client concerns is relevant to all your clients, regardless of their background. *Issue* is another term that can be used instead of *problem*. This further removes the pathology from the person and tends to put the person in a situational context. It may be a more empowering word for some clients. Carrying this idea further, *challenge* may be defined as a call to our strengths. Some might even talk about *an opening for change*.

Beyond that, the concepts of wellness and the positive asset search make good sense for the youth with whom I have worked. Change is most easily made from a position of strength—criticism and problem-oriented language can weaken. However, don't be afraid to challenge people to grow. Empathic confrontation can help your clients develop in positive ways.

(see Chapter 2). Cultural intentionality also means being aware that we cannot expect to solve all our clients' issues and challenges in a few sessions, but in the short time we have with them, we can make a difference. First think of what intentionality and flexibility mean for you as an interviewer or counselor. You can think of the interview as a "mini-culture" in which the client can learn new skills, attitudes, and behaviors that he or she can take into the "real world."

Cultural **intentionality** is acting with a sense of capability and choosing from among a range of alternative actions, interviewing skills, and helping theories, always considering the cultural and ethnic characteristics of the client. The intentional interviewer or counselor can generate alternatives in a given situation and approach a client's concerns from different vantage points, using a variety of skills and personal qualities, adapting styles to meet the needs of different individuals and cultures.

As you become competent in using skills and structuring an effective interview, you will have more flexibility to meet the changing needs of clients. The culturally intentional interviewer remembers a basic rule of helping: *If a helping lead or microskill doesn't work— try another approach!* A critical issue in interviewing is that the same comment may have different effects on individuals who have unique life experiences and multicultural backgrounds. Intentional interviewing requires awareness that cultural groups and individuals each have their own patterns of communication.

What does cultural intentionality means for the client? Clients will benefit and become stronger as they feel heard and respected and as they discover new ways to resolve their concerns. Making intentional decisions on specific immediate issues, such as choosing a college major, making a career change, deciding whether to break up a long-term relationship, or handling mild depression after a significant loss, will help them feel empowered and facilitate further action. Ultimately, our goal is to help clients develop skills that will empower them to face and effectively cope with current and future situations and challenges. This is a process we call self-healing or resilience.

▶ Resilience and Self-Actualization

> Resilience is a dynamic process whereby individuals exhibit positive behavioral adaptation when they encounter significant adversity, trauma, tragedy, threats, or even significant sources of stress.
>
> —S. Luthar, D. Cicchetti, & B. Becker

The development of client intentionality is another way to talk about client **resilience**—a major goal of interviewing and counseling. As counselors, we want to be flexible and move with changing and surprising events, but clients need the same abilities. Helping a client resolve an issue is our contribution to increasing client resilience. You have helped the client move from immobility to action, from indecision to decision, or from muddling around to clarity of vision. Pointing out to clients who change that they are demonstrating resilience and ability facilitates longer-term success.

The development of intentionality encourages a stronger sense of self (self-control), a major aspect of resilience. Neuroscience points out that self-control relies primarily on cognitive skills in the prefrontal lobes of the brain. Here working memory, reasoning, decision making, and the ability to put plans into action can work to modify stress and resolve issues.

Emotional regulation, basic to resilience, is defined by neuroscience as the ability to respond appropriately socially, but also as the ability to manage challenging situations without losing self-control. The primary emotional areas are deep in the brain's limbic system—and emotions and feelings are wired as our first reaction in any situation before a cognitive response, whether we know it or not. Thus emotional regulation becomes a key interviewing and counseling issue. For some, Chapter 6 on reflecting feelings becomes the most important chapter of all.

A good cognitive decision will be ineffective if emotional regulation and self-control are lost. Thus resilience and flexibility become all the more important. Emotional regulation enables us to inhibit impulsive behavior and focus on the task at hand, manage disagreements with others, and not let ourselves "lose it" in the face of major life challenges.

SELF-ACTUALIZATION AND SELF-IN-RELATION: KEYS TO RESILIENCE

Carl Rogers and Abraham Maslow have focused on self-actualization, which they define as follows:

> the curative force in psychotherapy—man's tendency to actualize himself, to become his potentialities . . . to express and activate all the capacities of the organism. (Carl Rogers)

> . . . intrinsic growth of what is already in the organism, or more accurately of what is the organism itself. . . . self-actualization is growth-motivated rather than deficiency-motivated. (Abraham Maslow)

Janet Surrey, a feminist theorist, might agree with Rogers and Maslow, but she points out:

> the self-in-relation involves the recognition . . . [that] the primary experience of self is relational; that is, the self is organized and developed in the context of important relationships. (Janet Surrey)

Regardless of the situation in which our clients find themselves, we ultimately want them to feel good about themselves and actualize the potential that is in all of us. Empathy demands that we understand and be with the client, but this does not mean approval of inappropriate, unwise, and harmful actions toward the client or others.

Both Rogers and Maslow had immense faith in the ability of individuals to overcome challenges and take charge of their lives. Many would call them the founders of the positive psychology movement. You want to be there for the client to facilitate the action of their potential selves.

Surrey and others add to this that the self does not arise alone. We can only actualize ourselves in relation to those around us. Those around us—family, community, school, multiculture—have had much to do with who we are. In short, we only actualize in relationships.

Counseling and psychotherapy sessions are indeed for the individual client, but let's not forget that the client exists in a multidimensional, multicultural, social context. Putting the client in context is all too often missing in counseling theory and practice. Rogers and Maslow both were interested in self-actualizing groups, communities, and nations. Surrey reminds us that as we build intentional, resilient clients, we need also to remember that self-actualizing, intentional clients are *selves-in-relationship*.

EXERCISE 1.3 Reflection: What Are the Goals of Counseling and Therapy?

Self-actualization and resilience are both challenging and important concepts.

How does self-in-relationship relate to self-actualization?

What experience and supports have led you to become resilient, more yourself, what you really are and want to be?

How have your bounced back (resilience) from major challenges you have faced?

What personal qualities or social supports helped you grow?

What does this say to your own approach to counseling and psychotherapy?

▶ Neuroscience: Counseling Changes the Brain

> Our interaction with clients changes their brain (and ours). In a not too distant future, counseling will be regarded as ideal for nurturing nature.
> —Óscar Gonçalves, Northeastern University and University of Minho, Portugal

> Psychotherapy is a biological treatment, a brain therapy. It produces lasting, detectable physical changes in our brain, much as learning does.
> —Eric Kandel, Nobel Prize Winner 2013

You will make a significant difference in your clients' lives. There is no longer any doubt as to the potential impact and long-term effectiveness of quality interviewing, counseling, and psychotherapy. Neuroscience research tells us that what we are doing in our field has basically been correct—that what we do "works."

We now know not only that your interviewing and counseling skills facilitate client cognitive and emotional growth but also, through research using functional magnetic imaging (fMRI), that your relationship and conversations can affect and observably increase the brain's neural connections. Your work with clients can enable them to modify memories in the hippocampus and to reframe past difficulties as they recognize the strengths they manifested during adversity (Goldin et al., 2013; Karlsson, 2011).

In effect, "brains can be rewired" (for example, Hölzel et al., 2011; Logothetis, 2008). Our brain's ability to change and grow is termed *neuroplasticity*. Throughout the lifespan, we and our clients can generate new neural networks of information and even new neurons (neurogenesis). Only a short time ago, it was believed that we lose neurons daily as we age, and you will find clients who are delighted to learn that they can increase the power and flexibility of their brains. Important in this process will be your relationship skills and your competence in helping skills and theories.

We also need to be aware of negative neuroplasticity. The first example one thinks of are the dangers of drug and alcohol addiction, and numerous studies and clinical examples show that these can do permanent damage. Developmental issues abound in that the depressed or addicted mother can damage the fetal brain. Poverty, neglect, and abuse injure the brain. Oppression from racism, sexism, bullying, and other forms of prejudice can be harmful. Out of all these can come reduced cognitive and emotional skills. Some are harmed by ineffective, thoughtless counselors and therapists, who may give inaccurate diagnoses or prescribe the wrong medication.

We may even become *neurocounselors* in the future. The label "neurocounselor" was used for the first time in a cover story in the American Counseling Association magazine *Counseling Today* (Montes, 2013). You read about neuroscience or see some new discovery about the brain on television almost daily. Your clients have access to the same information, and discussion of how the interviewing relationship involves potential changes in the brain may become a necessary part of your sessions.

One place to begin discussion with your clients is *brain plasticity*. It can be useful to point out to clients that brain plasticity means they cannot only change the way they think, feel, and behave but they can add new and permanent neural connections and even change memories as the two of you work together.

Increasingly, counselors are sharing with clients pictures of the brain from a book, or even mounting a poster on the wall. As they discuss counseling goals with the client, they point out that by working together they can achieve lasting changes in the brain. This egalitarian approach appears to increase client interest and increases effort on their part to take new ideas and learning home from the interview.

Neuroscience has brought us to a deeper awareness that stress underlies virtually all issues that clients bring to us. Severe stress can be damaging to the brain and also to the body. Some 80% of medical issues involve the brain and stress (Ratey, 2008). Stress management is becoming a central strategy for prevention and treatment of both mental and physical illness, regardless of our theoretical orientation.

When we listen empathically to clients and their stressors, clients can look at their situation and understand what is going on more clearly. Through being listened to, clients understand themselves better and are better able to make their own decisions. As one client said, "I had so much running around in my head that I was totally stressed, confused, and upset. Now that you have listened to my story, I can organize it, make sense, and decide what I want to do."

Those of you who are interested neuroscience, now or later, can search for "Allen and Mary Bradford Ivey" on YouTube and see a relatively recent presentation on the basics of neuroscience and counseling. Get acquainted with two of the authors.

► Office, Community, Phone, and Internet: Where Do We Meet Clients?

> Regardless of physical setting, you as a person can light up the room, street corner, even the Internet. Smiling and a warm, friendly voice make up for many challenging situations. It is the *how* you are, rather than *where* you are.
>
> —Mary Bradford Ivey

First, let's recognize that interviewing and counseling occur in many places other than a formal office. There are street counselors who work with youth organizations, homeless shelters, and the schools, as well as those who work for the courts, who go out into the community and get to know groups of clients. Counseling, interviewing, and therapy can be very informal, taking place in clients' homes, a neighborhood coffee shop or nearby park, and while they play basketball or just hang out on the street corner. The "office" may not exist, or it may be merely a cubicle in a public agency where the counselor can make phone calls, receive mail, and work at a computer, but not necessarily a place where he or she will meet and talk with clients. The office is really a metaphor for your physical bearing and dress—smiling, culturally appropriate eye contact, a relaxed and friendly body style.

As a school counselor, Mary Bradford Ivey learned early on that if she wanted to counsel recent immigrant Cambodian families, home visits were essential. She sat on the floor as the family did. She attended cultural events, ate and cooked Cambodian food, and attended weddings. She brought the Cambodian priest into the school to bless the opening ceremonies. She provided translators for the parents so they could communicate with the teachers. She worked with school and community officials to advocate for the special needs of these immigrants. The place of counseling and developing your reputation as a helper varies widely. Maintaining a pleasant office is important, but not enough.

Another approach, used by Mary, is to consider the clientele likely to come to your setting. Working in a school setting, she sought to display objects and artwork representing various races and ethnicities. The brightness of the artwork worked with well with children, and many parents commented favorably on seeing their culture represented. But most important, make sure that nothing in your office can be considered objectionable by any of those whom you serve.

PHONE, SKYPE, AND INTERNET COUNSELING

Historically, the emphasis has been on keeping the boundaries between counselor and client as clear and separate as possible, but this seems to be changing. Where once the therapist was opaque and psychologically unseen, you as a person have become more important.

As the importance of that fifth stage of the interview (action and follow-up) is recognized as increasingly critical to client change, many counselors are now using smartphones so that clients can follow up with them or ask questions. With smoking, alcohol, or drug cessation, being available can make a significant difference. However, this is also fraught with practical and ethical issues. You and the client both lose privacy, nonverbal communication will be missed, and confidentiality may be endangered. Skype and other visual phone services partially answer these questions, but they are still not the same as a face-to-face relationship.

Enter "online counseling" in your search engine, and a number of services will appear. Following is a composite result from Allen Ivey's visits to several online services:

Meeting life's challenges is difficult.

[Internet counseling center] enables you to talk with real-life professional counselors 24/7 in full confidentiality.

Choose a counselor.

First session is free.

Easy payments arranged.

We expect that more and more of these services will appear on the Internet, particularly now that limited face-to-face interaction is available on the phone and Internet. Some of these services may be quite helpful to clients at a reasonable cost, but others may be risky. Now search for "coaching services," and you will find an amazing array of possibilities. However, view all these sites with some attention as to how ethical and professional standards can be met on the Internet.

Distance Credentialed Counselor (DCC) is a national credential currently offered by the Center for Credentialing and Education (CCE). Holders of this credential adhere to the National Board for Certified Counselors' Code of Ethics and the Ethical Requirements for the Practice of Internet Counseling. These professionals adapt their counseling services for delivery to clients via technology-assisted methods, including telecounseling (telephone), secure email communication, chat, videoconferencing, and other appropriate software (CCE, n.d.). Professional associations are still working on these new approaches to counseling delivery. As with all types of counseling and therapy, ethics is the first concern. In Chapter 2 we present some beginning issues in ethics. You will also find Internet links to the professional ethics of the main national helping organizations.

 These websites can be accessed in the web resources or through a search engine.
Website: Bureau of Labor: Counselors. Counselors work in diverse community settings designed to provide a variety of counseling, rehabilitation, and support services.
Website: Bureau of Labor: Psychologists.
Website: Bureau of Labor: Social Workers.

▶ Your Natural Style and Beginning Expertise: An Important Audio or Video Activity

The microskills learned through this text will provide you with additional alternatives for culturally intentional responses to the client. However, these responses must be genuinely your own. If you use a skill or strategy simply because it is recommended, it is likely to be ineffective for both you and your client. Not all parts of the microskills framework are appropriate for everyone. You have a natural style of communicating, and these concepts must enhance your natural style, not detract from it.

Self-study of your own interviewing style is a valuable part of building a **portfolio of competencies**. Throughout this text and accompanying web resources are an array of exercises that are well worth recording. We have found that the portfolio you develop can be useful when you apply later for a practicum, internship, or even a job. You will be able to show the decision makers what you can do in the session and the results that come from your conversations with clients.

You will need varying patterns of helping skills with your clients. Couple your natural style with awareness and knowledge of individual and multicultural differences. How will clients respond to your natural style? You will need to be able to "flex" and be intentional as you encounter diversity among clients.

You may work more effectively with some clients than with others. Developing trust and a working alliance takes more time with some clients than with others. Many clients lack trust with interviewers who come from a cultural background different from their own. You may be less comfortable with teenagers than you are with children or adults. Some have difficulty with elders.

You are about to engage in a systematic study of the interviewing process. By the end of the book, you will have experienced many ideas for analyzing your interviewing style and skill usage. Along the way, it will be helpful to have a record of where you were before you began this training.

It is invaluable to identify your personal style and current skill level before you begin systematic training. Eventually, it is YOU who will integrate these ideas into your own practice. Let's start with your own work. Please read Box 1.3 and make plans to record your natural style *before* continuing further in this text.

 Interactive Exercise: Your Natural Helping Style: An Important Audio or Video Activity

---

BOX 1.3	Discovering Your Natural Style of Interviewing

Many of you now have cell phones, cameras, and computers with video and sound capabilities. All these provide the opportunity for effective feedback. There is nothing like seeing yourself with a client as it "really happened." You will want to make transcripts of some of your practice sessions. Reviewing key parts of an interview several times is immensely valuable. We also find that asking clients to watch or listen to a session just completed can be helpful for them and a great learning experience for you.

Guidelines for Audio or Video Recording

1. Find a volunteer client willing to role-play a concern, challenge, opportunity, or issue.
2. Interview the volunteer client for at least 15 minutes.
3. Use your own natural communication style.
4. Ask the volunteer client, "May I record this interview?"
5. Inform the volunteer client that the recorder can be turned off any time he or she wishes.
6. Select a topic. You and the client may choose interpersonal conflict or a specific issue selected by the client. As part of this session, begin with a discussion of client strengths, things that they have done right, and/or past successes at resolving challenges and issues.

7. Follow the ethical guidelines discussed in Chapter 2 (page 29). Common sense demands ethical practice and respect for the client.
8. Obtain feedback. You will find it very helpful to get immediate feedback from your client. As you practice the microskills, use the Client Feedback Form (Box 1.4). You may even find it helpful to continue using this form, or some adaptation of it, in your work as an interviewing professional.
9. Compare this baseline with subsequent recordings of your work throughout this text.

It will be more valuable if you conduct this exercise as soon as possible. Try to complete this interview before you read far into this book or hear too many instructor presentations. Your audio or video recording will document an accurate baseline of your natural style and skill level. As you progress through the text, compare this session with your later work.

BOX 1.4	Client Feedback Form

———————————————————————————————————— (DATE)

_____   _____
(NAME OF INTERVIEWER)   (NAME OF PERSON COMPLETING FORM)

Instructions: Rate each statement on a 7-point scale with 1 representing "strongly agree (SA)," 7 representing "strongly disagree (SD)," and N as the midpoint "neutral." You and your instructor may wish to change and adapt this form to meet the needs of varying clients, agencies, and situations.

	Strongly Agree SA			Neutral N		Strongly Disagree SD	
1. (Awareness) The session helped you understand the issue, opportunity, or problem more fully.	1	2	3	4	5	6	7
2. (Awareness) The interviewer listened to you. You felt heard.	1	2	3	4	5	6	7
3. (Knowledge) You gained a better understanding of yourself today.	1	2	3	4	5	6	7
4. (Knowledge) You learned about different ways to address your issue, opportunity, or problem.	1	2	3	4	5	6	7
5. (Skills) This interview helped you identify specific strengths and resources you have to help you work through your concerns and issues.	1	2	3	4	5	6	7
6. (Skills) The interview allowed you to identify specific areas in need of further development to cope more effectively with your concerns and issues.	1	2	3	4	5	6	7
7. (Action) You will take action and do something in terms of changing your thinking, feeling, or behavior after this session.	1	2	3	4	5	6	7
8. (Action) You will create a plan of action to facilitate change after this session.	1	2	3	4	5	6	7

What did you find helpful? What did the interviewer do that was right? Be specific—for example, not "You did great," but rather, "You listened to me carefully when I talked about _____."

What, if anything, did the interviewer miss that you would have liked to explore today or in another session? What might you have liked to have happen that didn't?

Use this space or the other side for additional comments or suggestions.

Summarizing Key Points of "Introduction: Foundations of Interviewing and Counseling"

You as Helper, Your Goals, Your Competencies

▶ Counseling is an art and a science.

▶ Interviewing and counseling have a solid research and scientific base.

▶ Love is listening. The identified listening skills are central to competent helping.

▶ Ask yourself: What brought you here? What natural talents do you bring? What do you need to learn to grow further?

Interviewing, Counseling, Psychotherapy, and Related Fields

▶ Interviewing is a basic process for gathering data, solving problems, and providing information and advice.

▶ Counseling is more comprehensive and is generally concerned with helping people cope with life challenges and develop new opportunities for further growth, whereas psychotherapy is focused on more deep-seated personality or behavioral difficulties.

▶ Currently, there is a high degree of overlap among interviewing, counseling, and psychotherapy.

▶ Multiple fields that use the skills of this text range from business interviewing through in-depth work in clinical social work, counseling or clinical psychology, and psychiatry.

▶ You will use the microskills presented in this text in any working or social context, but with varying emphasis and objectives.

The Microskills Approach

▶ Microskills are communication skill units that help you develop the ability to interact more intentionally with a client.

▶ The microskills hierarchy provides a visual picture of the skills and strategies.

▶ Natural talent in many activities is enhanced by the study and practice of single skills.

▶ You can integrate these skills into the session to make the relationship work and enable client growth and change.

▶ Applications in counseling and interviewing differ in how they use the microskills.

▶ The microskills model has been tested in more than 450 data-based studies.

▶ The microskills model is used in more than 1,000 settings throughout the world.

Cultural Intentionality

▶ Cultural intentionality is defined as acting with a sense of capability and choosing from among a range of alternative actions.

▶ When you use a specific microskill or strategy, you can expect how the client will respond. Be flexible; if one skill isn't working, try another.

▶ The use of microskills will increase your cultural intentionality and flexibility by showing you many ways to respond to a single client statement or concern.

▶ Clients will benefit, feeling heard and respected as they discover new ways to resolve their concerns.

Resilience and Self-Actualization

► Resilience is the capacity to achieve successful adaptation in the face of major challenges.
► Helping the client resolve an issue is your contribution to increasing client resilience.
► Self-actualization is at the heart of counseling and therapy; both help clients achieve their potential in positive ways.

Neuroscience: Counseling Changes the Brain

► The brain is flexible and can add new neurons and connections throughout the lifespan, a process known as neuroplasticity.
► Counseling and therapy change the brain.
► Counselor's and client's brain functioning changes through the interview interaction.
► Stress can reduce life quality and damage the brain.
► Stress is involved in the majority of health and mental health concerns.
► Stress management and reduction are a central goal of counseling and therapy.

Office, Community, Phone, and Internet: Where Do We Meet Clients?

► Interviewing and counseling occur in many places other than a formal office, and interviewers and counselors may spend much of their time outside the formal office.
► When working in an office, keep comfort and safety in mind and provide a sense of warmth.
► Use decorations that embrace diversity and games, objects, and books that are developmentally appropriate.
► Online is another setting where counseling is being provided. As with all type of counseling and therapy, ethics is the first concern.

Your Natural Style and Beginning Expertise: An Important Audio or Video Activity

► The microskills are meant to add to and enhance your natural style of communicating.
► Your ability to vary the use of microskills allows you to flex as you encounter client diversity.
► It is critical to document a baseline of your natural style and skill level before you begin systematic study of the microskills.
► Using the resources and portfolio of competencies can help you become an effective listening and helper.

Assess your current level of knowledge and competence as you complete the chapter:

1. Flashcards: Use the flashcards to check your understanding of key concepts and facilitate memorization of key information.

2. Self-Assessment Quiz: The quiz will help you assess your current knowledge and prepare for course examinations.

3. Portfolio of Competencies: Evaluate your present level of competence on the ideas and concepts presented in this chapter using the Self-Evaluation Checklist. Self-assessment of your competencies demonstrates what you can do in the real world.

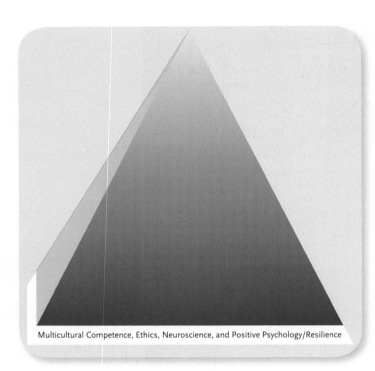

Multicultural Competence, Ethics, Neuroscience, and Positive Psychology/Resilience

Chapter 2
Multicultural Competence, Ethics, Positive Psychology, and Resilience

I am (and you also)

Derived from family

Embedded in a community

Not isolated from prevailing values

Though having unique experiences

In certain roles and statuses

Taught, socialized, gendered, and sanctioned

Yet with freedom to change myself and society.

—Ruth Jacobs*

This chapter has three distinct sections that are designed to be considered one by one. Effective interviews build on multicultural awareness, a solid sense of ethical practice, and a positive psychology approach. All are basic to encouraging and supportive resilience in our clients.

Chapter Goals

Awareness, knowledge, skills, and actions developed through the concepts of this chapter and this book will enable you to:

▶ Examine your identity as a multicultural being, how dimensions of diversity and privilege may affect the session, and the central importance of multicultural competence.

▶ Identify multicultural strengths in clients as a path toward wellness and resilience.

▶ Develop an awareness of ethics and the place of social justice in your practice.

▶ Define and apply positive psychology as a basis for fostering and building client resilience.

▶ Use therapeutic lifestyle changes (TLC) as positive wellness strategies in the session for physical and mental health.

*Based on R. Jacobs, "Be An Outrageous Older Woman" 1991, p. 37., Knowledge, Trends, and Ideas, Manchester, CT.

▶ Multicultural Competence

All interviewing and counseling are multicultural.

—Paul Pedersen

Every session has a cultural context that underlies the way clients and counselors think, feel, and behave.

—Carlos Zalaquett

Multicultural competence is imperative in the interview process. Awareness of our clients' multicultural background enables us to understand their uniqueness more fully. We live in a multicultural world where every client you encounter will be different from the last and different from you in some major way. Without a basic understanding of and sensitivity to a client's uniqueness, the interviewer will fail to establish a relationship and true grasp of a client's issues.

The following are responses you can expect when you take the broad array of multicultural issues into consideration. This is important for developing cultural intentionality.

MULTICULTURAL COMPETENCE	ANTICIPATED CLIENT RESPONSE
Interviewing and counseling rest on an ethical foundation of multicultural awareness, knowledge, skills, and action. Many issues of diversity have implications for the session. Important among them is awareness of your own cultural identity, intersecting multicultural identities, and privilege.	Anticipate that both you and your clients will appreciate, gain respect, and learn from increasing knowledge in intersecting identities, the nature of privilege, and multicultural competence. You, the interviewer, will have a solid foundation for a lifetime of personal and professional growth.

Interviewing and counseling have become global phenomena. The early history of interviewing, counseling, and therapy is populated primarily by famous White male European and U.S. figures such as Sigmund Freud, Carl Rogers, Viktor Frankl, Albert Ellis, and Aaron Beck. While their contributions are legion, they all give at best only minor attention to cultural difference or to women. The rise of the multicultural movement in the United States can be traced to the Civil Rights Act, followed by the growth of awareness in activists from groups such as African Americans, women, the disabled, lesbian/gay/bisexual/transgendered (LGBT) individuals, Vietnam War veterans, and others. All these identified and named oppression as a root cause of human distress. Counseling was slow to respond to this movement, but gradually it has become a central force in what is termed "psychological liberation."

Psychological liberation occurs when clients discover that what they saw as a personal issue is not just "their problem." With the counselor's help, clients begin to see that external racism, sexism, heterosexism, or other form of **oppression** is the underlying cause of many of their concerns. For example, veterans returning from Vietnam were often hospitalized for what we now call posttraumatic stress disorder (PTSD). But at the time there was no such term, so they were given diagnoses such as depressed, manic, or schizophrenic. The Veterans Administration asked therapists to search for malingers who were faking their symptoms just to obtain benefits.

It was veterans themselves, gathering in discussion groups without the "benefit" of professionals, who discovered the root underlying cause of their issues. Through this group work, many relieved themselves of guilt and moved on to health. As psychiatrists observed this phenomenon, they came up with the label for PTSD. We believe that this term is inaccurate, because external stressors are the real cause of the internalized issues. Consider deleting the word *disorder* and using just *posttraumatic stress (PTS)*. By labeling stress symptomology as a "disorder," psychiatry pathologized what, in truth, is a logical result of living in an insane environment.

Similarly, the helping professions have all too often failed to see that the issues that clients bring to us are deeply involved with societal dysfunction, harassment, and oppression. The multicultural movement shows that we need to examine external causes of personal concerns, whether clients are People of Color, those affected by a disability, those with mental issues, or children and adolescents who are bullied.

▶ Respectful Interviewing and Counseling

If we cannot end now our differences, at least we can help make the world safe for diversity.

—John F. Kennedy

The **RESPECTFUL model** (D'Andrea & Daniels, 2001) takes us more deeply into multicultural understanding. Please review the list in Box 2.1 and identify your multicultural self. It is possible that you have not thought of yourself as a multicultural being. As you consider the issues of multiculturality, we ask that you also examine your beliefs and attitudes toward those who are similar to and multiculturally different from you.

As you review your multicultural identity, what stands out for you? What might be surprising? What is most meaningful or salient in the way you think about yourself? Often our identity is most affected by significant experiences that shape our being. For example, certain givens in life such as being a man or woman or of a certain race or ethnicity affect how we see and experience ourselves. But other dimensions can be as important or more important in our identity. For some of you, it may be spiritual or religious values or where you lived when you were growing up; for others, it may be your education or being raised in a lower income situation. If you happen to be an older person or one who has been affected by physical or mental disability, that could be the most salient factor when you think of yourself.

Do not view the RESPECTFUL model as just a list of difficult issues and concerns. Rather, look at this list as a source of information about client resilience. Clients can draw amazing strength from their religious or spiritual background, the positive pride associated with racial/ethnic identity, their family background, or the community in which they grew up.

Nonetheless, there is also the possibility of **cultural trauma** in each of the RESPECTFUL dimensions. Large historical events and daily microaggressions in the form of insults to one's color, ethnicity, gender, sexual orientation, religious beliefs, or disability result in personal and group trauma, frustration, anger, hopelessness, and depression.

Multiculturalism needs to consider present and past histories of oppression, but we need also to look to each multicultural group for its positive strengths and what gives each

BOX 2.1 | The RESPECTFUL Model

Identify yourself on each of the dimensions of the RESPECTFUL model in Box 2.1. Then ask yourself how you might work with those who are different from you? And then identify strengths and positives that can be associated with each.

The 10 Dimensions		Identify yourself as a multicultural being.	What personal and group strengths can you develop for each multicultural dimension?	How effective will you be with individuals who differ from you?
R	Religion/spirituality			
E	Economic/social class background			
S	Sexual identity			
P	Personal style and education			
E	Ethnic/racial identity			
C	Chronical/lifespan status and challenges			
T	Trauma/crisis (estimated 90% experience; may be single trauma or repeated racism, sexism, bullying, etc.)			
F	Family background and history (single or two-parent, extended family, etc.)			
U	Unique physical characteristics (including disabilities, false standards of appearance, skills and abilities)			
L	Location of residence, language differences			

positive resilience. This, in turn, can be used by the individual client to work more comfortably with difficult and challenging situations.

Intersections among multicultural factors are also critical. For example, consider the biracial family (e.g., a child who is both Chinese and White, African descent and Latina/o). Both children and parents are deeply affected, and categorizing an individual into just one multicultural category is inappropriate. Or think of the Catholic lesbian woman who may be economically advantaged (or disadvantaged). Or the South Asian gay male with a Ph.D. For many clients, sorting out the impact of their multiculturality may be a major issue in counseling.

▶ Privilege as a Multicultural Interviewing Issue

Privilege is the greatest enemy of right.

—Marie von Ebner-Eschenbach

Move out of your comfort zone. You can only grow if you are willing to feel awkward and uncomfortable when you try something new.

—Brian Tracy

Privilege is power given to people through cultural assumptions and stereotypes. McIntosh (1988) comments on the "invisibility of Whiteness." European Americans tend to be unaware of the advantages they have because of the color of their skin. The idea of special privilege has been extended to include men, those of middle- or upper-class economic status, and others in our society who have power and privilege. For those with White privilege comes the awareness that White people are a minority in the global population. Within the United States, there are presently more infants and preschoolers of so-called "minorities" than Whites, so that during your work life in helping, Whites will become the new minority.

Income inequality and rigid class structures are now recognized as a central multicultural issue, perhaps even more important than other RESPECTFUL issues. In most nations, a small group holds the bulk of the wealth and believes that those of lower economic and social status are at fault for their own condition—they just need to study and work harder. Sociologists speak of educational and social systems producing social reproduction—the fact that over generations, class and income levels change very little. As a counselor, you will see students accumulate immense amounts of debt—and many of these same students fail to complete their education. Connections and personal influence are the way that most people find internships and later good jobs.

In short, societal structures are such that moving from one social class to another is well-nigh impossible. This is not true just in the United States. Researchers have found that the children of politicians, physicians, lawyers, and top business executives fill the universities not only of the United States but also of France, Germany, Great Britain, and even socialist Sweden. Moreover, even during full communism in Russia, the elite followed the same pattern as in capitalist countries (Bourdieu & Passeron, 1990).

What does this type of privilege mean for you as you work with students and others who face these obstacles? Listening to stories is obviously important, but a more activist teaching, consulting, and coaching role is needed. Consider becoming a mentor and helping your clients understand and cope with a challenging system. This is an area for social justice action.

Quite a few Whites, males, heterosexuals, middle-class people, and others currently enjoy the convenience of not being aware of their privileged state. The physically able see themselves as "normal," with little awareness that they are only "temporarily able" until old age or a trauma occurs. Out of privilege comes stereotyping of less dominant groups, thus further reinforcing the privileged status. Research has found that rich people have less empathy and generosity (Goleman, 2013).

However, avoid stereotyping anyone or any group's cultural identities. To say that all White rich males are insensitive or that all those who have experienced serious trauma are deeply troubled is just another form of stereotyping. The well-off often seek to help others, and many trauma victims are extremely resilient. Look for individual uniqueness, strength, and openness to change. Multicultural awareness enriches uniqueness only when it allows us to become more aware of how much each person is different from the other.

You, the interviewer, face challenges. For example, if you are a middle-class European American heterosexual male and the client is a working-class female of a different ethnicity or race, it can become more difficult to gain trust and rapport. If you are a young Person of Color and the client is older, White, and of a markedly different spiritual orientation, again it will take time to develop a relationship and working alliance.

Remember that the issues the client brings to you are the ones he or she currently sees as most important. Although these concerns often relate to multicultural identity, it is generally best to keep them in your awareness and only discuss them in the session if it seems potentially helpful to the client. However, there are areas where a much stronger stand needs to be taken. For example, if you a working with a woman who is trying to please her husband and accepts being beaten now and then, naming this as an issue related to trauma and sexism is often essential. However, this still must be done carefully to ensure safety, even to the point of taking the client to a safe house.

Clients may be talking about an academic issue and, in this process, may mention frustration with university facilities that don't meet their physical needs. They still need emotional and cognitive support to resolve their immediate issues, but they can also benefit from awareness of societal privilege that works against them. And you, as a counselor, have a responsibility to work with the campus. This may lead to a discussion of disability rights and/or what occurs in clients' daily lives around handling their issues.

▶ Awareness, Knowledge, Skills, and Action for Multicultural Competence

The American Counseling Association and the American Psychological Association have developed multicultural guidelines and specific competencies for practice. The concept of multicultural competency was originated by Derald Wing Sue and a supporting committee, who worked over many years to bring them to the attention of the helping professions (Sue et al., 1998; Sue & Sue, 2013). In these statements, the words ***multiculturalism*** and *diversity* are defined broadly to include many dimensions, as represented in the RESPECTFUL model.

The multicultural competencies include awareness, knowledge, and skills. Cultural competency training is now a requirement for medical licensure in several states and has become a standard in many helping professions. In addition, it has had increasing influence in business management and other fields.

Let us examine the multicultural guidelines and competencies in more detail.

AWARENESS: BE AWARE OF YOUR OWN ASSUMPTIONS, VALUES, AND BIASES

Awareness of yourself as a cultural being is a vital beginning. Unless you see yourself as a cultural being, you will have difficulty developing awareness of others. We will not elaborate further in this area, as the RESPECTFUL discussion above focuses on awareness.

The competency guidelines also speak to how contextual issues beyond a person's control affect the way a person discusses issues and problems. Oppression, discrimination, sexism, racism, and failure to recognize and take disability into account may deeply affect clients without their conscious awareness. Is the problem "in the individual" or "in the environment"? For example, you may need to help clients become aware that issues such as tension, headaches, and high blood pressure may be results of the stress caused by harassment and oppression. Many issues are not just client problems, but rather problems of a larger society.

> ### EXERCISE 2.1 Self-Awareness Practice
> To increase self-awareness of your own background, please return to the RESPECTFUL framework and identify yourself on each of the components of the model. You may wish to add other issues as well. Then write answers to the questions provided there. In addition, list where you stand within each area of privilege. What are your areas of privilege? Less privilege?
>
> Have you experienced cultural trauma and microaggressions, and how has this affected your sense of trust in others, both individuals and groups?
>
> Perhaps most challenging, where might you have biases of favoritism toward others or toward your own group? How might that affect your interviewing practice?
>
> End this self-examination with stories of strength and resilience that come from your life experience. What are you proud of? Build on what you can do, rather than what you can't do.

KNOWLEDGE: UNDERSTAND THE WORLDVIEW OF THE CULTURALLY DIFFERENT CLIENT

Worldview is formally defined as the way you and your client interpret humanity and the world. People of different historical, religious, and cultural backgrounds worldwide often have vastly different philosophic views on the meaning of life, right and wrong, and personal responsibility versus control by fate. Because of varying multicultural backgrounds, we all view people and the larger world differently. Multicultural competence stresses the importance of being aware of possible negative emotional reactions and biases toward those who are different from us. If people have learned to view certain groups through inaccurate stereotypes, we especially need to listen and learn respect for the worldview of the client; be careful not to impose your own ideas.

All of us need to develop knowledge about various cultural groups, their history, and their present concerns. If you work with Spanish-speaking groups, it is critical to learn the different history and issues faced by those from Mexico, Puerto Rico, the Caribbean, and Central and South America. What is the role of immigration? How do the experiences of Latinas/Latinos who have been in Colorado for several centuries differ from those of newly arrived immigrants? Note that diversity is endemic to the broad group we often term "Hispanic."

The same holds true for European Americans and all other races and ethnicities. Old-time New England Yankees in Hadley, Massachusetts, once chained Polish immigrants in barns to keep hired hands from running away. The tables are now turned and it is the Poles who control the town, but quiet tensions between the two groups still remain. Older gay males who once hid their identity are very different from young activists. Whether it is race or religion, ability or disability, we are constantly required to learn more about our widely diverse populations.

Traditional approaches to counseling theory and skills may be inappropriate and/or ineffective with some groups. We also need to give special attention to how socioeconomic factors, racism, sexism, heterosexism, and other oppressive forces may influence a client's worldview.

Despite the recent presidency of Barack Obama, racial disparities remain. Racial minorities still have more school dropouts and at all levels tend to be more dissatisfied with the educational system. While college attendance has nearly doubled, recent court decisions have resulted in fewer minorities at "top" state universities. Beyond schooling, People of Color encounter more poverty, violence, income disparities, and a variety of other discriminatory situations. Just getting a taxi in downtown New York can be difficult for an African American, even in business attire.

These large and small insults and slights, termed **microaggressions**, can result in physical and mental health problems (Sue, 2010). Repeated racial harassment (or bullying) can literally result in posttraumatic stress. What seems small at first is damaging through repetition. This is not a disorder, but a logical result of external oppression. As you work with minority clients, it is vital that you remain aware of the impact that these external system pressures and oppression may be having. One useful intervention in such situations is encouraging and helping these clients to examine how forms of oppression may relate to their present headaches, stomach upsets, high blood pressure, and an array of psychological stressors.

> **EXERCISE 2.2 Developing Awareness of Another Person's Experience of the RESPECTFUL Multicultural Framework**
> Developing awareness is seen as a first step toward cultural expertise. We are all culturally different, even if we think "we are the same." It will be most helpful to work with a person who culturally different from you or who comes from a different background.
>
> Use the RESPECTFUL model as the basis for discussion. Try to go through all 10 dimensions, noting where you are similar and where you are different. Ask the questions of awareness of each other. As you both become comfortable, search out biases and life experiences. Honesty will

likely be a challenge here, so don't "push it," but seek to learn that even those who seem similar are indeed different. Many (most?) of us have experienced some form of prejudice. It may not be racism or socioeconomic oppression, but discrimination exists in many subtle forms.

End this exercise with a focus on strengths in the various cultures that can lead to resilience and mental health.

SKILLS AND ACTION: AWARENESS AND KNOWLEDGE ARE MEANINGLESS UNLESS WE ACT

A classic study found that 50% of minority clients did not return to counseling after the first session (cited in Sue & Sue, 2013). This book seeks to address cultural intentionality by providing you with ideas for multiple responses to your clients. If your first response doesn't work, be ready with another.

Traditional counseling strategies are being adapted for use in a more culturally respectful manner (Ivey, D'Andrea, & Ivey, 2012). It is important to be mindful of cultural bias in assessment and testing instruments and the impact of discrimination on clients. Box 2.2 depicts the ongoing process of becoming multiculturally aware.

BOX 2.2 | **National and International Perspectives on Counseling Skills**

Multiculturalism Belongs to All of Us

Mark Pope, Cherokee Nation and Past President of the American Counseling Association

Multiculturalism is a movement that has changed the soul of our profession. It represents a reintegration of our social work roots with our interests and work in individual psychology.

Now, I know that there are some of you out there who are tired of culture and discussions about culture. You are the more conservative elements of us, and you have just had it with multicultural this and multicultural that. And, further, you don't want to hear about the "truth" one more time.

There is another group of you that can't get enough of all this talk about culture, context, and environmental influences. You are part of the more progressive and liberal elements of the profession. You may be a member of a "minority group" or you have become a committed ally. You may see the world in terms of oppressor and oppressed. I'll admit it is more complex than these brief paragraphs allow, but I think you get my point.

Here are some things that perhaps can join us together for the future:

1. We are all committed to the helping professions and the dignity and value of each individual.
2. The more we understand that we are part of multiple cultures, the more we can understand the multicultural frame of reference and enhance individuality.
3. Multicultural means just that—many cultures. Racial and ethnic issues have tended to predominate, but diversity also includes gender, sexual orientation,

age, geographic location, physical ability, religion/spirituality, socioeconomic status, and other factors.
4. Each of us is a multicultural being and thus all interviewing and counseling involve multicultural issues. It is not a competition as to which multicultural dimension is the most important. It is time to think of a "win/win" approach.
5. We need to address our own issues of prejudice—racism, sexism, ageism, heterosexism, ableism, classism, and others. Without looking at yourself, you cannot see and appreciate the multicultural differences you will encounter.
6. That said, we must always remember that the race issue in Western society is central. Yes, I know that we have made "great progress," but each progressive step we make reminds me how very far we have to go.

All of us have a legacy of prejudice that we need to work against for the liberation of all, including ourselves. This requires constantly examining honestly and at times this self-examination can be challenging, even painful. You are going to make mistakes as you grow multiculturally; but see these errors as an opportunity to grow further.

Avoid saying, "Oh, I'm not prejudiced." We need a little discomfort to move on. If we realize that we have a joint goal in facilitating client development and continue to grow, our lifetime work will make a significant difference in the world.

As you work your way through this book and the microskills hierarchy, you will discover ideas to increase your action skills for clients represented in the RESPECTFUL model. Becoming multiculturally competent requires a lifetime of learning and a willingness to act on that knowledge.

EXERCISE 2.3 Diversity: Experiential Project

This exercise in taking awareness, knowledge, and skills into action is adapted from the syllabi of Dr. Zalaquett's multicultural courses. It will help you expand your awareness, knowledge, skills, and actions regarding different cultural groups.

Select a diverse group and setting that is different from your own (e.g., attend a predominantly African American church, a mosque, a GLBT event, or a Deaf community Silent Dinner). Attend more than one event to gain a deeper cultural understanding of your chosen group.

▶ What did you see or hear that was consistent or inconsistent with your personal beliefs about this cultural/ethnic group?

▶ What did you learn about this group? What did you learn about yourself?

▶ What were your personal reactions in terms of your levels of comfort and feelings of acceptance and/or belonging?

▶ What experiences, beliefs, and values from your own upbringing do you think contributed to those personal reactions?

▶ How will this experience assist you in your professional development?

POLITICAL CORRECTNESS: OR IS THE ISSUE RESPECT FOR DIFFERENCES?

Political correctness (PC) is a term used to describe language that is calculated to provide a minimum of offense, particularly to the racial, cultural, or other identity groups being described. The existence of PC has been denounced by conservative, liberal, and other commentators. The term and its usage are hotly contested.

Given this controversy, what is the appropriate way to name and discuss cultural diversity? We argue that interviewers and counselors should use language empathically, and we urge that you use terms that the client prefers. Let the client define the name that is to be used, as it is respect of the client's point of view that counts in the issue of naming.

A woman is unlikely to enjoy being called a girl or a lady, but you may find some who use these terms. Some people in their 70s resent being called elderly or old, whereas others embrace and prefer this language. At the same time, you may find that the client is using language in a way that is self-deprecating. A woman struggling for her identity may use the word *girl* in a way that indicates a lack of self-confidence. The older person may benefit from a more positive view of the language of aging. A person struggling with sexual identity may find the word *gay* or *lesbian* difficult to deal with at first. You can help clients by exploring names and social identifiers in a more positive fashion.

Race and ethnicity are often central topics in discussions of multicultural issues. *African American* is considered the preferred term, but some clients prefer *Afro-Canadian* or *Black*. Others may feel more comfortable being called Haitian, Puerto Rican, or Nigerian. A person from a Hispanic background may well prefer Chicano, Mexican, Mexican-American, Cuban, Puerto Rican, Chilean, or Salvadorian. Some American Indians prefer Native American, but most prefer to be called by the name of their tribe or nation—Lakota, Navajo, Swinomish. Some Caucasians would rather be called British Australians, Irish Americans, Ukrainian Canadians, or Pakistani English. These people are racially White but also have an ethnic background.

Knowledge of the language of nationalism and regional characteristics are also useful. American, Irish, Brazilian, or New Zealander (or "Kiwi") may be the most salient self-identification. Yankee is a word of pride to those from New England and a word of derision for many Southerners. Midwesterners, those in Outback Australia, and Scots, Cornish, and Welsh in Great Britain often identify more with their region than with their nationality. Many in Great Britain resent the more powerful region called the Home Counties. And we must recognize that the Canadian culture of Alberta is very different from the cultures of Ontario, Quebec, and the Maritime Provinces.

Capitalization is another issue. The *New York Times* style manual does not capitalize *Black* and several other multicultural terms, but the capital has become a standard in counseling and psychology. *White* is also not capitalized by the *Times*, but it is helpful for White people to discover that they, too, have a general racial cultural identity. Capitalization of the major cultural groupings of Black and White is becoming more and more the standardized usage in counseling, human relations, psychology, and social work.

EXERCISE 2.4 Political Correctness
What occurs to you when you explore the importance of respectful naming? What are your thoughts about capitalization?

▶ Ethics and Morals: Professional and Personal

Ethics is nothing else than reverence for life.

—Albert Schweitzer

Action indeed is the sole medium of expression for ethics.

—Jane Addams

Albert Schweitzer was awarded the Nobel Peace Prize in 1952. A renowned philosopher and musician, he earned his medical degree and started practice in a hut in Africa. Jane Addams founded Hull House in Chicago 1889, which resulted in the formation of social work as a profession.

Ethics are thoughtful professional lists of do's and don'ts for our profession. Morals are the way we apply ethics through our commitment to excellence, reverence for others, and willingness to take action to engage life for ourselves and others.

Ethical codes can be summarized with the following statement: "Do no harm to your clients; treat them responsibly with full awareness of the social context of helping." As interviewers and counselors, we are morally responsible for our clients

ETHICS AND MORALS	ANTICIPATED CLIENT RESPONSE
Ethics are rules, typically prescribed by social systems and, in counseling, as professional standards. They define how things are to be done.	Following professional ethics results in client trust and provides us with guidelines for action in complex situations.
Morals are individual principles we live by that define our beliefs about right and wrong.	Morals represent our individual efforts and actions to follow ethical principles. A moral approach to interviewing and counseling helps us to remember that our personal actions count, both inside and outside the session.
A moral approach to interviewing and counseling allows us to apply ethical principles respectfully to our clients and ourselves.	Furthermore, a moral approach to the session may ask you to help clients examine their own moral and ethical decisions.

and for society as well. At times these responsibilities conflict, and you may need to seek detailed guidance from documented ethical codes, your supervisor, or other professionals.

CONFIDENTIALITY: OUR MORAL FOUNDATION

Confidentiality is the cornerstone of our tool kits.

—Robert Blum

We have said that the empathic relationship is central to developing a working alliance with clients. But without **confidentiality** and trust as our basis, we will have no relationship. Thus, from the very beginning, the amount of confidentiality you can provide your client needs to be crystal clear. Your clients need to know that absolute confidentiality is legally impossible (see Box 2.3), and thus it is essential to spell out the limits of confidentiality in your setting at the beginning.

As a student taking this course, you are a beginning professional; you usually do not have legal confidentiality. Nonetheless, you need to keep to yourself what you hear in class role-plays or practice sessions. Trust is built on your ability to keep confidences. Be aware that state laws on confidentiality vary. Informed consent is an ethical issue discussed later in the chapter, along with some ways to share the concept of confidentiality with clients.

| **BOX 2.3** | Confidentiality and Its Limits |

The following provisions regarding confidentiality are from the Code of Ethics and Practice of the Australian Counselling Association (2012).

3.4 Confidentiality

(a) Confidentiality is a means of providing the client with safety and privacy and thus protects client autonomy. For this reason any limitation on the degree of confidentiality is likely to diminish the effectiveness of counselling.

(b) The counselling contract will include any agreement about the level and limits of the confidentiality offered. This agreement can be reviewed and changed by negotiation between the counsellor and the client. Agreements about confidentiality continue after the client's death unless there are overriding legal or ethical considerations. In cases where the client's safety is in jeopardy any confidentially agreements that may interfere with this safety are to be considered void (see 3.6 'Exceptional circumstances').

3.6 Exceptional Circumstances

(a) Exceptional circumstances may arise which give the counsellor good grounds for believing that serious harm may occur to the client or to other people. In such circumstance the client's consent to change in the agreement about confidentiality should be sought whenever possible unless there are also good grounds for believing the client is no longer willing or able to take responsibility for his/her actions. Normally, the decision to break confidentiality should be discussed with the client and should be made only after consultation with the counselling supervisor or if he/she is not available, an experienced counsellor.

(b) Any disclosure of confidential information should be restricted to relevant information, conveyed only to appropriate people and for appropriate reasons likely to alleviate the exceptional circumstances. The ethical considerations include achieving a balance between acting in the best interests of the client and the counsellor's responsibilities under the law and to the wider community.

(c) While counsellors hold different views about grounds for breaking confidentiality, such as potential self-harm, suicide, and harm to others they must also consider those put forward in this Code, as they too should imbue their practice. These views should be communicated to both clients and significant others, e.g., supervisor, agency, etc.

Australian Counselling Association (2012), *Code of Ethics and Practice.*

Professionals encounter many challenges to confidentiality. Some states require you to inform parents before counseling a child and to share information from interviews with them if they ask. If issues of abuse should arise, you must report this to the authorities. If the client is a danger to self or others, then rules of confidentiality change; the issue of reporting such information needs to be discussed with your supervisor. As a beginning interviewer, you will likely have limited, if any, legal protection, so limits to confidentiality must be included in your approach to informed consent.

Dual relationships can present challenging ethical and moral issues. They occur when you have more than one relationship with a client. Another way to think of this is the concept of conflict of interest.

If your client is a classmate or friend, you are engaged in a dual relationship. If you live in a small town, you are likely to encounter some of your clients at the grocery store or elsewhere. These situations may also occur when you counsel a member of your church or school. Personal, economic, and other privacy matters can become complex issues. You can examine statements on dual relationships in more detail in professional ethical codes.

DIVERSITY, MULTICULTURALISM, ETHICS, AND MORALITY

We need to help students and parents cherish and preserve the ethnic and cultural diversity that nourishes and strengthens this community—and this nation.

—Cesar Chavez

The American Counseling Association (2014) focuses the Preamble to its Code of Ethics on diversity as a central ethical issue:

> The American Counseling Association (ACA) is an educational, scientific, and professional organization whose members work in a variety of settings and serve in multiple capacities. Counseling is a professional relationship that empowers diverse individuals, families, and groups to accomplish mental health, wellness, education, and career goals.

Human services professionals have added a moral personal dimension to this ethical statement (National Organization for Human Services, 1996; Neukrug, 2012). They first speak of the need for advocacy for the rights of others, particularly groups that have been disadvantaged or oppressed. For us, this means that following ethical principles in the office often is not enough. Morally, we are asked to move to the community and work to prevent discrimination and oppression. It also means that we need to look at our own RESPECTFUL identity and consider the morality of how we have been treated in the past and how we might want to treat our clients with moral respect.

You will work with clients who have made mistaken moral judgments about themselves. Some clients may be judging themselves too harshly for what they have done or left undone. Other clients may fail to recognize their own moral failures. You will have an interesting challenge as you face the moral dilemmas of clients, both those of which they are aware and those that they deny, ignore, or may not even be aware of.

ETHICS, MORALITY, AND COMPETENCE

> Don't judge each day by the harvest you reap but by the seeds that you plant.
> —Robert Louis Stevenson

Awareness of what we can and cannot do is basic to moral competence.

> **C.2.a. Boundaries of Competence.** Counselors practice only within the boundaries of their competence, based on their education, training, supervised experience, state and national professional credentials, and appropriate professional experience. Whereas multicultural counseling competency is required across all counseling specialties, counselors gain knowledge, personal awareness, sensitivity, dispositions, and skills pertinent to being a culturally competent counselor in working with a diverse client population. (American Counseling Association, 2014)

We all need to constantly monitor whether we are competent to counsel clients around the issues that they present to us. For example, you may be able to help a client work out difficulties occurring at work, but you discover a more complex underlying issue of serious depression undercutting the client's ability to find or keep a job. You may be competent to help this client with career and vocational issues but may not have had enough experience with depression. While maintaining a supportive attitude, you **refer** the client to another counselor for therapy while you continuing to work with the job issues. Although it is essential that you not work beyond your competence, the morality of ethics demands that you continue studying to expand your competence through reading, inservice training, and supervision.

INFORMED CONSENT

Counseling is an international profession. The Canadian Counselling and Psychotherapy Association (2007) approach to **informed consent** is particularly clear:

> **B4. Client's Rights and Informed Consent.** When counselling is initiated, and throughout the counselling process as necessary, counsellors inform clients of the purposes, goals, techniques, procedures, limitations, potential risks and benefits of services to be performed, and other such pertinent information. Counsellors make sure that clients understand the implications of diagnosis, fees and fee collection arrangements, record keeping, and limits of confidentiality. Clients have the right to participate in the ongoing counselling plans, to refuse any recommended services, and to be advised of the consequences of such refusal.

The American Psychological Association (2010) stresses that psychologists should inform clients if the interview is to be supervised and provides additional specifics:

> **Standard 10.01 (c)** When the therapist is a trainee and the legal responsibility for the treatment provided resides with the supervisor, the client/patient, as part of the informed consent procedure, is informed that the therapist is in training and is being supervised and is given the name of the supervisor.
>
> **Standard 4.03 (c)** Before recording the voices or images of individuals to whom they provide services, psychologists obtain permission from all such persons or their legal representatives.

When you work with children, the ethical issues around informed consent become especially important. Depending on state laws and practices, it is often necessary to obtain written parental permission before interviewing a child or before sharing information about the interview with others. The child and family should know exactly how any information is to be shared, and interviewing records should be available to them for their comments and evaluation. An essential part of informed consent is stating that both child and parents have the right to withdraw their permission at any point. Needless to say, these same principles apply to all clients—the main difference is parental awareness and consent.

When you enter into role-plays and practice sessions, inform your volunteer "clients" about their rights, your own background, and what clients can expect from the session. For example, you might say:

> I'm taking an interviewing course, and I appreciate your being willing to help me. I am a beginner, so only talk about things that you want to talk about. I would like to [audio or video] record the interview, but I'll stop immediately if you become uncomfortable and delete it as soon as possible. I may share the recording in a practicum class or I may produce a written transcript of this session, removing anything that could identify you personally. I'll share any written material with you before passing it in to the instructor. Remember, we can stop any time you wish. Do you have any questions?

You can use this statement as a starting point and eventually develop your own approach to this critical issue. The sample practice contract in Box 2.4 may be helpful as you begin.

BOX 2.4 Sample Practice Contract

The following is a sample contract for you to adapt for practice sessions with volunteer clients. (If you are working with a minor, the form must be signed by a parent as appropriate under HIPPA standards.)

Dear Friend,

I am a student in interviewing skills at [insert name of class and college/university]. I am required to practice counseling skills with volunteers. I appreciate your willingness to work with me on my class assignments.

You may choose to talk about topics of real concern to you, or you may prefer to role-play an issue that does not necessarily relate to you. Please let me know before we start whether you are talking about yourself or role-playing.

Here are some dimensions of our work together:

Confidentiality. As a student, I cannot offer any form of legal confidentiality. However, anything you say to me in the practice session will remain confidential, except for certain exceptions that state law requires me to report.

Even as a student, I must report (1) a serious issue of harm to yourself; (2) indications of child abuse or neglect; (3) other special conditions as required by our state [insert as appropriate].

Audio and/or video recording. I will be recording our sessions for my personal listening and learning. If you become uncomfortable at any time, we can turn off the recorder. The recording(s) may be shared with my supervisor [insert name and phone number of professor or supervisor] and/or students in my class. You'll find that recording does not affect our practice session so long as you and I are comfortable. Without additional permission, recordings and any written transcripts are destroyed at the end of the course.

Boundaries of competence. I am an inexperienced interviewer; I cannot do formal counseling. This practice session helps me learn interview skills. I need feedback from you about my performance and what you find helpful. I may give you a form that asks you to evaluate how helpful I was.

_____   _____
Volunteer Client   Interviewer

Date _____

EXERCISE 2.5 Informed Consent

Box 2.4 presents a sample informed consent form, or practice contract. In a small group, develop your own informed consent form that is appropriate for your practice sessions, school, or agency.

 Group Practice Exercise: Develop an Informed Consent Form

PRIVACY RULES

The Health Insurance Portability and Accountability Act (HIPAA) took effect in 1996. We include it here because, among other functions, it requires the protection and confidential handling of protected health information.

Following is a summary of some key elements of the Privacy Rule, including who is covered, what information is protected, and how protected health information may be used and disclosed. For a complete outline of **HIPAA privacy** requirements, visit the website of the U.S. Department of Health and Human Services, Office for Civil Rights, at http:// www.hhs.gov.

1. Protected health information. The Privacy Rule defines "protected health information (PHI)" as all individually identifiable health information held or transmitted by a covered entity or its business associate, in any form or media, including electronic, paper, or oral. "Individually identifiable health information" is information, including demographic data and personal identifiers such as name, address, birth date, and social security number, that identifies the individual, or could reasonably be used to identify the individual, and that relates to:

▶ The individual's past, present, or future physical or mental health or condition
▶ The provision of health care to the individual
▶ The past, present, or future payment for the provision of health care to the individual

The Privacy Rule does not include protected health information from employment records that a covered entity maintains in its capacity as an employer as well as education and certain other records subject to, or defined in, the Family Educational Rights and Privacy Act, 20 U.S.C. §1232g.

2. De-identified health information. This is information that makes it impossible for others to identify a client, and there are no restrictions on its use or disclosure. Information can be de-identified in two ways: (1) a formal determination by a qualified statistician or (2) the removal of specified identifiers of the individual and of the individual's relatives, household members, and employers. De-identification is adequate only if the covered entity has no actual knowledge that the remaining information could be used to identify the individual.

When you visit a physician, you are asked to sign a version of the privacy statement. Mental health agencies make their privacy statements clearly available to clients and often post them in the office.

SOCIAL JUSTICE AS MORALITY AND ETHICS IN ACTION

I've become even more convinced that the type of stress that is toxic has more to do with social status, social isolation, and social rejection. It's not just having a hard life that seems to be toxic, but it's some of the social poisons that can go along with the stigma of poverty.

—Kelly McGonigal

Jane Addams, the founder of social work, has infused her thinking throughout the National Association of Social Workers. Their code of ethics is strongest on social justice and emphasizes that action beyond the interview in the community may be needed to address unfairness of many types.

> **Ethical Principle:** *Social workers challenge social injustice.* Social workers pursue social change, particularly with and on behalf of vulnerable and oppressed individuals and groups of people. Social workers' social change efforts are focused primarily on issues of poverty, unemployment, discrimination, and other forms of social injustice. These activities seek to promote sensitivity to and knowledge about oppression and cultural and ethnic diversity. Social workers strive to ensure access to needed information, services, and resources; equality of opportunity; and meaningful participation in decision making for all people.

We now know that childhood poverty, adversity, and stress produce lifelong damage to the brain. These changes are visible in cells and neurons and include permanent changes in DNA (Marshall, 2010). Therefore, just treating children of poverty through supportive counseling is not enough; for significant change to occur, prevention and social justice action are critical.

The incidence of hypertension among African Americans is reported to be 4 to 7 times higher than that of Whites, and the stress and impact of long-term racism may be a high-ranking contributor (Hall, 2007). Social justice action has both mental and physical implications.

There are two major types of social justice action. The first and most commonly discussed is action in the community to work against the destructive influences of poverty, racism, and all forms of discrimination. These preventive strategies are now considered a vital dimension of the "complete" counselor or therapist. Getting out of the office and understanding society's influence on client issues is central. Clients who have suffered social injustice of virtually any type (poverty, bullying, sexism, heterosexism) will also benefit from joining groups that work in some way to prevent or alleviate the impact of oppression. One route toward healing is working with others, or even by oneself, for those who have experienced injustice. This can range from work in soup kitchens to participating in a protest, joining a "Take Back the Night" walk, or simply writing letters or articles in the local paper.

The second type of social justice action occurs in the interview. When a female client discusses mistreatment and harassment by her supervisor, the issue of oppression of women should be named as such. The social justice perspective requires you to help her understand that the problem is not caused by her behavior or how she dresses. By naming the problem as sexism and harassment, you often free the client from self-blame and empower her for action. You can also support her in efforts to bring about change in the workplace.

You will sometimes be challenged in the session by clients who hold moral and ethical values that differ from yours. The question always comes up—should you confront and challenge them? First and foremost, how fragile is the client and will your comments be disrupting? If so, it is best to hold your tongue and seek to support the client where he or she needs help. When the difficult topics come up (religion, politics, oppression of many types), be prepared with your own moral values, but apply them carefully. If it becomes too difficult, refer; do not impose your values.

Box 2.5 lists websites of some key ethical codes in English-speaking areas of the globe. All codes provide guidelines on competence, informed consent, confidentiality, and

Professional Organizations With Ethical Codes

American Academy of Child and Adolescent Psychiatry (AACAP)
Code of Ethics

http://www.aacap.org

American Association for Marriage and Family Therapy (AAMFT)
Code of Ethics

http://www.aamft.org

American Counseling Association (ACA) Code of Ethics

http://www.counseling.org

American Psychological Association (APA)
Ethical Principles of Psychologists and Code of Conduct

http://www.apa.org

American School Counselor Association (ASCA)
Ethical Standards for School Counselors

http://www.schoolcounselor.org

Australian Counselling Association (ACA) Code of Ethics

http://www.theaca.net.au/

Australian Psychological Society (APS) Code of Ethics

http://www.psychology.org.au

British Association for Counselling and Psychotherapy (BACP)
Ethical Framework for Good Practice in Counselling &
Psychotherapy

http://www.bacp.co.uk/

Canadian Counselling Association (CCA) Codes of Ethics

http://www.ccacc.ca

Canadian Counselling and Psychotherapy Association

http://www.ccpa-accp.ca/

Commission on Rehabilitation Counselor Certification (CRCC)
Code of Professional Ethics for Rehabilitation Counselors

http://www.crccertification.com/

International Union of Psychological Science (IUPsyS)
Universal Declaration of Ethical Principles for Psychologists

http://www.iupsys.net

National Association of School Nurses (NASN)
Code of Ethics

http://www.nasn.org

National Association of School Psychologists (NASP)
Principles for Professional Ethics

http://www.nasponline.org

National Association of Social Workers (NASW)
Code of Ethics

http://www.naswdc.org

National Career Development Association (NCDA)
Code of Ethics

http://www.ncda.org

New Zealand Association of Counsellors (NZAC)
Code of Ethics

http://www.nzac.org.nz

International Coach Federation (ICF) Code of Ethics

http://www.coachfederation.org

Ethics Updates provides updates on current literature, both
popular and professional, that relate to ethics.

http://ethics.sandiego.edu

Ethical codes of Latin American countries can be found at the
Society of Interamerican Psychology's Grupo de Trabajo de
Ética y Deontología Profesional webpage under "Informacion
del Grupo y Documentos."

http://www.sipsych.org

diversity. Issues of power and social justice are explicit in social work and human services and implicit in other codes.

> **EXERCISE 2.6 Examining Professional Ethical Codes**
> In Box 2.5, select the organization that is most relevant to your interests and examine its ethical code in detail. Also select the code of one other helping profession, and look for similarities and differences between the two codes. Website addresses are correct at the time of printing but can change. For a keyword web search, use the name of the professional association and the words *ethics* or *ethical code*.

▶ Positive Psychology: Building Client Resilience

Your mind is a powerful thing. When you fill it with positive thoughts, your life will start to change.

—Anonymous

Optimists literally don't give up as easily and this links to greater success in life.

—Elaine Fox

If you help clients recognize their strengths and resources, you can expect them to use these positives as a basis for resolving their issues. There is a second story behind the first story of client difficulties and concerns. Where have they done things right and succeeded? Of course, our first task is to draw out the worrisome story and issues that brought them to the interview or counseling session. Searching for the positive psychology/wellness story does not deny client concerns and serious challenges. But our aim is constantly to watch and listen for strengths that will eventually be part of the solution.

RESILIENCE AND OPTIMISM

Positive psychology's central aim is to encourage and develop optimism and resilience. *Optimism* is defined in various dictionaries with many affirmative words—among them, *hope, confidence,* and *cheerfulness.* It also includes a trust in which we expect things to work out and to get better, a sense of personal power, and a belief in the future. Optimism is a key dimension of resilience and the ability to recover and learn from one's difficulties and challenges.

Research has useful lessons for us here. People who are more optimistic have an increased ability to eliminate, reduce, or manage stressors and negative emotions. Furthermore, they are better able to approach and face their difficulties. Optimists tend to live healthier lives, suffer less from physical illness, and, of course, feel better about themselves and their abilities (Kim, Park, & Peterson, 2011; Nes & Segerstrom, 2006).

Out of this research has come an effective six-point scale to measure optimism (see Box 2.6). We suggest that you use this scale to assess your own level of optimism. At times you may want to share this scale with clients, or ask them questions to discover their level of optimism. One of our goals in counseling and therapy is to increase resilience, and helping clients become more optimistic and hopeful is part of this process.

Using the scale in Box 2.6, you can develop an optimism score by adding items 2, 3, 4 and a pessimism score by adding 1, 5, and 6. These six items have proven both reliable and predictive. We suggest that you use them as an indication of what a positive attitude toward life means. A positive attitude appears critical for the development of resilience.

BOX 2.6 A Six-Point Optimism Scale

Please say how much you agree or disagree with the following statements: 1 = Strongly disagree, 2 = Somewhat disagree, 3 = Slightly disagree, 4 = Slightly agree, 5 = Somewhat agree, 6 = Strongly agree.

1. _____If something can go wrong for me, it will.

2. _____I'm always optimistic about my future.

3. _____In uncertain times, I usually expect the best.

4. _____Overall, I expect more good things to happen to me than bad.

5. _____I hardly ever expect things to go my way.

6. _____I rarely count on good things happening to me.

Source: Based on Scheier, M., Carver, C., & Bridges, M. (1994). Distinguishing optimism from neuroticism (and trait anxiety self-mastery, and self-esteem): A reevaluation of the Life Orientation Test. *Journal of Personality and Social Psychology, 67*(6), 1063–1078.

POSITIVE PSYCHOLOGY AND RESILIENCE	ANTICIPATED CLIENT RESPONSE
Help clients discover and rediscover their strengths. Find strengths and positive assets in clients and in their support system. Identify multiple dimensions of wellness. In addition to listening, actively encourage clients to learn new actions that will increase their resilience.	Clients who are aware of their strengths and resources can face their difficulties and resolve issues from a positive foundation. They become resilient and can bounce back from obstacles and defeat.

 Complete Personal Wellness Form

BUILDING RESILIENCE THROUGH STRENGTHS AND RESOURCES

Positive psychology brings together a long tradition of emphasis on positives within counseling, human services, psychology, and social work.

—Martin Seligman

In recent years, the field of counseling has developed an extensive body of knowledge and research supporting the importance of **positive psychology**, a strength-based approach. Psychology has overemphasized the disease model and all too often places a self-defeating and almost total focus on difficulties, ignoring the client's own strengths in the resolution of issues. Nonetheless, the positive psychology movement is well aware that happiness is not possible in the midst of excessive stress.

In addition to drawing out the problematic negative stories, we also need to search for stories of strength and success. Seek out and listen for times when clients have succeeded in overcoming obstacles. Listen for and be "curious about their competencies—the heroic stories that reflect their part in surmounting obstacles, initiating action, and maintaining positive change" (Duncan, Miller, & Sparks, 2004, p. 53). In the microskills, we call this the "positive asset search." Find something that the client is doing right and discover the client's resources and supports. Use these strengths and positive assets as a foundation for personal growth and resolving issues.

Encouraging and teaching clients to becoming fully engaged in life is basic to positive psychology. Your client may have "retired" from life to be safe. In terms of life satisfaction, engagement was found to be more central than happiness. Find areas in which your clients have been involved, activities that they care about. Often they have stopped exercising, meditating, or playing tennis or golf. They may have cut themselves off from friends, church, or other groups that provide interest and support. How much time do they spend on passively watching TV or other screen time taking them away from daily life?

"Life—be in it" is a classic Australian saying. Australia is known as one of the happiest countries in the world.

Therapeutic lifestyle changes (TLCs), discussed in the following section, are a key route to identifying and encouraging an engaged lifestyle. Examples of lifestyle changes include increased exercise, better nutrition, meditation, and helping others, which in turn helps us feel better about ourselves.

At the same time, when we view engagement from a broad multicultural frame, many clients, regardless of race or ethnicity, are hungry, abused, or suffering from trauma. Perhaps their unemployment benefits are about to run out or a family member may be seriously ill. These clients may consider it a luxury to find the time to study better nutrition, to exercise, or, particularly, to meditate. But you can help the engagement process and lead the client to greater resilience by encouraging a short walk—even around the block. You can provide information on how to make wiser food decisions. Meditation may make no sense to some, but deep breathing, visual imaging of family strengths, or short relaxation training can help alleviate stress. A client might be reminded to draw on spiritual traditions or help with a church project. As always, contact with other humans makes a difference.

Seligman was surprised to find in his research that meaning was the most essential aspect of a satisfying life. This would not surprise Viktor Frankl, famous survivor of the Nazi concentration camp at Auschwitz, who stated that those of us who find a *why*, a meaning for our lives, can live with virtually any *how*. Meaning and a life vision carry many people through the most difficult times in their lives.

Chapter 10 focuses on the skill of reflection of meaning, which, as Seligman observes, needs much more attention in interviewing and counseling practice. With some clients, your work with life meaning and vision may be important in ensuring that new thoughts, behaviors, and action actually happen. For some clients, the only route toward health may be the will to meaning.

EXERCISE 2.7 Story Telling

Engage a friend, a classmate, or a family member, asking them to tell you about a time they faced a challenge or an issue of some difficulty, but something not serious. First, ask them to tell you the story of what went wrong and why. Then, ask them to tell you the same story focusing on what they did right, what worked, and what strengths they showed.

What did you learn? This exercise in storytelling may be valuable in the long run. Clients usually come to you worrying about issues that concern them. Listen fully, but later couple this with stories of strength and resilience. If we face our challenges with our strengths, we are better prepared for a resilient resolution.

▶ Therapeutic Lifestyle Changes: Specifics on Which to Build Wellness for Lifetime Resilience

Three quarters of health costs are related to lifestyle.

—Anna Matthews

Some 20,000,000 new neural connections (synapses) are made and lost daily. We need to make new ones to compensate for those we lose. Therapeutic lifestyle changes (TLCs) facilitate the development of new connections and brain health. They can increase cognitive reserves, along with self-esteem, mental and physical health—even happiness and longevity. TLCs are cost effective, backed up by extensive research in neuroscience, psychology, and medicine. TLCs now need to become central in interviewing and counseling practice, as useful in the session as counseling and therapy theories.

As we move to a positive approach to mental and physical health, we are recognizing a new role for interviewing and counseling. While all of the TLCs discussed below are supported by research, be aware that a team approach may be required. You cannot be an expert in all therapeutic lifestyle changes. As needed, refer clients to medical personnel, nutritionists, physical therapists, personal trainers, and other behavioral health professionals.

TLCs are closely related to the holistic Indivisible Self wellness model of Myers and Sweeney (2004, 2005), who point out that any new wellness activity will reverberate throughout the brain and body. Through extensive research, including factor analysis, they identified 17 aspects of wellness: spirituality, gender identity, cultural identity, self-care and self-worth, friendship, love, leisure, stress management, realistic beliefs, emotional awareness, creative thinking, a feeling of control, work, problem solving, positive humor, nutrition, and exercise.

Each of the Myers and Sweeney findings is an important part of the fabric of positive psychology. The TLCs described below further validate their innovative, breakthrough model. These TLCs are derived from and named for early work in preventive medicine. Now validated by research findings in medicine, neuroscience, and psychology, they have found a significant place in positive psychology and counseling practice.

As shown in Chapter 1, stress underlies most mental health issues, and stress management has become a vital part of the helping process. Stress sends damaging cortisol to the brain and body. Stress impairs our natural resilience and makes us vulnerable to losing personal control, leads us to become unnecessarily angry or depressed, and can harm interpersonal relationships.

Stress also shows as anxiety, which we typically assign as an issue to be resolved as soon as possible. But short-term anxiety, according to prize-winning research by Alison Brooks (2013), can best be described as excitement. Brooks asked people to perform an anxiety-producing task such as giving a speech or solving a difficult math problem. One group was asked to say "I am excited," while others were told to say "I am calm" or "I am anxious." Saying "I am excited" before giving a speech resulted in better performance ratings. In short, defining that surge in body feeling as being "pumped up" better prepares us for action. This study found that the very act of making a statement to oneself resulted in the thought turning into action. Helping your clients change their language around opportunities and challenges they face can be facilitative.

What does this research have to say about positive psychology, stress management, and therapeutic lifestyle changes (TLC)? First, approached with an optimistic and "can do" attitude, your clients will be able to deal with stressors, small or large, more effectively. Second, as you help clients examine their lifestyle choices, you can empower them with a more joyful and energetic approach to life.

All the therapeutic lifestyle changes discussed below are central to recognizing stress as a normal event and handling stress effectively. Stress management is considered the central aspect of wellness and is an important strategy in cognitive behavioral therapy. Throughout this book, we will also show how microskills and counseling theory are used to help clients manage cognitive and emotional stress. TLCs build brain and body health, resulting in a longer lifespan with more joy and zest. TLCs help prevent Alzheimer's, so an early start is wise.

However, let us recall the most important message of this book—the empathic relationship and listening are what make a significant difference. Bringing up the TLCs in the session could lead to giving advice on how the client "should" live. As we all know, advice that is given before we are ready to hear it does little good; it often builds resistance, even anger.

Thus, avoid advice, as suggested in Box 2.7. Some agencies request information on lifestyle as part of intake. It would be natural to follow up and discuss what the client has already mentioned. In the session, listen for stress, possible depression, and/or lifestyle issues. Sensitively phrased questions can bring out whether or not a client is exercising, sleeping well, and eating reasonably. Others lifestyle areas such as social relations can lead to the opportunity to discuss possible interest in the TLCs. Also review the tables below as they list positive and negative consequences of each TLC.

TLCS FOR STRESS MANAGEMENT, BUILDING MENTAL AND PHYSICAL HEALTH, BRAIN RESERVE, AND RESILIENCY[†]

It is unethical for a physician (or counselor) to meet with a patient and fail to prescribe exercise.

—John Ratey, M.D.

Physical exercise. The above quote from John Ratey, who teaches exercise science at Harvard Medical School, is significant for those of us who interview, counsel, or engage in psychotherapy. We have many complex theories and methods, but the helping fields have forgotten to make the simple prescription of exercise central to the helping session. Key to stress management and behavioral health is getting blood flowing to the brain and body. Exercise increases brain volume; it is also valuable in obesity prevention, may help prevent cancer, and may slow the onset of Alzheimer's disease.

Two studies offer examples of how exercise affects mental health. Duke University found that after one year, those with major depressive disorder who did regular exercise received as much benefit as they would have gained from medications. In addition, regular exercisers were less likely to relapse and return into depression. Another research study found that women with generalized anxiety disorder who did two weeks

---

[†]By permission of Allen E. Ivey © 2013, 2016. Permission is granted for duplication with the request that this copyright information remain on the document.

TABLE 2.1 Therapeutic Lifestyle Changes: Physical Exercise

Meaningful Exercise	Lack of Exercise
Enhances sleep	Increases wakefulness
Increases dopamine, gray matter	Obesity
Reduces depression, anxiety	Ill health
Increases lifespan	Reduces lifespan Increases likelihood of heart disease (4 hours TV = 80% increase in heart death

of resistance training or aerobic exercise decreased their worry symptoms by 60% (Herring et al., 2012).

Potential activities in the session. Be sure to check on the physical activities of each client. Watch for changes that indicate less exercise. At appropriate moments, point out that exercise has positive benefits for physical and mental health. Follow up regularly to see if an exercise plan has been implemented. Refer to a physician in cases where there is any history of significant illness or present signs of ill health. Recommendations for clients who are beginning to exercise for the first time require a visit to a health provider.

Nutrition, weight, and supplements. Relatively new to our field is the importance of helping clients deal with the central health and brain issue of nutrition. *Avoid the whites* (pasta, sugar, salt) and *snack only on healthy food.* Vegan, vegetarian, and Mediterranean diets have proven to be effective. Fat activates genes that cause apoptosis (cell death). Obesity facilitates diabetes and other illnesses, including cancer, and it encourages the development of Alzheimer's disease. Supplements can be valuable, but encourage clients to consult their physician before taking supplements.

Potential activities in the session. As this can be an area of sensitivity for some clients, issues of eating need to be approached with care. Those who are concerned about prejudice against weight need special attention and support. Intake forms before interviewing or counseling begins can ask for information on eating style and nutrition, thus opening the way for you to bring up the topic later. What can you share with your clients in a way that they might listen? Basically, we suggest initiating a brief discussion of one of the issues mentioned here and observing client response. Too much information may overwhelm the client and result in no action. Those who want to think seriously about better nutrition will

TABLE 2.2 Therapeutic Lifestyle Changes: Nutrition

Good Nutrition	Poor Nutrition
Organic food	Junk food, causing inflammation
Healthy snacks	
Low-fat, complex carbs	Sugar/pasta/white bread
Olive oil	Palm oil, cottonseed oil, etc.
Richly colored fruits/vegetables	"Dirty dozen" fruits and vegetables with most pesticide residue
Wild salmon, fish oil, flaxseed	Meat hormones and antibiotics
Walnuts and other nuts	Processed food/snacks
Pure water	BHP plastic water bottles

TABLE 2.3 Therapeutic Lifestyle Changes: Social Relations

Productive Relationships	Destructive Relationships
Love, sex	Isolation, living alone
Joyful relationships	Negativism, criticism
Extended lifespan	Brain cell death
Higher levels of oxytocin ("love hormone")	Anger, fear, stress
Helping others	Ignoring or harming others

likely need referral to a nutritionist, as most medical personnel do not have sufficient time or interest to explore this issue. David Katz, M.D., of Yale, who was nominated in 2009 for U.S. Surgeon General, talks of "forks, fingers, and feet." ("Fingers" refer to the importance of not smoking.) The "3-F" framework is a possible way to start the conversation.

Social relations. Being with people in a positive way makes a significant difference in wellness. Interpersonal relationships, of course, are often the central issue in interviewing and counseling. We want our clients to engage socially as fully as possible, as this not only builds mental health but also builds the brain and body. Love and close relationships build health. People with negative or ambivalent relationships have been found to have shorter protective telomeres, thus predicting age-related disease (Uchino et al., 2012).

Potential activities in the session. Naturally, this is where careful listening skills become essential, but in addition to negative stories, we need to search for the positive stories and strengths that can lead to better relationships. There is a long history in counseling and psychotherapy of searching for the "problem," identifying weaknesses that need to be corrected, and diagnosing and labeling before searching out what the client "can do" rather than what he or she "can't do." Of course, clients don't come to us looking for what they did right. Rather, they want solutions to alleviate their issues—and as soon as possible. Thus, we need to draw out the problematic story and the issues behind it, but simultaneously listen for strengths and supports, then use these as a foundation for positive change.

Albert Ellis, the famed originator of rational emotive behavioral therapy (REBT), is also considered by many to be the founder of the cognitive behavioral therapy (CBT) tradition. As a teen, he found himself awkward, with little in the way of social skills. He decided that he wanted a girlfriend, so he started asking girls on dates, both those he knew and those he didn't. He was virtually always turned down, but he kept taking the risk. He did not let failure get him down, and eventually he succeeded. Many of your clients have difficulty taking social risks, meeting new people and relating to them. One of your tasks is to listen to their stories and, working with them, to find new ways of relating. One of these is role-playing social tasks that your client finds challenging.

Cognitive challenge. Take a course, learn a language, learn to play an instrument—basically do something different for growth and the creation of new neural networks. Uncertainty can be growth producing, challenging the assumptions that we have worked so hard to accumulate while young. With a brain already full of well-connected pathways, adult learners should "jiggle their synapses a bit" by confronting thoughts that are contrary to their own.

Potential activities in the session. Check out the degree of stimulating and cognitive involvement of your clients. Generally, we want to encourage and support more.

TABLE 2.4 Therapeutic Lifestyle Changes: Cognitive Challenges

Productive Cognitive Challenges	Limiting Cognitive Challenges
Any cognitive challenge	TV, too much screen time
Change of any type	Repetitious routine tasks
Learning a musical instrument	Being alone
Learning a new language or skill	Vegetating, sitting too much
Playing card games and word games	Boredom
Playing brain games	Taking the easy way out

The destructive factors all too easily lead to depression. However, an overly active schedule, both cognitively and socially, can lead to anxiety and sleeplessness. Our goal is to find cognitive/emotional balance.

Sleep. A full rest is critical for brain functioning and development of new neural networks. Lack of sleep is one of the indications for depression or anxiety.

Potential activities in the session. Your office intake form needs to include questions about how well clients sleep. Your task in the session is to follow through and learn more. If clients are not sleeping well, you need to make frequent contact with them and follow their sleep patterns. However, counseling around sleep issues belongs primarily to experts and medical professionals. If you sense serious problems such as sleep apnea or continued inability to sleep, referral to medical professionals and possibly a formal sleep study is essential.

Meditation and relaxation. If you meditate 10–20 minutes or more each day, it will make a significant difference and calm you throughout the day. Research and clinical evidence is clear that meditation not only provides better mental health—it also changes the brain and increases gray matter (Kabat-Zinn & Davidson, 2012). A relaxed focus in extended prayer, the lighting of candles, saying the Rosary, or attending healing services may function similarly to meditation and help the immune system. You can download Jon Kabat-Zinn's *Guided Mindfulness Meditation* at the Apple Store. An app called *GPS for the Soul* provides an excellent introduction to meditation and daily exercises to help us slow down and "be here now."

Potential activities in the session. The first thing to do is assess what areas of stress management, meditation, and relaxation may be relevant to your client. Then, contract with the client for what is most meaningful. You can teach many basic forms of meditation as you talk with a client in almost any situation.

TABLE 2.5 Therapeutic Lifestyle Changes: Sleep

Sufficient Sleep	Insufficient Sleep
7–9 hours of sleep	Sleep deprived
Reading, meditating, quiet, no TV	Screen time
Increases metabolism, hormones	Sleepy at school/job
Consolidates learning/memory	Parts of brain turn off
	Increased risk of accidents
Positive mood, increase in motivation	Loss of emotional control
Critical for physical health	Increased eating

TABLE 2.6   Therapeutic Lifestyle Changes: Meditation and Relaxation

Productive Use of Meditation/Relaxation	Failure to Calm Mind
20 minutes meditation (or more)	Lack of awareness of what stress does to mind and body
10 minutes systematic muscle relaxation	Failing to take time for oneself
Yoga	Overinvolvement in too many activities
Tai Chi	Continued worry and anxiety
Prayer	Focus on money, achievement, status
Spiritual meditation	Always needing to be busy
Listening to soft or culturally appropriate music	Multitasking
Focus on a mountain, hill, or stream	
Silent walking meditation	
Deep breathing	

Multicultural pride and cultural identity. Our personal identity as multicultural beings affects both our mental and physical health. The harassment that comes with racism, ethnic prejudice, and lack of opportunity deeply affects People of Color. Regardless of our race or ethnicity, we still face issues of religious discrimination and favoritism, economic injustice, ableism, sexism, heterosexism, and other forms of oppression.

You were asked at the beginning of this chapter to identify your multicultural identity through the RESPECTFUL model. A positive psychology/wellness approach recognizes that multiple forms of oppression do exist, but finding personal and cultural strengths can increase one's self-respect and strengthen identity.

Potential activities in the session. If clients show interest, introduce them to the RESPECTFUL model and ask them for examples of strengths and positive assets that will lead to resiliency. All aspects of the RESPECTFUL model have heroes who have made a difference. Who are the cultural heroes and models? What have been their strengths for action, success, or survival that the client can draw on? Look for strength stories in each RESPECTFUL area. What is the client's own personal commitment to the values that he or she has discussed, and how might they be acted upon?

TABLE 2.7   Therapeutic Lifestyle Changes: Cultural Identity

Multicultural Identity/Pride	Limited Multicultural Awareness
Loud and proud about one's multicultural identity its full RESPECTFUL dimensions	Denial of one's culture and/or lack of awareness that one has a culture
Awareness of family strengths	Alienated from family
Awareness of positives in one's school, community, region, or nation, but also able to identify and seek to correct weaknesses	Only focuses on negatives in the community, region, or nation, or ignores or denies weaknesses or limitations
Free and able to change self and work toward involvement, action, and change	Passive and oblivious to responsibility for change in the world

While the seven factors above are fundamental to mental and physical health, there are other research-based TLCs that may be as or more useful to many of your clients (Ivey, Ivey, & Zalaquett, 2012). They are summarized briefly in Box 2.8.

As always, TLCs should be considered in light of the possible need for referral to other sources of assistance. At the same time, you can make a significant difference in helping clients to think through their own mental and physical health plans.

BOX 2.8 Additional Therapeutic Lifestyle Changes

Drugs and Alcohol

Drugs seem only to harm or destroy brain cells. Drugs are especially dangerous for the critical ages 13–18. Teen marijuana use increases the risk of psychosis by 8–10% and causes cognitive decline. Research shows that one or two glasses of alcohol for men and one for women helps the brain and heart and delays Alzheimer's. But alcohol is also associated with increases in breast and other cancers. Moderation is the key.

Medication and Supplements

Interactions are often missed by professionals. More than one antidepressant is no better than one, and exercise and meditation often as good as meds. Look for USP or DSVP verification on supplements, and check with a physician. For those over 45, a daily baby aspirin is highly recommended. Watch blood pressure, body mass index (BMI), and dental hygiene. Dangerous inflammation can often not be seen, so try to control with appropriate diet. There are strong data linking omega-3 with mental and physical health.

Positive Thinking/Optimism/Happiness

Positive thoughts and emotions rest primarily in the executive frontal cortex, while negative emotions of sadness, anger, fear, disgust, and surprise lie in the deeper limbic parts of the brain. Research shows that positive thinking and therapy can build new neural structures and actually reduce the power and influence of negative neural nets.

Beliefs, Values, and Spirituality

Dartmouth Hospital found years ago that those with a strong spiritual orientation recovered more quickly and got out of the hospital sooner. This has been replicated in several studies. Research on cognitive therapy found that using spiritual imagery with depressed clients was highly effective. Imagery and metaphor can affect several areas of the brain simultaneously. A strong positive faith or belief system can make a significant difference in one's life. One can visualize positive examples of strength models of faith, such as Jesus, Moses, Buddha, Mohammed, Viktor Frankl, Gandhi, and Martin Luther King. Visualizing positive strengths of family members such as grandparents can remind one of basic value systems. Your belief system and spiritual orientation contribute to healthy, positive thinking and an optimistic attitude.

Take a Nature Break, Rather Than a Coffee Break

The Wall Street Journal (Wang, 2012) has summarized research showing that memory, attention, and mood increase by 20% if one goes for a 10-minute quiet walk in nature, or even spends time viewing nature scenes in a quiet room. Japanese research validates the value of nature; even looking at or visualizing the color green seems is helpful. Small breaks can be an important part of stress management—and even improve studying and exams.

No Smoking

Smoking narrows blood vessels, reducing blood flow to the brain. It is our central addictive and most dangerous drug in terms of early death. An issue is that smoking often relaxes and increases one's ability to concentrate. Successful no-smoking counseling typically requires a committed client and close follow-up by phone or email—continuous support and reminders appear to help.

Control Screen Time

The many types of screens are changing our brains, resulting in short-term attention and increasing both hyperactivity and obesity. The constant flicker both activates and tires the brain. A study by the National Heart, Lung, and Blood Institute (2013) revealed that children ages 8–18 average 4.5 hours a day watching TV, 1.5 hours on a computer, and 1 hour playing video games. As a result, they go to bed later and have more difficulty getting up.

Art, Music, Dance, Literature

What is the client's passion? Returning to and becoming involved with the arts can be life changing. The relaxation and here-and-now emotions of music and dance bring satisfaction and a needed break in routine. Many of us find peace and oneness through participating in or viewing the arts.

Relaxation and Having Fun

This will increase dopamine release to the nucleus accumbens. Dance, tennis, enjoy a sunset, you name it—it is good for the brain.

More Education

It is clear that the farther you go in the educational system, the less your chance of Alzheimer's and the better your health. Ensure that child raising provides richness and encourages reading. Continuing education as we age helps develop new neural networks. Take courses; learn a musical instrument or a new language. However, education is often oriented to the privileged. Those from economically disadvantaged backgrounds receive less. Even when they go to college, they drop out more frequently, often with high levels of debt. Support networks are needed. Work for better and more just schooling.

Money and Privilege

Not having to worry about funds, being able to get the best medical care, buying the best food are not options available for all. Most of us who read this page, regardless of race/ethnicity, enjoy some form of privilege. Regardless of background, many benefit from being educated and reasonably affluent. White privilege brings consequential benefits to this group. Coming into the world with privilege also brings responsibility.

Helping Others and Social Justice Action

Working for justice and volunteering make a positive difference to both helper and helpee. Stress hormones are reduced, and telomeres lengthen. Fetal development depends on a healthy mother. Poverty, abuse, and trauma produce damaging cortisol, destroying brain cells permanently. Racial and oppressive harassment of all types causes stress and raises cortisol. Neuroscience and genetic/epigenetic research clearly shows that environmental conditions are key to health and that social justice action will enable children and adults to function more fully, effectively, and joyfully.

Joy, Humor, Zest for Living, and Keep It Simple

Client: Doctor, I can't sleep. I keep thinking someone is under my bed and I get up several times a night to look.

Psychiatrist: I can fix that. Four years of therapy twice a week.

Client: How much will it cost?

Psychiatrist: $100. An hour.

Client: I can't afford it, sorry.

Two years later, they meet on the street.

Psychiatrist: Nice to see you. How are things going?

Client: Great, no more problems. I told my neighbor, and he came over and cut off the legs of the bed.

We often make things too complex with fancy theorizing. TLCs are a shortcut to health that will change our practice. Learned optimism heals and can change a life. Focus on strengths and what clients *can do* rather than on what they can't. The positive approach builds cognitive reserve that will enable them to meet challenges more effectively. Keep it simple, use humor and laughter, and increase zest for life.

Summarizing Key Points of "Multicultural Competence, Ethics, Positive Psychology, and Resilience"

Multicultural Competence

▶ Interviewing and counseling are global phenomena used with individuals from many different cultures and customs.
▶ It is an ethical imperative that interviewers and counselors be multiculturally competent and continually increase their awareness, knowledge, skills, and action in multicultural areas.

Respectful Interviewing and Counseling

▶ The RESPECTFUL model lists 10 key multicultural dimensions, thus showing that cultural issues inevitably will be part of the interviewing and client relationship.

Privilege as a Multicultural Interviewing Issue

▶ Race and ethnicity are especially important multicultural issues. We need to be aware that being White, male, and economically advantaged often puts a person in a privileged group.

▶ The interviewer needs to be client centered rather than directed by "politically correct" terminology. This is an issue of respect, and clients need to say what is comfortable and appropriate for them.

Awareness, Knowledge, Skills, and Action for Multicultural Competence

▶ Developing multicultural awareness, knowledge, skills, and actions is a lifelong process of continuing learning.

▶ Awareness of yourself as a cultural being is a vital beginning.

▶ There are many ways in which you and your client interpret humanity and the world. Learn about your own and other worldviews.

▶ Use a culturally and diversity sensitive approach to interviewing and counseling. Adapt your strategies in a culturally respectful manner.

Ethics and Morals: Professional and Personal

▶ Counselors must practice within boundaries of their competence, based on education, training, supervised experience, state and national credentials, and appropriate professional experience.

▶ Confidentiality of information is addressed in the ethics codes of all helping professions and provides the basis for trust and relationship building. Informed consent requires us to tell clients of their rights. When recording sessions, we need permission from the client.

▶ Interviewers and counselors have more perceived power than clients. Efforts need to be made to equalize power in the relationship.

▶ Helping professionals are asked to work outside the interview to improve society and are called upon to act on social justice issues.

Positive Psychology: Building Client Resilience

▶ Positive psychology promotes the development of optimism and resilience.

▶ Optimism can be measured with scales such as the Six-Point Optimism Scale.

▶ Search for positive assets, which are the resources and strengths that clients bring with them.

▶ Clients aware of their strengths and resources are more resilient and can face challenges in a more effective manner.

Therapeutic Lifestyle Changes: Specifics on Which to Build Wellness for Lifetime Resilience

▶ Most of us can benefit from the therapeutic lifestyle changes (TLCs) for managing stress and building mental and physical health, brain reserve, and resilience. An effective plan to implement the TLCs can improve the quality of your life and work.

▶ Review the TLCs and select one to three as a beginning step toward improving your own mental and physical health.

► You can use the TLCs in every interview to help clients increase well-being and effective coping.

► As with your own health program, it does little good to change all the TLCs at once. The client can select those that are most immediately meaningful.

Assess your current level of knowledge and competence as you complete the chapter:

1. Flashcards: Use the flashcards to check your understanding of key concepts and facilitate memorization of key information.

2. Self-Assessment Quiz: The quiz will help you assess your current knowledge and prepare for course examinations.

3. Portfolio of Competencies: Evaluate your present level of competence on the ideas and concepts presented in this chapter using the Self-Evaluation Checklist. Self-assessment of your competencies demonstrates what you can do in the real world.

Attending, Empathy, and Observation Skills

Multicultural Competence, Ethics, Neuroscience, and Positive Psychology/Resilience

Chapter 3
Attending, Empathy, and Observation Skills
Fundamentals of All Interviewing and Counseling Approaches

I love the feeling of the fresh air on my face and the wind blowing through my hair.
> —Evel Knievel, world-famous motorcyclist who jumped over the Grand Canyon

Attending behavior is like air. We seldom notice it, but when we discover others listening to us, like Evel, we love the feeling. When we are not heard, we are stultified and may even gasp for understanding.

Your first task in a helping relationship is to listen empathically. The second is to listen and observe with individual and multicultural sensitivity.

> —Allen Ivey

Attending behavior encourages client talk. You will use attending behavior to help clients tell their stories and to reduce your own talk time. Through empathic understanding, selective attention, and sometimes silence, you can help clients refocus their stories in more positive ways.

Observation skills will help you understand what is going on between you and your clients, both verbally and nonverbally. Moving beyond just words is vital. Observation skills will enable you to respond appropriately to the wide variety of individual and multicultural differences.

Chapter Goals

Awareness, knowledge, skills, and actions developed through the concepts of this chapter and this book will enable you to:

▶ Communicate to clients that you understand and are interested in their stories.

▶ Become aware of verbal and nonverbal attending behavior styles in the interview— including the styles of both you and your client.

► Identify the importance of empathy and how to assess your own levels of empathic understanding.

► Modify your patterns of attending to establish empathic relationships with the widely differing styles and needs of individuals, including those of many different multicultural backgrounds.

Assess your current level of knowledge and competence as you begin the chapter:

1. **Self-Assessment Quiz:** The chapter quiz will help you determine your current level of knowledge. You can take it before and after reading the chapter.

2. **Portfolio of Competencies:** Before you read the chapter, please fill out the Self-Evaluation Checklist to assess your existing knowledge and competence on the ideas and concepts presented in this chapter. Then, at the end of the chapter, complete the checklist again to summarize your competencies after study and practice.

► Defining Attending Behavior

Listening is the core of developing a relationship and making real contact with our clients. How can we define effective listening more precisely? The following personal case study may help you to identify listening in terms of clearly observable behaviors.

> One of the best ways to identify and define listening skills is to experience the opposite—poor listening. Think of a time when someone failed to listen to you. Perhaps a family member or friend failed to hear your concerns, a teacher or employer misunderstood your actions and treated you unfairly, or you had that all-too-familiar experience of calling a computer helpline and never getting anyone who comprehended your issues. Remember the frustration, hurt, or even anger when you were not heard or understood?

The following exercise was actually used at the very start of the microskills approach. A volunteer beginning counselor was videotaped. Out of this exercise, the most fundamental aspects of listening were identified for the first time.

EXERCISE 3.1 Identifying Listening Skills
Please take time to think through your own definition of what observable behaviors are involved in listening.

Find a partner to role-play an interview in which one of you plays the part of a poor listener. The poor listener should feel free to exaggerate in order to identify concrete behaviors of the ineffective interviewer. If no partner is available, think of a specific time when you felt that you were not heard. What feelings and thoughts occur to you when you recall that someone you cared about did not listen?

Now write down the specific behaviors of poor listening that you identified. Later in this chapter, compare your observations with what is said here.

Examples of poor listening and other ineffective interviewing behaviors are numerous—and instructive. Doing the exaggerated role-play is often humorous, but your strongest memory of poor listening may be of disappointment and even anger at not being heard. If you are to be effective and competent, do the opposite of the ineffective interviewer: Attend and listen!

Attending behavior is not just for interviewing and counseling. It is fundamental to interpersonal communication in all cultures, although the style of listening will vary. Attending is key to building a working relationship or therapeutic alliance with your clients. All theories and methods of interviewing, counseling, and psychotherapy rest on a base of attending behavior.

Neuroscience brain imaging has proven the importance of attending in another way. When a person attends to a stimulus such as the client's story, many areas of the brain of both interviewer and client show activity. In effect, attending and listening "light up" the brain. So does ineffective and hurtful nonlistening, but involving different areas of the brain.

THE FOUR KEY DIMENSIONS OF ATTENDING BEHAVIOR

Attending behavior is defined as supporting your client with individually and culturally appropriate verbal following, visuals, vocal quality, and body language (the 3 V's + B). When you use each of the microskills, you can anticipate how the client may respond. These predictions are never 100% perfect, but research has shown what you can usually expect (Daniels, 2010). If your first attempt at listening is not received, you can intentionally flex and show your interest and involvement by going back to something the client said earlier.

ATTENDING BEHAVIOR	ANTICIPATED CLIENT RESPONSE
Support your client with individually and culturally appropriate visuals, vocal quality, verbal tracking, and body language (3 V's + B).*	Clients will talk more freely and respond openly, particularly about topics to which attention is given. Depending on the individual client and culture, anticipate fewer breaks in eye contact, a smoother vocal tone, a more complete story (with fewer topic jumps), and more comfortable body language.

Your interview style changes the brain for good or ill. You are actually rewiring the brain and facilitating rewriting of old, ineffective memories to new and more useful ones. If your client comes to you with a career decision, a marital concern, or a traumatic fire or flood experience, you have the potential to help focus on strengths for resilience and help the person move from indecision to positive action.

Now let us turn to further discussion of how skills and competence can "light up" the interview.

▶ Discerning: Basic Techniques and Strategies of Attending Behavior

Discerning: Ability to see and understand people, things, or situations clearly and intelligently.

—Merriam-Webster Dictionary Online

Attending behavior is the first critical skill of listening; it is a necessary part of all interviewing and counseling. To communicate that you are indeed listening or attending to the client, you need the following "three V's + B." Of course, each of these needs to be modified to meet individual and multicultural differences.

---

*Norma Gluckstern is thanked and recognized for suggesting the acronym 3 V's + B.

1. **Visual/eye contact.** Look at people when you speak to them.
2. **Vocal qualities.** Communicate warmth and interest with your voice. Think of how many ways you can say, "I am really interested in what you have to say," just by altering your vocal tone and speech rate.
3. **Verbal tracking.** Stay with the client's story. Don't change the subject.
4. **Body language.** Be yourself—authenticity is essential to building trust. To show interest, face the client squarely, lean slightly forward with an expressive face, and use encouraging gestures. Smile to show warmth and interest in the client.

The three V's + B reduce interviewer talk time and provide clients with an opportunity to tell their stories in as much detail as needed. You will find it helpful to observe your clients' verbal and nonverbal behavior. Note the topics to which your clients attend and those they avoid. You will find that clients who are culturally different from you may have differing patterns of attending and listening. Use client observation skills and adapt your style to meet the needs of the unique person before you.

Each person you work with will have a unique style of communicating. Furthermore, the multicultural background of each client may modify her or his communication style both verbally and nonverbally. People with various disabilities also represent a cultural group with whom you may need to vary your attending style, as illustrated in Box 3.1.

BOX 3.1 Attending Behavior and People With Disabilities

Attending behaviors may require modification if you are working with people with disabilities. It is your role to learn clients' unique ways of thinking and being and how they deal with important issues. Focus on the person, not the disability. For example, think of a person with hearing loss rather than "hearing impaired," a person with AIDS rather than "AIDS victim," a person with a physical disability rather than "physically handicapped." So-called handicaps are often societal and environmental rather than personal.

People who are blind or have limited vision

Clients who are blind or partially sighted may not look at you when they speak. Expect clients with limited vision to be more aware of and sensitive to your vocal tone. People who are blind from birth may have unique patterns of body language. It may be helpful to teach clients attending skills, such as orienting their face and body to other people, that enable them to communicate more easily with the sighted.

People who are deaf or have hearing loss

Many people who are deaf do not consider themselves impaired in any way. People who were born deaf have their own language (signing) and their own culture that often excludes the hearing. You are unlikely to work with this type of client unless you are skilled in sign language and are trusted among the deaf community.

You may be skilled in sign language or you may counsel a deaf person through an interpreter. To be effective, you will need specific training in the use of an interpreter and a basic understanding of deaf culture. Eye contact is vital in counseling a deaf client or while using an interpreter. It will isolate the client when you speak to the interpreter instead of to the client or use phrases such as "Tell him. . . ."

(continued)

BOX 3.1 (continued)

For those with moderate to severe hearing loss, speak in a natural way, but not fast. Extensively paraphrase the client's words. Speaking loudly is often ineffective, as ear mechanisms may not equalize for loud sounds. In turn, teaching those with hearing loss to paraphrase what others say to them can help them to communicate with others.

People with physical disabilities

We cannot place people with physical disabilities in any one group; each person is unique. Consider a person who uses a wheelchair, an individual with cerebral palsy or Parkinson's disease, someone who may have lost a limb, or a person who is physically disfigured by a serious burn. They all have the common problem of physical disability, but their body language and speaking style will vary. You must attend to each individual as a complete person from her or his unique perspective.

People who are temporarily able

We suggest that you consider yourself one of the many who are *temporarily able*. Age and life experience will bring most of us some variation of the challenges previously described. For older individuals, the issues discussed here may become the norm rather than the exception. Approach disabled clients with humility and respect.

The National Council on Disability has a searchable website for further details: http://www.ncd.gov.

Case Study: Is Attending Enough?
Video Activity: Sharpening Your Observation Skills
Video Activity: A Negative and a Positive Example of Attending.

▶ Discerning: Individual and Multicultural Issues in Attending Behavior

Listen before you leap! A frequent tendency of the beginning counselor or interviewer is to try to solve the client's difficulties in the first 5 minutes. It is critical that you slow down, relax, and attend to client narratives. Think about it—the client most likely developed her or his concern over a period of time. Attending and giving clients talk time demonstrates that you truly want to hear their story and major challenges. In addition, clients have varying attending and interacting styles. By observing their patterns of conversation, you are also learning about how they interact with others outside of the session.

VISUAL/EYE CONTACT

Direct eye contact is considered a sign of interest in European–North American middle-class culture, with more eye contact while listening and less while talking. However, research indicates that some African Americans in the United States may have reverse patterns—looking more when talking and slightly less when listening. Among some traditional Native American and Latin groups, eye contact by the young is a sign of disrespect. Imagine the problems this

BOX 3.2

National and International Perspectives on Counseling Skills

Use With Care: Culturally Incorrect Attending Can Be Rude

Weijun Zhang

The visiting counselor from North America got his first exposure to cross-cultural counseling differences at one of the counseling centers in Shanghai. His client was a female college student. I was invited to serve as an interpreter. As the session went on, I noticed that the client seemed increasingly uncomfortable. What had happened? Since I was translating, I took the liberty of modifying what was said to fit each other's culture, and I had confidence in my ability to do so. I could not figure out what was wrong until the session was over and I reviewed the videotape with the counselor and some of my colleagues. The counselor had noticed the same problem and wanted to understand what was going on. What we found amazed us all.

First, the counselor's way of looking at the client—his eye contact—was improper. When two traditional Chinese talk to one another, we use much less eye contact, especially with a person of the opposite sex. The counselor's gaze at the Chinese woman could be considered rude or seductive in Chinese culture.

Although his nods were acceptable, they were too frequent by Chinese standards. The student client,

probably believing one good nod deserved another, nodded in harmony with the counselor. That unusual head bobbing must have contributed to the student's discomfort.

The counselor would mutter "uh-huh" when there was a pause in the woman's speech. While "uh-huh" is a good minimal encouragement in North America, it happens to convey a kind of arrogance in China. A self-respecting Chinese would say the more personal and active *er* (oh), or *shi* (yes) to show he or she is listening. How could the woman feel comfortable when she thought she was being slighted?

He shook her hand and touched her shoulder. I told our respected visiting counselor afterward, "If you don't care about the details, simply remember this rule: in China, a man is not supposed to touch any part of a woman's body unless she seems to be above 65 years old and displays difficulty in moving around."

"Though I have worked in the field for more than 20 years, I am still a lay person here in a different culture," the counselor commented as we finished our discussion.

may cause when the teacher or counselor contradicts basic cultural values by saying, "Look at me!" Some traditional Native American, Inuit, and Aboriginal Australian groups generally avoid eye contact, especially when talking about serious subjects. Cultural differences in eye contact are numerous (see Box 3.2), but we must recall that individual differences are often even greater and avoid stereotyping any client or group with so-called "normative" patterns.

You want to look at clients and notice quick shifts and breaks in eye contact. Clients often tend to look away when thinking about a complex issue or discussing distressing topics. It may be wise to avoid direct and solid eye contact when the client is uncomfortable. Also, think about your own breaks in eye contact. You may find yourself avoiding eye contact while discussing certain topics. There are counselors who say their clients talk about "nothing but sex" and others who say their clients never bring up the topic. Both types indicate to their clients whether the topic is appropriate and interesting through their style of eye contact. If you find that your clients are avoiding a topic, look at your own behavior, not just that of the client.

VOCAL QUALITIES

Your voice is an instrument that communicates much of the feeling you have toward another person or situation. Changes in its pitch, volume, or speech rate often convey the emotions underlying your words.

Keep in mind that different people are likely to respond to your voice differently. Try the following exercise with a group of three or more people.*

---

*This exercise was developed by Robert Marx, School of Management, University of Massachusetts, Amherst.

Ask the group to close their eyes and note your vocal qualities as you speak. Talk for 2 or 3 minutes in your normal tone of voice on any subject of interest to you. How do they react to your tone, your volume, your speech rate, and perhaps even your accent? Ask the group for feedback. What does this feedback say to you?

This exercise reveals a central point of attending: People differ in their reactions to the same stimulus. Some people find a particular voice interesting; others may find that same voice boring or even threatening. People differ, and what is successful with one person or client may not work with another.

Accent is a particularly good example of how different people will react differently to the same voice. Obviously we need to avoid stereotyping people because their accents are different from ours. How do you react to the following accents: Australian, BBC English, Canadian, French, Chinese, New England U.S., and Southern U.S.? Are you aware that the person who speaks two or more languages is advantaged? How many languages can you speak?

If you think about the way you tell a story, you will likely find yourself giving increased vocal emphasis to certain words and phrases; this is known as **verbal underlining**. Clients, of course, do the same. The key words that a person underlines by means of volume and emphasis are often ideas of particular importance.

Awareness of your voice and observation of the changes in others' vocal qualities will enhance your attending skills. Speech hesitations and breaks and the timing of vocal changes can signal distress, anxiety, or discomfort. Clearing one's throat may indicate that words are not coming easily.

VERBAL TRACKING

Verbal tracking is staying with your client's topic to encourage full elaboration of the narrative. Just as people make sudden shifts in nonverbal communication, they change topics when they aren't comfortable. In middle-class U.S. communication, direct tracking is appropriate, but in some Asian cultures such direct verbal follow-up may be considered rude and intrusive.

Verbal tracking is especially helpful to both the beginning and the experienced interviewer who is lost or puzzled about what to say next in response to a client. *Relax*; you don't need to introduce a new topic. Ask the client a question or make a brief comment about something the client has said in the immediate or near past. Build on the client's topics, and you will come to know the client very well over time.

Observe yourself and use **selective attention**. Clients tend to talk about what interviewers are willing to hear. A famous training film (Shostrum, 1966) shows three eminent counselors (Albert Ellis, Fritz Perls, and Carl Rogers) all counseling the same client, Gloria. Gloria changes the way she talks and responds very differently as she works with each counselor. Research on verbal behavior in the film revealed that Gloria tended to match the language of the three different counselors (Meara, Pepinsky, Shannon, & Murray, 1981; Meara, Shannon, & Pepinsky, 1979). Each expert indicated, by his nonverbal and verbal behavior, what he wanted Gloria to talk about.

Should clients match the language of the interviewer, or should you, the interviewer, learn to match your language and style to that of the client? Most likely, both approaches are relevant, but in the beginning, you want to draw out client stories from their own language perspective, not yours. What do you consider most essential in the interview? Are there topics with which you are less comfortable? Some interviewers are excellent at helping clients talk about vocational issues but shy away from interpersonal conflict and sexuality. Others may find their clients constantly talking about interpersonal issues, excluding critical practical issues such as getting a job.

Consider the following abbreviated case.

Titanya: [Speaks slowly, seems to be sad and depressed] I'm so confused and mixed up right now. The first term went well and I passed all my courses. But this term, I am really having trouble with chemistry. It's hard to get around the lab in my wheelchair and I still don't have a textbook yet. [An angry spark appears in her eyes, and she clenches her fist.] By the time I got to the bookstore, they were all gone. It takes a long time to get to that class because the elevator is on the wrong side of the building for me. [looks down at floor] Almost as bad, my car broke down and I missed two days of school because I couldn't get there. [The sad look returns to her eyes.] In high school, I had lots of friends, but somehow I just don't fit in here. It seems that I just sit and study, sit and study. Some days it just doesn't seem worth the effort.

There are several different directions an interviewer could follow to help Titanya unpack and understand her confused and frustrated situation. Where would you go, given the multiple possibilities? List at least three possibilities for follow-up to this client statement.

One way to respond is to reflect on the main theme of the client's story. For example, "You must feel like you're being hit from all directions. Which would you like to talk about first?" Different interviewers place emphasis on different issues. Some interviewers consistently listen attentively to only a few key topics while ignoring other possibilities. Be alert to your own potential patterning of responses. It is important to see that no issue gets lost, but equally important to avoid confusion by not trying to solve everything at once.

Howard Busby, a deaf professor at Gallaudet College, has commented on this case in a personal email to Allen and Mary. He points out that the client's problem could be related to the disability, the issues at school, or both:

The wheelchair might be the issue as much as how this client is dealing with it. The client obviously has a disability, but it does not have to be disabling unless the client makes it so. Although I am categorized as having a disability due to deafness, I have never allowed it to be disabling.

My interpretation of the client is that this depression, on the surface, could be the result of mobility restriction. However, there are other factors that might have caused the depression, even if there were no need for a wheelchair. It is easy for the client to blame problems on the disability and thus distract from personal issues. The concern could be a poor and unsupportive campus environment, learned helplessness on the part of the client, or a combination of these and other multiple factors.

All of us would do well to consider Dr. Busby's comments. He is pointing out the importance of taking a broad and comprehensive view of all cases. We must avoid stereotyping and be sensitive to individual and cultural differences.

ATTENTIVE AND AUTHENTIC BODY LANGUAGE

The anthropologist Edward Hall once examined film clips of Southwestern Native Americans and European North Americans and found more than 20 different variations in the way they walked. Just as cultural differences in eye contact exist, body language patterns also differ.

A comfortable conversational distance for many North Americans is slightly more than arm's length, and the English prefer even greater distances. Many Latin people prefer half that distance, and some people from the Middle East may talk standing only inches apart. As a result, the slightly forward lean we recommend for attending is not appropriate all the time.

What determines a comfortable interpersonal distance is influenced by multiple factors (Hargie, Dickson, & Tourish, 2004, p. 45).

Gender: Women tend to feel more comfortable with closer distances than men.

Personality: Introverts need more distance than extraverts.

Age: Children and the young tend to adopt closer distances.

Topic of conversation: Difficult topics such as sexual worries or personal misbehavior may lead a person to more distance.

Personal relationships: Harmonious friends or couples tend to be closer. When disagreements occur, observe how harmony disappears. (This is also a clue when you find a client suddenly crossing the arms, looking away, or fidgeting.)

A person may move forward when interested and away when bored or frightened. As you talk, notice people's movements in relation to you. How do you affect them? Note your own behavior patterns in the interview. When do you markedly change body posture? A natural, authentic, relaxed body style is likely to be most effective, but be prepared to adapt and be flexible according to the individual client and multicultural background.

ATTENTION, CONSCIOUSNESS, AND ATTENDING BEHAVIOR

> Attending behavior is the *act of listening*.
> Attention is where we and others direct our attending behavior perceptual systems.
> Where we direct our attention is considered by many as the definition of consciousness.
>
> —Allen Ivey

The importance of attending behavior and attention cannot be overstated. At one level, the actions of attending behavior are simple and straightforward. But there is a significant difference between one's ability to attend and truly understanding the mind or consciousness of the other person, the client.

Being attentive, giving your attention to the client, means surrendering yourself in the other. It means that the client is the key person in your conscious field of awareness. Speaking more broadly, the famed philosopher William James said, "attention is the behavioral and cognitive process of selectively concentrating on one aspect of the environment while ignoring other things."

What, then, is the meaning of the word *consciousness*, a concept that philosophers have pondered for centuries. A new definition proposed by Michael Graziano (2013) is simple and straightforward: Consciousness is a perceptual process—what we see, hear, feel, taste, touch, think, feel in the here-and-now moment. What we perceive, even in our mind's eye as we think internally, is the closest we ever get to conscious experience. Consciousness is all that we tell ourselves about what we observe and perceive. It is our stories.

Neuroscience notes that consciousness involves the brain's attentional system of orienting, executive control, and resting state areas of the brain (Peterson & Posner, 2012).

What does this mean for you, the counselor, therapist, or coach? First, drawing out client stories and facilitating the rewriting of ineffective stories are basic to the nature of helping. A key route to becoming conscious of clients is empathic understanding. Interestingly, at the same time, your own self-awareness—your attention to your own experience—is part of your consciousness. Some attention to yourself with empathy will help you become more attuned to others, including your clients.

High proficiency in attending behavior is required if you are to be truly conscious of your clients. Attending behavior and attention at first seem simple, but achieving professional excellence is a lifetime process that requires you to become conscious of the client's consciousness. Needless to say, beginning and advanced professionals will always need to increase and practice their attending, attention, and understanding of others.

▶ Defining Empathy and Empathic Understanding

To sense the client's world as if it were your own, but without ever losing the "as if" quality—this is empathy and this seems essential to therapy.

—Carl Rogers

Putting yourself into "another person's shoes" or viewing the world through "someone else's eyes and ears" is another way to describe empathy. It was Carl Rogers (1957, 1961) who brought the importance of empathy to our attention. It is vital to listen carefully, enter the world of the client, and communicate that we understand the client's world as the client sees and experiences it.

EMPATHY	ANTICIPATED CLIENT RESPONSE
Experiencing the client's world and the client's story as if you were that person; understanding his or her key issues and saying them back accurately, without adding your own thoughts, feelings, or meanings. This requires attending and observation skills plus using the important key words of the client, but distilling and shortening the main ideas.	Clients will feel understood and engage in more depth in exploring their issues. Empathy is best assessed by clients' reaction to a statement and their ability to continue discussion in more depth, and eventually with better self-understanding.
Additive empathy, also known as positive empathy, used carefully, accesses strengths and positive emotions, communicating that the client has within the self the potential for growth.	Clients who have first been heard accurately with basic, interchangeable empathic responses often bloom and grow faster with positive, additive empathy.

Rogers's influential definition of empathy has been elaborated by many others (cf. Carkhuff, 2000; Egan, 2014; Ivey, D'Andrea, & Ivey, 2012). Empathy is central to building the "working alliance" and the relationship factors (the common factors that explain 30% of change) that make for successful interviewing, counseling, and psychotherapy (Miller, Duncan, & Hubble, 2005).

▶ Discerning: Levels of Empathy

Here we describe three types of empathic understanding that we will use in this book to analyze interview transcripts.

Subtractive empathy: In **subtractive empathy**, interviewer responses give back to the client less than what the client stated or distorts what has been said. The listening or influencing skills are used inappropriately.

Basic, interchangeable empathy: In **basic empathy**, often called **interchangeable empathy**, interviewer responses are roughly interchangeable with those of the client. The interviewer is able to say back accurately what the client has said. It is the most common and important counselor comment level. Rogers pointed out that listening itself is necessary and sufficient to produce client change. The basic listening sequence (see the later chapters of this book) focuses on basic empathy.

Additive, positive empathy: Based on skilled and insightful attending, **additive empathy** focuses on strengths and positives in the client, leading to a "can do" rather than a "can't do" attitude. But basic empathy needs to come first. The influencing skills of the second half of the book require a positive, additive empathic approach.

These three anchor points are often expanded to classify and rate the quality of empathy shown in a session. You can use empathy rating in your practice with microskills. Later, in your professional work, keep checking whether you have maintained interest in your clients and are fully empathic.

Client: I don't know what to do. I've gone over this problem again and again. My husband just doesn't seem to understand that I don't really care any longer. He just keeps trying in the same boring way—but it doesn't seem worth bothering with him anymore.

Level 1 Empathy	(subtractive) That's not a very good way to talk. I think you ought to consider his feelings, too.
	(lightly subtractive) Seems like you've just about given up on him. You don't want to try anymore. (interpreting the negative)
Level 2 Empathy	(basic empathy or interchangeable response) You're discouraged and confused. You've worked over the issues with your husband, but he just doesn't seem to understand. At the moment, you feel he's not worth bothering with. You don't really care. (Hearing the client accurately is the place to start all empathic understanding. Level 2 is always central.)
Level 3 Empathy	(slightly additive) You've gone over the problem with him again and again to the point that you don't really care right now. You've tried hard. What does this mean to you? (The question adds the possibility of the client's thinking in new ways, but the client still is in charge of the conversation.) (additive and perhaps transformational) I sense your hurt and confusion and that right now you really don't care anymore. Given what you've told me, your thoughts and feelings make a lot of sense to me. At the same time, you've had a reason for trying so hard. You've talked about some deep feelings of caring for him in the past. How do you put that together right now with what you are feeling? (A summary with a mild self-disclosure. The question helps the client develop her own integration and meanings of the issues at the moment.)

▶ Refining: The Power of the Positive

Every word you say counts, not every gram of medication.
—Ted Kaptchuk, Harvard Medical School

It has been clear for years that the words we say produce expectations that have a major impact on clients' bodies and minds. Your empathic authenticity and caring as a person are powerful agents of change.

Historically, a classic study broke new ground when it was discovered that if teachers were told that certain elementary school students were expected to do well that year, their performance results were superior to those of other students about whom nothing had been said (Rosenthal & Jacobson, 1968). Teachers who expected certain students to improve brought about significant change. Since then, numerous studies have proven the importance of positive expectations in changing behavior.

Negative expectations also affect performance. African Americans and women sometimes suffer "stereotype threat affect." More than 300 papers have examined this issue. It has been found that lowered expectation for African Americans reduces performance on tests (Steele, 1997). Similarly, women have been shown to perform less effectively as a result of lowered expectations and stereotype threat (Stoet & Geary, 2012).

It should be patently clear that a positive psychology approach with expectation of client success will be most helpful to your clients. If you focus on the negative, you will reinforce negative thoughts, feelings, and even neural networks in the brain. Positive expectations produce positive results.

THE NEUROSCIENCE OF EMPATHY: MIRROR NEURONS

Well-established data from neuroscience have confirmed the importance of empathy and located related structures in specific areas of the brain. Many believe that mirror neurons are one of the most significant discoveries in recent science. Empathy is identifiable through functional magnetic resonance imagining (fMRI) and other technologies. Key to this process are the mirror neurons, which fire both when a human or animal acts and when it observes actions by another.

One of the earliest studies measuring the importance of mirror neurons and empathy asked closely attached partners to watch each other (through a one-way mirror) receive a mild shock. It was found that the emotional areas of both the shocked partner and the observing partner fired simultaneously (Singer et al., 2004), although the pain areas were activated only in the shocked partner. This basic finding has been consistently replicated. We have been able measure the similarity in brain activity in interviewing; the more similarity, the more empathy and interviewer comprehension of the client's message (Stephens, Silbert, & Hasson, 2010). Conversely, this "neural coupling" disappears when story comprehension is not effective; when listening skills are not successfully implemented (subtractive), empathy falls apart.

As you become a more experienced and empathic listener, your mirror neurons are very likely to fire in a pattern similar to those of your client. When they hurt, you may notice tight or sad feelings in your body. When they smile, you may return the smile automatically. Equally important, if you are pleasant, smile, and care for them, their self-concept may change and their mirror neurons may start following you. If you respect them, they are more likely to respect themselves.

A different thing occurs with children or teenagers who are diagnosed with conduct disorder or the sociopathic adult. Their mirror neurons fire significantly less when they see pain in others. They do not sense the sadness or joy that they see in others the same we that you do. It has been shown that the pleasurable neurotransmitter dopamine is produced in some sociopaths when they see others harmed (Buckholtz & Marois, 2012). We still need to be empathic with these challenging cases, but we also need to add a bit of toughness in ourselves. As you listen, you will begin to understand where they are coming from; through this, you *may* be able to help them.

Empathic listening is a vital part of helping your clients feel supported, grow and change. Empathy and listening are not just abstract concepts—they are clearly measureable and make a difference in other people's lives.

▶ Defining Observation Skills

Some authorities say that 85% or more of communication is nonverbal. Observing the client's and your own verbal and nonverbal behavior is a critical dimension of effective interviewing. Emotional tone is often conveyed through nonverbals, which may overrule and be more important than what you are saying to the client. Think about how many ways you can say "I understand," with vocal tones varying from warm and supportive to neutral to sarcastic.

Observation skills help you understand how the client likely is responding. If a client breaks eye contact, "something" is happening, but the reasons for looking away will vary. It may be that the topic is uncomfortable, or a recollection from the past is distracting, or it may be a reaction to you or something you said. Do not overanalyze, however—the client may have heard a noise outside the window. And, of course, be careful to watch for individual and cultural differences. Watch for small behaviors. A casual nose wipe can indicate a client who is unsure and feels uncomfortable, or perhaps one who is not giving you a fully honest story.

CLIENT OBSERVATION SKILLS	ANTICIPATED CLIENT RESPONSE
Observe your own and the client's verbal and nonverbal behavior. Anticipate individual and multicultural differences in nonverbal and verbal behavior. Carefully and selectively feed back observations to the client as topics for discussion.	Observations provide specific data validating or invalidating what is happening in the session. Also, they provide guidance for the use of various microskills and strategies. The smoothly flowing interview will often demonstrate movement symmetry or complementarity. Movement dissynchrony provides a clear clue that you are not "in tune" with the client.

▶ Discerning: Specifics of Observation

Clients may break eye contact, shift their bodies, and change vocal qualities as their comfort level changes when they talk about various topics. You may observe clients crossing their arms or legs when they want to close off a topic, using rapid alterations of eye contact during periods of confusion, or exhibiting increased stammering or speech hesitations when topics are difficult. If you watch yourself carefully on tape, you, as interviewer, will also exhibit many of these same behaviors at times.

Jiggling legs, making complete body shifts, or suddenly closing one's arms most often indicates discomfort. Random, discrepant gestures may indicate confusion, whereas a person seeking to control or organize a situation may move hands and arms in straight lines and point fingers authoritatively. Smooth, flowing gestures, particularly those in harmony with the gestures of others such as family members, friends, or the interviewer, may suggest openness.

Often people who are communicating well "mirror" each other's body language. They may unconsciously sit in identical positions and make complex hand movements together as if in a ballet. This is termed **movement synchrony**. **Movement complementarity** is paired movements that may not be identical but are still harmonious. For instance, one person talks and the other nods in agreement. You may observe a hand movement at the end of one person's statement that is answered by a related hand movement as the other takes the conversational "ball" and starts talking.

You may also observe discrepancies in nonverbal behavior. **Movement dissynchrony** occurs when a client is talking casually about a friend, for example, with one hand tightly clenched in a fist and the other relaxed and open, possibly indicating mixed feelings toward the friend. Lack of harmony in movement is common between people who disagree markedly or even between those who may not be aware they have subtle conflicts. You have likely seen this type of behavior in couples that you suspect have problems in communicating.

Some expert counselors and therapists deliberately "mirror" their clients. Experience shows that matching body language, breathing rates, and key words of the client can heighten interviewer understanding of how the client perceives and experiences the world. But be careful with deliberate mirroring. A practicum student reported that a client's nonverbal behavior seemed especially unusual. Near the end of the session, the client commented, "I know you guys; you try to mirror my nonverbal behavior. So I keep moving to make it difficult for you." You can expect that some clients will know as much about observation skills and nonverbal behavior as you do. What should you do in such situations? Use the skills and concepts in this book with *honesty* and *authenticity*. Talk with your clients about their observations of you without being defensive. Openness works!

▶ Observing Attending Behavior and Empathy in Action: Do I Want to Become a Counselor?

In the example interviews that follow, the client, Jared, is a first-semester sophomore exploring career choices. Like many reading this book, he is considering the helping professions. Jared has already stated that he is considering psychology as a major and counseling as a career field. The first example is designed to illustrate ineffective interviewing; it provides a sharp contrast with the second, more positive example. In addition to the verbal content, note that the observation of nonverbal client reactions provides a good indication of the level of empathic understanding.

Jerome does several things wrong in the first session, with clear impact on the client. Compare that with the more effective second session.

NEGATIVE EXAMPLE

INTERVIEWER AND CLIENT CONVERSATION	PROCESS COMMENTS
1. *Jerome:* The next thing on my questionnaire is your job history. Tell me a little bit about it, will ya?	The vocal tone is overly casual, almost uninterested. The interviewer looks at the form, not at the client. He is slouching in a chair. (Subtractive empathy)
2. *Jared:* Well, I guess the job that, uh . . .	Client stammers and appears a bit hesitant and unsure as to what to say or do.
3. *Jerome:* Hold it! There's the phone. [Long pause while Jerome talks to a colleague.] Uh, okay, okay, where were we?	The interview is for the client. Avoid phone calls and other interruptions during the session. Such behavior shows disrespect for Jared. If a break is necessary, remember what occurred just before you were called away. (Subtractive empathy)
4. *Jared:* The job that really comes to mind is my work as a camp counselor during my senior year of high school. I really liked counseling kids and . . .	The client's eyes brighten and he leans forward as he starts to talk about something he likes. Notice that nonverbal behavior tells where the client is emotionally and how he is responding in the here and now.

(continued)

INTERVIEWER AND CLIENT CONVERSATION	PROCESS COMMENTS
5. *Jerome:* [interrupts] Oh, yeah, I did a camp counselor job myself. It was at Camp Itasca in Minnesota. I wasn't that crazy about it, though. But I did manage to have fun. What else have you been doing?	When he is interrupted by the topic jump, Jared looks up with surprise, then casts his eyes downward, as if he realizes he is no longer the focal point of attention. It would have been better if Jerome had asked Jared for specifics about what he liked in his work as a camp counselor. (Subtractive empathy)
6. *Jared:* Ah . . . well, I . . . ah . . . wanted to tell you a little bit more about this counseling job, I . . .	Jared is interested in this topic and tries again, but notice the speech hesitations.
7. *Jerome:* [interrupts] I got that down on the form already, so tell me about something else. A lot of counseling types have done peer counseling in school. Have you?	The leading closed question puts Jared on the spot. (Subtractive empathy)
8. *Jared:* Ah . . . no, I didn't.	The topic jump now has Jared at a complete standstill. He stammers, hesitates, and looks puzzled and confused.
9. *Jerome:* Many effective counselors were peer helpers in school . . .	For the first time, Jerome sits up with direct eye contact, but he is now working on his agenda. One senses that he is not fond of his job and would rather be doing something else. (Subtractive empathy)
10. *Jared:* [interrupts] No, I lived in a small town. I never learned about peer counseling until I got to college.	Jared is now interrupting and sits forward and talks a little more loudly. His anger is beginning to show.

This sample interview is extreme, but not as rare as we might wish. It illustrates some of the many ineffective things an interviewer can do. Jerome clearly had poor visuals, vocals, verbal tracking, and body language.

Now let's give Jerome another chance.

POSITIVE EXAMPLE

INTERVIEWER AND CLIENT CONVERSATION	PROCESS COMMENTS
1. *Jerome:* Jared, so far, I've heard that you are interested in counseling as a career. You've liked your psychology courses and you find that friends come to you to talk about their concerns. In this next phase of the session, I'd like to review a form with you that may help us plan together. The next item concerns your job history. Could you tell me a little bit about the jobs you've had in the past?	Jerome leans forward slightly, maintains good eye contact, and his vocal tone is warm and friendly. He personalizes the interview by using the client's name. He summarizes what he has heard and tells the client what is going to happen next. He asks an open question to obtain information from Jared's point of view. (Interchangeable empathy)
2. *Jared:* Sure, the job that comes to mind is camp counselor at the YMCA during my senior year of high school and my first two years of college.	Jared's vocal tone is confident and relaxed. He seems eager to explore this area.
3. *Jerome:* Sounds like you really liked it. Tell me more.	Jerome observes Jared's enthusiastic words and nonverbals and encourages him to say more. (Additive, positive empathy)

(continued)

INTERVIEWER AND CLIENT CONVERSATION	PROCESS COMMENTS
4. *Jared:* Well, . . . when I was a kid people always asked me what I wanted to be when I grew up. I had no idea, but when I worked at this camp, I realized that I truly liked working with kids, and I was good at it.	Jared smiles and continues.
5. *Jerome:* Uh-huh.	Jerome smiles and continues good eye contact and encourages Jared to go on. (Additive empathy)
6. *Jared:* I like to work with people—maybe that's why I'm thinking of majoring in psychology or maybe human services or social work.	Jerome begins to see the relationship between past jobs Jared has held and possible future majors in college and career choices.
7. *Jerome:* I see. So you really liked helping kids in the camp. You think it might be a good career direction. Could you give me a specific example of what you particularly liked at the camp?	Jerome summarizes Jared's comments. Asking for specific examples moves client talk from general to specifics. (Interchangeable reflection) This is followed by a question focusing on strengths. (Additive, positive empathy)
	Jerome is identifying client strengths, which can serve as a basis for both decision making and problem solving. Jared continues to talk smoothly, and his posture and body movement indicate comfort.
9. *Jerome:* Sounds like you really feel good about your work there.	Jerome focuses on Jared's underlying feeling tone. (Interchangeable empathy, but focused on positives)
10. *Jared:* I feel real good because I helped them and we had lots of fun at the same time. I learned how to play the ukulele, and we had some great singing sessions. I thought, you know, that this could be a profession for me.	Jared continues with enthusiasm and verifies good feelings. If feeling tones are correctly identified, the client will often acknowledge these feelings and continue discussion in more depth. Whether a positive or negative feeling, you can anticipate more discussion on that topic.
11. *Jerome:* You felt really good and had fun with the kids, making you think counseling might be a good way to go. What else leads you that way?	Jerome pinpoints Jared's enthusiasm, paraphrases some of Jared's important words, and then asks an open *what else* question to encourage more exploration on the topic. (Interchangeable and additive empathy)
12. *Jared:* Well, kids came to me a lot to talk about their issues with their families and friends. I might even want to be a camp director. I also thought I might want to become a counselor to help people with real personal problems. I just know I want to work with people, but I'm not exactly sure how.	Throughout all this discussion Jerome and Jared have a comfortable relaxed relationship with a solid demonstration of the three V's + B. We now have a far better interviewer and client relationship and a basis for further discussion of majors and possible careers.

The focus in this second session is on attending to Jared and drawing out some early parts of his career story. We now have some positive ideas of where to go with this session. This time Jerome effectively demonstrates visuals, vocal qualities, verbal tracking, and culturally appropriate body language. The result is a comfortable conversation with free-flowing words and positive nonverbals.

EXERCISE 3.3 Reflection

This might be a good time for you to think about what led you to the study of interviewing and counseling.

 Interactive Exercise: Identification and Classification

► Practicing Attending, Empathy, and Observation Skills

EXERCISE 3.4 Obtaining Feedback on Your Vocal Tone

Ask friends to close their eyes and note your vocal qualities as you speak. Talk for a few minutes in your normal tone of voice on any subject of interest to you. How do they react to your tone, your volume, your speech rate, and perhaps even your accent? Ask them for feedback. What does this feedback say to you?

Then pick out a single sentence or phrase such as "I really like you" or "This dinner was terrific." Say the sentence using different tones of voice ranging from approval to disapproval. Tone determines meaning as much as or more than the actual words.

EXERCISE 3.5 Observing the Attending Behavior of Others

During the next few days, spend some time observing the 3 V's + B of your friends, family members, and others. Regarding others, it may be helpful to view a television show where you observe participants in action.

Visuals. Do they maintain eye contact more while talking or listening? What do the eyes communicate?

Vocal tone. Note changes in speech rate and volume. Where do they hesitate? Give special attention to the emotional tone underlying their words.

Verbals. Observe two people talking. How often does the topic change, and who leads the change? Who talks the most? What is their talk time ratio?

Body language. Note gestures, full body movements, patterns of breathing, and use of space. Pay special attention to the face—smiling, frowning. Sometimes you will see two emotions played out at one time.

EXERCISE 3.6 Observing Your Own Skills of Attending

Video record (using a camera, computer, tablet, or smartphone) an individual attending behavior practice session with a classmate, friend, or family member. It may be useful to demonstrate first negative attending and then positive, comparing the differences. What do you and the volunteer observe as you view your interview? You may wish to use the downloadable feedback form available on the web resources.

EXERCISE 3.7 Empathic Understanding

Review the same recorded interviews and assess the level of empathic understanding. Rate yourself and other interviewers you observe on the three-level empathy scale (subtractive, interchangeable, additive).

Feedback Form: Attending Behavior is available online. Use this form to obtain feedback and increase mastery of skills.

▶ Refining: Attending Behavior, Empathy, and Observation in a Multicultural Context

While there are many more dimensions that could be discussed here, we have selected a few key items that may be helpful in the future.

OPENING AND CLOSING CLIENT CONVERSATION

Skilled counselors and interviewers use attending skills to open and close client talk, making the most effective use of limited time in the interview. If a client talks insistently about the same topic over and over again, this can become an issue. Simply paying attention and staying on the topic may be nonproductive. Intentional nonattending may be useful to shift clients away from repetitive or negative topics.

The first and usually the most effective way to help a client move on further with her or his issues is to break in and paraphrase or summarize what has been said, thus indicating that the client has been heard. For example:

> Sondra, may I break in here? I'd like to summarize what I've heard you say and I want to make sure I have a full understanding. What I heard you say was . . . [here you continue with your summary].

If the client has told you the same story three times, at a minimum your repetition makes it four!

Clients need to tell their stories, and if the story is traumatic, extensive storytelling and repetition may be necessary. Be patient and continue to listen carefully in those cases. With children, the stories also can go on and on, and the same principles apply. Attending to their stories and repeating back to them key elements will enable them to feel heard and move on.

Assuming that the client has not been traumatized, after you summarize what you've heard, you can ask a question referring to another topic, usually one mentioned earlier in the interview. Or if you sense important specifics in the longer story, turn attention to those issues: "I heard . . . and that seems worth following up. Could you tell me more about . . . ?"

If you find yourself unable to help the client move on, look for your own failures. There is a real possibility that you are tired or even bored with the conversation. Are you maintaining eye contact? Are there subtle shifts in body posture? Perhaps your vocal tone is flat and uninterested—or even critical. The client, even though speaking, is still noticing your behavior and reactions. If you "tune out," the client will be aware.

In short, be patient, be kind, and listen fully!

THE VALUE OF SILENCE

You hear quite a bit by not saying a thing. Observing also made me a student of body language. I could look at people talking to each other and sense what was coming next, or if what they were saying was accurate. Listening and watching made me a rather patient and calm person.

—Piers Morgan, CNN

Sometimes the most useful thing you can do is to support your client silently. As a counselor, particularly as a beginner, you may find it hard to sit and wait for clients to think through what they want to say. Your client may be in tears, and you may want to give support through your words. However, sometimes the best support may be simply being with the person nonverbally and not saying anything. Consider offering a tissue, as even this small gesture shows you care without any need to say anything. In general, it's always good to have a box or two of tissues for clients to take even without asking or being offered.

Silence can be challenging and even frightening. After all, doesn't counseling mean talking about issues and resolving issues verbally? When you feel uncomfortable with silence, look at your client. If the client appears comfortable, draw from her or his ease and join in the silence. If the client seems disquieted by the silence, rely on your attending skills. Ask a question or make a comment about something relevant mentioned earlier in the session.

Finally, remember the obvious: *Clients can't talk while you do.* Review your interviews for talk time. Who talks more, you or your client? With most adult clients, the percentage of client talk time should be more than that of the interviewer. With less verbal clients or young children, however, the interviewer may need to talk slightly more than they do or tell stories to help clients verbalize. A 7-year-old child dealing with parental divorce may not say a word initially. But when you read a children's book on feelings about divorce, he or she may start to ask questions and talk more freely.

 Interactive Exercise: Basic Competence: Shera Looks at Careers

---

INDIVIDUAL AND MULTICULTURAL ISSUES IN NONVERBAL BEHAVIOR

Learning about multicultural difference requires respect for the individual and also for the critical importance of all aspects of the RESPECTFUL framework.

As you engage in observation, recall that each culture has a different style of nonverbal communication. For example, Russians say yes by shaking the head from side to side and no by moving the head up and down. Most Europeans do exactly the opposite. An old, but still highly relevant, study counted the average number of times in an hour friends of different cultural groups touched each other while talking in a coffee shop. The results showed that British friends in London did not touch each other at all, French friends touched 110 times, and Puerto Rican friends touched 180 times (cited in Asbell & Wynn, 1991).

BOX 3.3

BOX 3.3 Einstein, Little Square Things, and Idiots

Gary Wharton

Albert Einstein said, "I fear the day that technology will surpass our human interaction. The world will have a generation of idiots." When I read this quote, I wasn't quite sure what he meant.

But, I started to observe people of all ages, sizes, and shapes walking around holding little square things in their faces. So fascinated were they that they seemed to be oblivious to their surroundings. I saw people walking into other people. (I soon learned to get out of the way of these square-thing-watching creatures). I saw people walking into walls, trees, moving vehicles, and sometimes into places they shouldn't be. I observed a woman, concentrating mightily on her square thing, walk into the men's room. Within seconds, she was out of said restroom, sans square thing in her face, which was by now quite red.

CBS Sunday Morning showed a cafeteria where 99% of the customers were staring into square things. No one was talking to anyone. They were just staring into square things. Frightening.

I am a "talker." My family was full of talkers. When we got together, we talked. Today, no one talks. I watch a couple walk into a restaurant, sit down, turn on the square things, and speak nary a word to each other. The only time they talk is when they give the waiter their order. Sad.

Perhaps Einstein was a bit severe using the word "idiots." I don't think things will go that far, but I do have a concern when I see my children, my granddaughter, and my son-in-law sitting at the table staring at their square things. I would like to have a conversation with them, but I guess they find their square things more interesting than a conversation with me. Can you believe that?

The other night we were with a group of friends at a restaurant. Someone asked for the time. Instantly, many hands were grabbing for purses, square things on belts and square things in pockets. I calmly looked at my watch and said, "It's twenty minutes till eight." The rest of the evening I was viewed with looks of hostility.

However, both verbal and nonverbal communication appears to be changing due to the culture of cell phones and other electronic media. Touching may not be as common in any culture as it used to be. People who talk on cell phones are less likely to communicate in public. "Cell phones do have the potential to make us more focused on what's going on in our personal lives, and the personal lives of the people that we're close to. Maybe we're not paying enough attention to things outside that realm" (Campbell & Kwak, 2013). People on cell phones were found considerably less willing to break off the call and help others (Banjo, Hu, & Sundar, 2008). There are also indications that smiling at others and other positive nonverbal indicators may be decreasing.

Box 3.3 presents a more humorous view of the impact of cell phones on verbal and nonverbal communication.

There are many differences in nonverbal cultural behaviors. To name just two, smiling is a sign of warmth in most cultures, but in some situations in Japan, smiling may indicate discomfort. Eye contact may be inappropriate for traditional Navajo but highly appropriate and expected for a Navajo official who interacts frequently with European Americans.

Be careful not to assign your own ideas about what is "standard" and appropriate nonverbal communication. You will want to begin a lifetime of studying nonverbal communication patterns and their variations. In terms of counseling sessions, you will find that changes in style may be as important as, or more important than, finding specific meanings in communication style. Edward Hall's *The Silent Language* (1959) remains a classic reference for nonverbal communication, while Paul Ekman (2013) is the current leading authority.

You can also visit several useful websites devoted to nonverbal communication. Insert in your search engine the key words "nonverbal communication." The visuals available on the Internet will provide clearer examples of nonverbal communication than we can provide through the written word.

PRACTICING ATTENDING, EMPATHY, AND OBSERVATION SKILLS

EXERCISE 3.9 Personal Reflections on Attending Behavior, Empathy, and Nonverbal Communication

Please spend some time reflecting on the 3 V's + B. How does this compare with your own original definition of listening skills at the beginning of this chapter? How do you rate on your ability to listen to others? How empathic are you? Most likely, you had times when you really did sense how others were feeling and thinking. What was going on? Nonverbal communication issues have become more prominent in the media. What stood out for you in this chapter? What prior knowledge did you bring to this book that is important to add in your thinking about interviewing and counseling practice?

EXERCISE 3.10 Observing How Those Around You Either Attend or Fail to Attend

Focus on one of the following at a time to sharpen your awareness and skills in observation. First, note talk time and how people sometimes dominant a conversation. Can you find a balanced conversation between two people. What is the difference? What about verbal following? Topic jumps? Interruptions? What does each of these do to the conversation? What happens to non-verbal behavior when topic jumps and interruptions occur? Then single out visuals—eye contact patterns, vocal tone, and body language. What do you notice about empathic understanding? Can you distinguish subtractive, interchangeable, or additive comments?

EXERCISE 3.11 Recording an Individual Practice Session and Obtaining Feedback

Join with a classmate, or ask a friend or family member to work with you as a volunteer. Request that the person tell you a story, ideally a story that has some meaning to the storyteller. Video record the session (using a camera, computer, tablet, or smartphone), obtaining written or verbal permission as appropriate. Some possible topics include an enjoyable experience, ranging from a hike to travel to a date; interpersonal conflict; or a difficult decision. Draw out the story using attending behavior. After 5 to 10 minutes, conclude and seek verbal feedback from the volunteer. If possible, obtain written feedback as well, using a downloaded form from the web resources. Watch the recorded session, making note of both your own and the other person's behavior, giving special attention to empathic understanding.

▶ "Magic," Samurai, Neuroscience, and the Importance of Practice to Mastery and Competence

Samurai skilled sword movements seem magical to many observers. However, the Japanese masters of the sword learn their skills through a complex set of highly detailed training exercises. The process is broken down into specific components that are studied carefully, one at a time. Extensive and intensive practice is basic to a samurai. In this process of mastery, the trainee often suffers and finds handling the sword awkward at times. The skilled individual may even find performance worsening during the practice of single skills. Being aware of what one is doing can interfere with coordination and smoothness in the early stages.

Once the individual skills have been practiced and learned to perfection, the samurai retire to a mountaintop to meditate. They deliberately forget what they have learned. When they return, they find the distinct skills have been naturally integrated into their style or way of being. The samurai then seldom have to think about skills at all; they have become samurai masters.

Why is the samurai "magic," you may ask? It is not really magic; it is extensive and continued intentional practice of basic skills that gradually become a natural part of the person.

Baseball fans still believe that Ted Williams and Joe DiMaggio "had it in their genes." It is a bit different from that. Giftedness is now recognized as a scientific error, but that error is still promoted in the popular media. Natural talent does exist, but it needs to be developed and nurtured through careful practice. Expertise across all fields depends on persistence, intentional practice, and the search for excellence (Ericsson, Charness, Feltovich, & Hoffman, 2006).

You are in this course because of your natural talent and caring for others. But we as counselors, like the samurai, baseball, tennis and golf players, and even the Beatles, can only develop our skills fully with extensive intentional directed practice, coupled with evaluation by clients, self-examination, and continuous feedback from supervisors and colleagues.

Intentional practice is the "magic"! This means that you need to recognize and enhance your natural talents, but greatness only happens with extensive practice. Practice is the breakfast, lunch, and dinner of champions. Skipping practice means mediocre performance.

Learn the skills of this book, but allow yourself time to integrate these ideas into your own natural authentic being. It does not take magic to make a superstar, but it does require systematic and intentional practice to achieve full competence in interviewing, counseling, and psychotherapy.

Interactive Exercise: Observation of Attending Patterns in Daily Life
Group Practice Exercise: Group Practice With Attending Behavior
Interactive Exercise: Examining Your Own Verbal and Nonverbal Styles

Summarizing Key Points of "Attending, Empathy, and Observation Skills: Fundamentals of All Interviewing and Counseling Approaches"

Defining Attending Behavior

▶ Listening is the core of developing a relationship and making real contact with clients.
▶ Attending behavior is fundamental to interpersonal communication in all cultures.
▶ All theories and methods of interviewing, counseling, and psychotherapy rest on a base of attending behavior, which is necessary to build a working relationship or therapeutic alliance with the client.
▶ Neuroscience demonstrates that your brain "lights up" when you are listening.
▶ Demonstrating poor listening skills through role-play is an effective way to identify the importance of listening and the specific skills of attending behavior.
▶ Attending behavior involves the 3 V's + B: visuals, vocal qualities, verbal tracking, and body language.

Discerning: Basic Techniques and Strategies of Attending Behavior

▶ It is helpful to observe clients' verbal and nonverbal behavior in a culturally sensitive way.
▶ The 3 V's + B reduce interviewer talk time and provide clients with an opportunity to tell their stories in as much detail as needed.

Discerning: Individual and Multicultural Issues in Attending Behavior

▶ Listen before you leap!
▶ By attending and giving clients talk time, you demonstrate that you truly want to hear their story and allow them time to unfold their issues.
▶ Vocal qualities express emotions, and each client may interpret your voice differently.
▶ Verbal tracking is the skill of focusing on client topics and identifying topics of concern.
▶ Respect multicultural and individual characteristics in the client's body language. Remain authentic to your own style.
▶ Nonattention may help clients shift from negative, nonproductive topics to more appropriate conversation. There are times when silence is the best approach.

Defining Empathy and Empathic Understanding

▶ Empathy and empathic listening will help clients feel understood, engage in the counseling process, and gain a better understanding of themselves.

Discerning: Levels of Empathy

▶ Subtractive empathy takes away from client conversation.
▶ Interchangeable empathy catches the essence of what the client has said and usually encourages them to move on.
▶ Additive, positive empathy finds strengths and resources in the client. Questions are often labeled "potentially additive" because we must wait to see how the client responds to be sure how helpful the comment was.

Refining: The Power of the Positive

▶ The words we say produce expectations that have a major impact on clients' bodies and minds.
▶ Positive and negative expectations can improve or deteriorate clients' performance.

Defining Observation Skills

▶ Observing verbal and nonverbal behavior is critical to understanding the client. In addition, watch videos of your own patterns to assess your strengths and areas for improvement.

Discerning: Specifics of Observation

▶ Movement synchrony, movement complementarity, and mirroring may represent particularly effective moments and close relationships in the interview. Also, these movements provide information regarding the client's experience.
▶ Movement dissynchrony and noncommunicative body movements may indicate that there are issues not being discussed in the here and now of the interview or that the client is uncomfortable with you or the session.
▶ Multicultural variations in verbal and nonverbal behavior will require a lifetime of learning on your part.

Observing Attending Behavior and Empathy in Action: Do I Want to Become a Counselor?

▶ Be sensitive to clients' multicultural and individual characteristics. Never stereotype.
▶ When the interviewer demonstrates culturally appropriate attending skills, it shows interest in the client and promotes greater client talk time.
▶ Expect individual and cultural differences on the four key dimensions of attending.

Practicing Attending, Empathy, and Observation Skills

▶ Use the exercises in this section to improve your 3 V's + B and master observation and empathic understanding skills.

Refining: Attending Behavior, Empathy, and Observation in a Multicultural Context

▶ Use attending skills to open and close client talk.
▶ Value silence.
▶ Remember clients cannot speak while you are talking; give clients talk time.

"Magic," Samurai, Neuroscience, and the Importance of Practice to Mastery and Competence

▶ The samurai effect helps us understand the importance of practicing and assimilating each skill.
▶ Many highly demanding activities, including sports, driving, golf, dance, and music, are improved with single skills practice.
▶ Continued and intentional practice are needed to achieve mastery.
▶ The microskills approach breaks interviewer skills into single units. When learning single skills one by one, you may experience a temporary decrease in competence. However, continued intentional practice and experience develops competence and a natural integration of skills.

Assess your current level of knowledge and competence as you complete the chapter:

1. Flashcards: Use the flashcards to check your understanding of key concepts and facilitate memorization of key information.

2. Self-Assessment Quiz: The quiz will help you assess your current knowledge and prepare for course examinations.

3. Portfolio of Competencies: Evaluate your present level of competence on the ideas and concepts presented in this chapter using the Self-Evaluation Checklist. Self-assessment of your competencies demonstrates what you can do in the real world.

SECTION II
The Basic Listening Sequence and Organizing an Effective Interview

Without individually and culturally appropriate attending behavior, there can be no interviewing, counseling, or psychotherapy. Attending behavior is basic to all the communication skills of the Microskills hierarchy.

This section, which includes Chapters 4–7, presents the **basic listening sequence (BLS)** that will enable you to draw out the major facts and feelings central to client concerns. We suggest that you think of the BLS holistically. Used effectively and intentionally, these basic listening skills will, in many cases, be all that is necessary to facilitate client change. Furthermore, the influencing skills of the second half of this book will be more meaningful and effective if founded on the basic listening sequence, backed by appropriate cultural and individual attending.

Chapter 4, Questions: Opening Communication, provides specifics for drawing out client cognitions and emotions. Questions can help to "fill in the blanks" of client stories. Three key questions ensure that information is complete: "What else?" "Could you tell me more?" and, at the end of the session (or periodically), "What have we missed?"

Chapter 5, Encouraging, Paraphrasing, and Summarizing: Active Listening and Cognition, provides skills for helping clients bring out their stories concretely and specifically. By saying back to clients the essence of what they have said, you help them clarify their understanding and check out whether your listening was accurate. You will also learn their style of cognitive processing and be better able to match all other helping leads to individual differences. Effective use of these skills may be rewarded by enabling the client to continue the conversation at a new level. And you may receive a smile of recognition.

Chapter 6, Observing and Reflecting Feelings: The Heart of Empathic Understanding, may be considered a second anchor to attending. Emotions are experienced in the body and mind before cognitions. Cognitive decisions require an emotional base if they are to be implemented over the long run. The central role of the prefrontal cortex (PFC) is executive functioning, and its place in emotional regulation is key.

Chapter 7, How to Conduct a Five-Stage Counseling Session Using Only Listening Skills, focuses on holistic integration of the basic listening sequence. Now that you have become competent in the listening skills, you will be able to complete a full, well-structured interview. Being able to work through a full session using only listening skills may be sufficient in itself. If not, then you and the client have a joint understanding that will make the use of influencing skills and theory more meaningful and long lasting.

Neuroscience offers two key ideas to enrich the BLS: executive functioning and emotional regulation. **Executive functioning** refers to the ability of the prefrontal cortex to organize cognition and make decisions. **Emotional regulation** relates to reflection of feelings and the brain's limbic system, but it is also a key part of executive functioning. Emotions fuel and color decisions, and they are activated before cognitions (Collura, Zalaquett, Bonnstetter, & Chatters, 2014). Emotions need executive organization to either "tighten up" or "loosen" to help the client achieve either more controlled or more relaxed behavior.

With sufficient practice, you can reach the following competency objectives:

1. Master the basic listening sequence (using questions, encouraging, paraphrasing, summarizing, and reflecting feelings) to draw out the cognitions, emotions, and behaviors relevant to clients' concerns.
2. Observe clients' reactions to your skill usage and intentionally modify your listening skills so that you are with them more fully in the here and now.
3. Conduct a complete interview using only listening and observing skills.

When you have accomplished these tasks, you may find that your clients have a surprising ability to resolve their own issues and challenges without further intervention on your part. You may also gain a sense of confidence in your own ability as an interviewer.

When in doubt, listen!

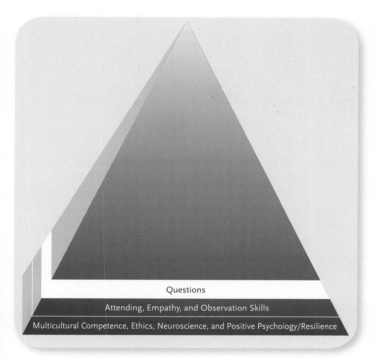

Questions

Attending, Empathy, and Observation Skills

Multicultural Competence, Ethics, Neuroscience, and Positive Psychology/Resilience

Chapter 4
Questions
Opening Communication

Asking the right questions takes as much skill as giving the right answers.

—Robert Half

Your mind will answer most questions if you learn to relax and wait for the answer.

—William S. Burroughs

Questions can make the interview flow and bring out client goals and other critical information that otherwise might be missed. With questions, however, the leadership comes mainly from the interviewer. The client often responds using your frame of reference. Questions potentially can take away from client self-direction.

Chapter Goals

Awareness, knowledge, skills, and actions developed through the concepts of this chapter and this book will enable you to:

▶ Enrich client stories by bringing out a more complete description, including important details.

▶ Choose questions that are most likely to enlarge and clarify the story.

▶ Open or close client talk intentionally, according to the individual needs of the interview.

▶ Adapt questioning style to meet the needs of clients with varying RESPECTFUL backgrounds.

Assess your current level of knowledge and competence as you begin the chapter:

1. Self-Assessment Quiz: The chapter quiz will help you determine your current level of knowledge. You can take it before and after reading the chapter.

2. Portfolio of Competencies: Before you read the chapter, please fill out the Self-Evaluation Checklist to assess your existing knowledge and competence on the ideas and concepts presented in this chapter. Then, at the end of the chapter, complete the checklist again to summarize your competencies after study and practice.

▶ Defining Questions

Questions are an essential part of most theories and styles of helping, particularly brief counseling, career counseling, and cognitive behavioral therapy. The employment counselor facilitating a job search, the social worker conducting an assessment interview, and the high school guidance counselor helping a student work on college admissions all need to use questions. Moreover, the diagnostic process, while not counseling, uses many questions.

Two basic types of questions are open and closed.

Open questions are those that can't be answered in a few words. They tend to facilitate deeper exploration of client issues. They encourage others to talk and provide you with maximum information. Typically, open questions begin with *what*, *how*, *why*, or *could*. For example, "Could you tell me what brings you here today?" However, you will find that a closed question that is especially significant will draw out extensive data as well.

Closed questions enable you to obtain specifics and can usually be answered in very few words. They may provide critical information, but the burden of guiding the talk remains even more on the interviewer. Use too many and clients may close up. Closed questions often begin with *is*, *are*, or *do*. For example, "Are you living with your family?"

> Keneki is nearing completion of his junior year in high school. The school requires that each student be interviewed about plans after graduation—work, technical school, the armed forces, or college. You are the high school counselor and have called Keneki in to check on his future plans. He is known as a "nice boy." His grades are average, placing him in the middle third of his class, and he is not particularly verbal or talkative.
>
> What are some questions that you could use to draw him out and help him think ahead to the future? What would you anticipate achieving by using these questions? If you ask too many questions, what potential problems will you face?
>
> You may want to compare your questions with our thoughts on page 90 at the end of this chapter.

Interactive Exercise: Open vs. Closed Questions
Interactive Exercise: Basic Competence

Following are some of the results you can expect when using open and closed questions.

OPEN QUESTIONS	ANTICIPATED CLIENT RESPONSE
Open questions often begin with *who, what, when, where,* or *why*. Closed questions may start with *do, is,* or *are*. *Could, can,* or *would* questions are considered the most open, and have the advantage of giving more power to the client, who can more easily say, "No, I can't" or provide a very brief comment.	Clients will give more detail and talk more in response to open questions. Effective questions encourage more focused client conversations with more pertinent detail and less wandering.

CLOSED QUESTIONS	ANTICIPATED CLIENT RESPONSE
Closed questions often start with *is, are, do*, but also can start with other stems that produce the same result.	Closed questions are effective for obtaining specific facts and information, but used too extensively, they may turn off the client. Rather than "Do you use drugs?" try "Could you share with me any experience with recreational drugs (or pain-killing medications)?"

▶ Discerning the How of Using Questions

Questions can be facilitative, or they can be so intrusive that clients want to say nothing. Use the ideas presented here to help you define your own questioning techniques and strategies and how questioning fits with your natural interviewing style.

QUESTIONS HELP BEGIN THE INTERVIEW

With verbal clients and a comfortable relationship, the open question facilitates free discussion and leaves plenty of room to talk. Here are some examples:

"What would you like to work on (or explore, discuss, talk about) today?"

"Could you tell me what prompted you to see me?"

"How have things been since we last talked together?"

"The last time we talked you planned to talk with your partner about your sexual difficulties. How did it go this week?"

The first three open questions provide room for the client to talk about virtually anything. The last question is open but provides some focus for the session, building on material from the preceding week. These types of questions will work well for a highly verbal client. However, such open questions may be more than a nontalkative client can handle. It may be best to start the session with more informal conversation—focusing on the weather, a positive part of last week's session, or a current event of interest to the client. You can turn to the issues for this session as the client becomes more comfortable.

THE FIRST WORD OF QUESTIONS MAY DETERMINE HOW A CLIENT RESPONDS

Question stems often, but not always, result in predictable outcomes. *What* questions most often lead to facts.

"What happened?"

"What are you going to do?"

How questions may lead to an exploration of process or feeling and emotion.

"How could that be explained?"

"How do you feel about that?"

Why questions can lead to a discussion of reasons.

"Why did you allow that to happen?"

"Why do you think that is so?"

Use *why* questions with care. While understanding reasons may have value, a discussion of reasons can also lead to sidetracks. In addition, many clients may not respond well because they associate *why* with a past experience of being grilled. You could also reword the *why* questions above: "*What* happened inside you that allowed that to happen?"; "Please tell me *how* you came to think that is so." Any of the *could*, *can*, or *would* questions are also alternatives.

Could, *can*, or *would* questions are considered maximally open and also contain some advantages of closed questions. Clients are free to say, "No, I don't want to talk about that." *Could* questions offer more control for the client.

"Could you tell me more about your situation?"

"Would you give me a specific example?"

"Can you tell me what you'd like to talk about today?"

Give it a try and you'll be surprised to see how effective these simple guidelines can be.

 Interactive Exercise: Observation of Questions in Your Daily Interactions

OPEN QUESTIONS HELP CLIENTS ELABORATE AND ENRICH THEIR STORY

A beginning interviewer often asks one or two questions and then wonders what to do next. Even more experienced interviewers can find themselves hard-pressed to know what to do next. To help the session start again and keep it moving, ask an open question on a topic the client presented earlier in the interview.

"Could you tell me more about that?"

"How did you feel when that happened?"

"Given what you've said, what would be your ideal solution to the problem?"

"What might we have missed so far?"

"What else comes to your mind?"

One of the following often accomplishes the same result.

"Tell me more . . ."

"And? . . ."

At times, silence with an open supportive stance.

QUESTIONS CAN REVEAL CONCRETE SPECIFICS FROM THE CLIENT'S WORLD

The model question "Could you give me a specific example?" is one of the most useful open questions available to any interviewer. Many clients tend to talk in vague generalities; specific, concrete examples enrich the interview and provide data for understanding action. Suppose, for example, that a client says, "Ricardo makes me so mad!" Some open questions that aim for concreteness and specifics might be:

"Could you give me a specific example of what Ricardo does?"

"What does Ricardo do, specifically, that brings out your anger?"

"What do you mean by 'makes me mad'?"

"Could you specify what you do before and after Ricardo makes you mad?"

Closed questions can bring out specifics as well, but even well-directed closed questions may take the initiative away from the client. However, at the discretion of the interviewer, closed questions may prove invaluable.

"Did Ricardo show his anger by striking you?"

"Does Ricardo tease you often?"

"Is Ricardo on drugs?"

If working with teens, a specific useful question, often more informative than those above, is to change the above to an open question, "Could you tell me if any of your friends are on drugs?" One of the best predictors that a client is using or abusing drugs or medications is to ask about what the peer group is doing. Questions like these may encourage clients to say out loud what they have only hinted at before.

THE BRAIN AND QUESTIONS: WHAT IS THE APPROPRIATE BALANCE OF FOCUSING ON DIFFICULTIES VERSUS STORIES OF STRENGTH?

Stories presented in the helping interview are often negative and full of serious challenges. Carl Rogers, the founder of client-centered counseling, was always able to find something positive in the client. He considered positive regard and respect for the client essential for future growth. This is illustrated clearly in neuroscience research. Negative emotions and feelings are located primarily in the amygdala, deep in the brain. Positive emotions (developed later in human evolution) are located in many areas, but the nucleus accumbens sends out signals to the prefrontal cortex, enabling focus on the positive (Ratey, 2008). Research suggests it takes at least five positive comments to balance one negative; it can take many more if the "neggie" hits deep (Schwartz Gottman & Gottman, 2014).

If it is the put-down experience that is more powerful, think of how long it might take to recover from these (some of these we never forget):

Parent: How could you do that? You are the worst child than I have seen. I won't speak to you until you apologize.

Peers at school bullying: You're a sissy, a freak, ugly, you're a whore, fatso, string bean, you act like a girl/boy, you can't play with us, now we are all going to beat you up.

Partner: I see a muffin top. Why are you so messy/smelly? Why can't you get along with anyone? Making love with you is like . . . (fill in the blank).

Simple, but early in the day from a friend: Your hair looks funny.

Our lives are full of challenges and the repeated hassle of thoughtless comments and put-downs can result in permanent memories that you will encounter in your counseling practice. This makes positive psychology, wellness, and therapeutic lifestyle changes all the more central to interviewing, counseling, and therapy.

Thus, we strongly recommend that every session include time spent on strengths and positives. "What's new and good?" is used in some counseling theories to start the session. Likely more helpful is your ability to search out hidden strengths that helped the person survive difficulties and trauma. Commenting periodically on client's abilities and drawing out stories of strength are important in counteracting the inevitable negatives we experience daily. Don't forget spiritual and family strengths.

POTENTIAL ISSUES WITH QUESTIONS

Questions can have immense value in the interview, but we must not forget how they could interfere with the relationship.

Bombardment/grilling. Too many questions may give too much control to the interviewer and tend to put clients on the defensive.

Multiple questions. Another form of bombardment, throwing out too many questions at once may confuse clients. However, it may enable clients to select which question they prefer to answer.

Questions as statements. Some interviewers may use questions to sell their own points of view. "Don't you think it would be helpful if you studied more?" This question clearly puts the client on the spot. On the other hand, "What occurs to you as you think about improving your grades?" might be helpful to get some clients thinking in new ways. Consider alternative and more direct routes of reaching the client. A useful standard is this: If you are going to make a statement, do not frame it as a question.

***Why* questions.** *Why* questions can put interviewees on the defensive and cause discomfort. As children, most of us experienced some form of "Why did you do that?" Any question that evokes a sense of being attacked can create client discomfort and defensiveness.

A CROSS-CULTURAL EXAMPLE OF HOW QUESTIONS MAY BE INAPPROPRIATE

If your life background and experience are similar to your client's, you may be able to use questions immediately and freely. If you come from a significantly different cultural background, your questions may be met by distrust and given only grudging answers. Questions place power with the interviewer. A poor client who is clearly in financial jeopardy may not come back for another interview after receiving a barrage of questions from a clearly middle-class interviewer. If you are African American or European American and working with an Asian American or a Latino/a, an extreme questioning style can produce mistrust. If the ethnicities are reversed, the same problem can occur.

Allen was conducting research and teaching in South Australia with Aboriginal social workers. He was seeking to understand their culture and their special needs for training. Allen is naturally inquisitive and sometimes asks many questions. Nonetheless, the relationship between him and the group seemed to be going well. But one day, Matt Rigney, whom Allen felt particularly close to, took him aside and gave some very useful corrective feedback:

> You White fellas! . . . Always asking questions! My culture considers many questions rude. But I know you, and that's what you do. But this is what goes on in my mind when you ask me a question. First, I wonder if I can trust you enough to give you an honest answer. Then, I realize that the question you asked is too complex to be answered in a few words. But I know you want an answer. So I chew on the question in my mind. Then, you know what? Before I can answer the first question, you've moved on to the next question!

Allen was lucky he had developed enough trust and rapport that Matt was willing to share his perceptions. Many People of Color have said that this kind of feedback represents how they feel about interactions with White people. People with disabilities, gays/lesbians/bisexuals/questioning/transgendered people, spiritually conservative persons, and many others—anyone, in fact—may be distrustful of the interviewer who uses too many questions.

On the other hand, questions can be useful in group discussions to help at-risk youth redefine themselves in a more positive way, as suggested in Box 4.1.

BOX 4.1 National and International Perspectives on Counseling Skills

Using Questions With At-Risk Youth

Courtland Lee, Past President, American Counseling Association, University of Malta

Malik is a 13-year-old African American male who is in the seventh grade at an urban junior high school. He lives in an apartment complex in a lower middle class neighborhood with his mother and 7-year-old sister. Malik's parents have been divorced since he was 6, and he sees his father very infrequently. His mother works two jobs to hold the family together, and she is not able to be there when they come home from school.

Throughout his elementary school years, Malik was an honor roll student. However, since he started junior high school, his grades have dropped dramatically and he expresses no interest in doing well academically. He spends his days at school in the company of a group of seventh- and eighth-grade boys who are frequently in trouble with school officials.

This case is one that is repeated among many African American early teens. But this problem occurs among other racial/ethnic groups as well, particularly those who are struggling economically. And the same pattern occurs frequently even in well-off homes. Many teens are at risk for getting in trouble or using drugs.

While still a boy, Malik has been asked to shoulder a man's responsibilities, picking up things his mother can't do. Simultaneously, his peer group discounts the importance of academic success and wants to challenge traditional authority. And Malik is making the difficult transition from childhood to manhood without a positive male model.

I've developed a counseling program designed to empower adolescent Black males that focuses on personal and cultural pride. The full program focuses on the central question, *"What is a strong Black man?"* While this question is designed for group discussion, it is an important one for adolescent males in general, who might be engaging in individual work. The idea is to use this question to help the youth redefine, in a more positive sense, what it means to be strong and powerful. Some of the related questions that I find helpful include:

What makes a man strong?

Who are some strong Black men that you know personally? What makes these men strong?

Do you think that you are strong? Why?

What makes a strong body?

Is abuse of your body a sign of strength?

Who are some African heroes or elders that are meaningful to you? What did they do that made them strong?

How is education strength?

What is a strong Black man?

What does a strong Black man do that makes a difference for his people?

What can you do to make a difference?

Needless to say, you can't ask an African American adolescent or a youth of any color these questions unless you and he are in a positive and open relationship. Developing sufficient trust to ask these challenging questions may take time. You may first have to get out of your office and into the school and community to become a person of trust.

My hope for you as a professional counselor is that you will have a positive attitude when you encounter challenging adolescents. They are seeking models for a successful life, and you may become one of those models yourself. I hope you think about establishing group programs to facilitate development and that you'll use some of these ideas with adolescents to help move them toward a more positive track.

► Observing Questions in Action: Interview Transcript, Conflict at Work

Virtually all of us have experienced conflict on the job—angry or difficult customers, insensitive supervisors, lazy colleagues, or challenges from those we supervise. In the following set of transcripts, we see an employee assistance counselor, Jamila, meeting with Kelly, a junior manager who has a conflict with Peter. The first session illustrates how closed questions can bring out specific facts but can sometimes end in leading the client, even to the point of putting the counselor's ideas into the client's mind.

CLOSED QUESTION EXAMPLE

INTERVIEWER AND CLIENT CONVERSATION	PROCESS COMMENTS
1. Jamila: Hi, Kelly. What's happening with you today?	Jamila has talked with Kelly once in the past about difficulties she has had in her early experiences supervising others for the first time. She begins the session with an open question that could also be seen as a standard social greeting.
2. Kelly: Well, I'm having problems with Peter again.	Jamila and Kelly have a good relationship. Not all clients are so ready to discuss their issues. More time for developing rapport and trust will be necessary for many clients, even on return visits. Her body is tense, what's going on?
3. Jamila: Is he arguing with you?	She asks a closed question, thus defining the issue without hearing Kelly's latest story about Peter. This is Level 1 subtractive, as Jamila is defining the situation, rather than "going with" the client.
4. Kelly: [hesitates] Not really; he's so difficult to work with.	Kelly sits back nervously in her chair and waits for the interviewer to take the lead.
5. Jamila: Is he getting his work in on time?	Jamila seeks to diagnose the situation by asking a series of closed questions. She is defining the story for her client—and, in this process may miss what is really happening. (Level 1 subtractive empathy)
6. Kelly: No, that's not the issue. He's even early.	Jamila looks puzzled. It is clear that Jamila lacks empathy at this point. Stress is building for the client.
7. Jamila: Is his work decent? Does he do a good job?	Jamila is starting to grill Kelly with more closed questions. (Subtractive empathy)

The interview continues and Kelly succumbs to Jamila's version of her story. *Asking the right questions takes as much skill as giving the right answers.*

Closed questions can overwhelm clients and can be used to force them to agree with the interviewer's ideas. Although this example seems extreme, encounters like this are too common in daily life and in interviewing and counseling sessions. Think about some of your supervisors or teachers—sound familiar? There is a power differential between clients and counselors, which we all too often forget.

OPEN QUESTION EXAMPLE

The interview is for the client, not the interviewer. Using open questions, Jamila learns about Kelly's situation. Paraphrasing William S. Burroughs statement at the chapter beginning: (The client's) mind will answer most questions if you learn to relax and wait for the answer.

INTERVIEWER AND CLIENT CONVERSATION	PROCESS COMMENTS
1. Jamila: Hi, Kelly. What's happening with you today?	Jamila uses the same easy beginning as in the closed question example. She has excellent attending skills and is good at relationship building.
2. Kelly: Well, I'm having problems with Peter again.	Kelly responds in the same way as in the first demonstration.
3. Jamila: More problems? Could you share more with me about what's been happening lately?	Open questions beginning with "could" provide some control to the client. Potentially a "could" question may be responded to as a closed question and answered with "yes" or "no." But in the United States, Canada, and other English-speaking countries, it usually functions as an open question. (Level 2 interchangeable empathy)
4. Kelly: This last week Peter has been going off in the corner with Daniel, and the two of them start laughing. He's ignoring most of our staff, and he's been getting under my skin even more lately. In the middle of all this, his work is fine, on time, and near perfect. But he is so impossible to deal with.	We are hearing Kelly's story. The anticipation is that that Kelly will respond to open questions with information. She provides an overview of the situation and shares how it is affecting her. Increasing body tension is apparent as Kelly looks down and tightens her fist.
5. Jamila: I hear you. Peter is getting even more difficult and seems to be affecting your team as well. It's really stressing you out and you look upset. Is this pretty much how you are feeling about things?	When clients provide lots of information, we need to ensure hearing them accurately. Jamila summarizes what has been said and acknowledges Kelly's emotions. The closed question at the end is termed a **perception check** or **checkout**. Periodically checking with your client can help you in two valuable ways: (1) It communicates to clients that you are listening and encourages them to continue. (2) It allows the client to correct any wrong assumptions you may have. (Level 3 additive empathy)
6. Kelly: That's right. I really need to calm down.	In the here and now of the empathic session, we see Kelly's body relaxing a bit and the stress level decreasing. Observing body changes in the session makes a difference in your effectiveness.
7. Jamila: Let's change the pace a bit. Could you give me a specific example of an exchange you had with Peter last week that didn't work well?	Jamila asks for a concrete example. Specific illustrations of client issues are often helpful in understanding what is really occurring. (Potentially additive empathy)

(continued)

INTERVIEWER AND CLIENT CONVERSATION	PROCESS COMMENTS
8. Kelly: Last week, I asked him to review a bookkeeping report prepared by Anne. It's pretty important that our team understand what's going on. He looked at me like "Who are you to tell *me* what to do?" But he sat down and did it that day. Friday, at the staff meeting, I asked him to summarize the report for everyone. In front of the whole group, he said he had to review this report for me and joked about me not understanding numbers. Daniel laughed, but the rest of the staff just sat there. He was obviously trying to get me. I just ignored it. But that's typical of what he does.	Specific, concrete examples can be representative of recurring problems. The concrete specifics from one or two detailed stories provide a better understanding of what is really happening. Now that Jamila has heard the specifics, she is better prepared to be helpful. However, as Kelly tells this story, we again see her showing more body tension. Remember the maxim: Keep calm and carry on. When tension and stress predominate, our thoughts, feelings, and behaviors can be dysfunctional.
9. Jamila: I see and understand your tension. Underneath it all, you're furious. Let's stop for just a moment. I know that you have many strengths. Could you tell me about a time when you have worked through some challenging issues?	Jamila's observation of nonverbal behavior is a dimension of additive empathy. The change of pace is to ensure that Kelly is aware that we often find resolution to our issues from our strengths and resources. (The change of pace is potentially additive.)
10. Kelly: Hummm (pause). Before he came aboard, all was smooth and the staff worked together very well. I used to go around several times a week talking to them individually. After I was put down by Peter, I kind of froze and retreated to my office.	Here we learn that talking with staff regularly is a skill that is not currently being used. We also see that Kelly has forgotten this relationship skill she already has.
11. Jamila: I see. . . . I hear you saying that things were *smooth* but you *froze* and stopped talking to the staff, which apparently was important in your *success*. What does that say to you?	This is a Level 2 interchangeable empathic paraphrased response with Kelly's main words. But the word *success* is Jamila's, which is potentially a Level 3 additive response. What happens next?
12. Kelly: Oh . . . I get it. Maybe I am paying too much attention to Peter and need to return my focus to my staff and perhaps this will be a beginning.	Here we see the client finding her own strength and positive assets. Listen and clients will often find their own solutions, although skilled questioning helps.

We see here the use of positive psychology and a strength-based approach. We resolve our issues best from what we can do, rather than what we can't do. The next several exchanges between Jamila and Kelly elaborate strengths. We see Jamila relax more fully, and her tone of voice shows less anxiety and tension.

13. Jamila: Could you tell me why you froze and didn't use the strengths that we know you have?	Here is the *why* question, potentially additive or subtractive, but it does put Kelly on the spot. It may or may not be helpful.
14. Kelly: (pause) Gee, I don't know why.	Sometimes the search for reasons via *why* works well, but not always.
15. Jamila: **Gender** can be an issue; men do put women down at times. Would you be willing to consider that possibility?	Jamila carefully presents her own hunch. But instead of expressing her own ideas as truth, she offers them tentatively with a "would" question and reframes the situation as a "possibility." (Potentially, this could be additive or subtractive empathy—we need to see what happens next.)

INTERVIEWER AND CLIENT CONVERSATION	PROCESS COMMENTS
16. Kelly: Jamila, it makes sense. I've halfway thought of it, but I didn't really want to acknowledge the possibility. But it is clear Peter has taken Daniel away from the team. Until Peter came aboard, we worked together beautifully. [pause] Yes, it makes sense for me. I think he's out to take care of himself. I see Peter going up to my supervisor all the time. He talks to the female staff members in a demeaning way. Somehow, I'd like to keep his great talent on the team, but how when he is so difficult?	With Jamila's help, Kelly is beginning to gain a broader perspective. She thinks of several situations indicating that Peter's ambition and sexist behavior need to be addressed. (Jamila's comment turns out to be additive. However, if Kelly had not accepted her idea, it would have been subtractive, taking the client off course.)
17. Jamila: So, the problem is becoming clearer. You want a working team, and you want Peter to be part of it. Well, let us continue. Let's focus on how you can use your strengths and talents, as well as some wonderful resources among your staff.	Jamila provides support for Kelly's new frame of reference and ideas for where the interview can go next. She suggests that time needs to be spent on using positive assets and wellness strengths. Kelly can best resolve these issues if she works from a base of resources and capabilities. (Level 3 additive empathy)

In this example, we see that Kelly has been given more talk time and empathic support. The questions focused on concrete specifics to clarify and expand the story. The positive asset search is a central part of successful questioning. Issues are best resolved when we bring in strengths as well as listen to difficulties.

Questioning is a most helpful skill, but do not forget the dangers of using too many.

► Practicing Questioning Skills

EXERCISE 4.1 Intentional Prediction
This exercise helps you examine intentional prediction. Your task is to read Jamila and Kelly's interview and check out what type of statement the client made following the interviewer's question.

EXERCISE 4.2 Practicing With Open Versus Closed Questions
Write open and closed questions to elicit further information from five different clients. Include in your questions a search for positive strengths and resources.

Suzanne, who has just discovered that she is pregnant.

Tabitha, a high school student, who has done poorly on college aptitude tests.

Alizon, 8 years old, who has been bullied on the playground.

Mr. and Mrs. Olson, who have just learned that Mr. Olson has cancer.

Martine, whose has a girlfriend from another cultural group and his parents are strongly objecting to the relationship.

Can you ask closed questions designed to bring out specifics of the situation? Can you use open questions to facilitate further elaboration of the topic, including the facts, feelings, and possible reasons? What special considerations might be important with each person as you consider age-related multicultural issues?

EXERCISE 4.3 Test Out Questions in Your Daily Interactions
Apply the basic question stems *what, how, why*, and *could* in your daily activities, and observe how clients respond differently to each.

EXERCISE 4.4 Record an Individual Practice Session and Obtain Feedback (audio, video, computer, or smartphone)
Join with a classmate, or ask a friend or family member to work with you as a volunteer. Regardless of gender or age, ask the volunteer to role-play the clients described above. Draw out the issues, concerns, or stories primarily through questioning. In addition, be sure to draw out positive strengths and stories, and notice how that affects the client and possibly changes nature of the session.

EXERCISE 4.5 Group Practice With Feedback (audio, video, computer, or smartphone)
Especially effective is meeting with a group of three or four, taking turns as interviewer and client. One member can run the recording equipment and observe. The second observer takes notes with an emphasis on the questioning skills of reflection of feeling. Both observers will be helpful in pointing out nonverbal behavior. In a group such as this, the downloaded forms from the web resources will be especially helpful.

 Feedback Form: Questions

▶ Refining: Extensions of the Questioning Skill

Some questions are essential at times: "What else?" "What might we have missed today?"

Clients do not always provide you with necessary information, and sometimes the only way to get at missing data is by asking questions. For example, the client may talk about being depressed and unable to act. As a helper, you could listen to the story carefully but still miss underlying issues relating to the depression. You could ask an open question: "What is happening in your life right now or with your family?" The client's answer might tell you that a separation or divorce is impending, that a job has been lost, or that there is some other significant dimension underlying the concern. What you first interpreted as depression becomes modified as a shorter-term issue, and the conversation takes a different direction.

An incident in Allen's life—when his father became blind after open heart surgery—illustrates the importance of questions. Was the blindness a result of the surgery? No; it was because the physicians failed to ask the basic open question "Is anything else happening physically or emotionally in your life at this time?" If that question had been asked, the physicians would have discovered that Allen's father had developed severe and unusual headaches the week before surgery was scheduled, and they could have diagnosed an eye infection that is easily treatable with medication.

In counseling, a client may speak of tension, anxiety, and sleeplessness. You listen carefully and believe the problem can be resolved by helping the client relax and plan changes in her work schedule. However, you ask the client, "What else is going on in your life?" In response the client shares a story of sexual harassment, and the goals of the session change.

Finally, at the close of a session, consider asking your client something like "What else should we have discussed today?" or "What have we missed today?" Another good question that will give you feedback on the session and some surprising answers is "What one thing stood out for you today from our conversation?" Often the answers to these questions open up new avenues that need to be considered in the future.

USING OPEN AND CLOSED QUESTIONS WITH LESS VERBAL CLIENTS

Generally in the interview, open questions are preferred over closed questions. Yet it must be recognized that open questions require a verbal client, one who is willing to share with you. Here are some suggestions to encourage clients to talk with you more freely.

Build trust at the client's pace. A central issue with hesitant clients is trust. Extensive questioning too early can make trust building a slow process with some clients. If the client is required to meet with you or is culturally different from you, he or she may be less willing to talk. Trust building and rapport need to come first, and your own natural openness and social skills are essential. With some clients, trust building may take a full session or more.

Search for concrete specifics. Some interviewers and many clients talk in vague generalities. We call this "talking high on the abstraction ladder." This may be contrasted with concrete and specific language, where what is said immediately makes sense. If your client is talking in very general terms and is hard to understand, it often helps to ask, "Could you give me a *specific, concrete* example?"

As the examples become clearer, ask even more specific questions: "You said that you are not getting along with your teacher. What specifically did your teacher say (or do)?" Your chances for helping the client talk will be greatly enhanced when you focus on concrete events in a nonjudgmental fashion, avoiding evaluation and opinion. Following are some examples of concrete questions focusing on specifics.

Draw out the linear sequence of the story: "What happened first? What happened next? What was the result?"

Focus on observable concrete actions: "What did the other person say? What did he or she do? What did you say or do?"

Help clients see the result of an event: "What happened afterward? What did you do afterward? What did he or she do afterward?" Sometimes clients are so focused on the event that they don't yet realize it is over.

Focus on emotions: "What did you feel or think just before it happened? During? After? What do you think the other person felt?"

Note that each of these questions requires a relatively short answer. These types of open questions are more focused and can be balanced with some closed questions. Do not expect your less verbal client to give you full answers to these questions. You may need to ask closed questions to fill in the details and obtain specific information. "Did he say anything?" "Where was she?" "Is your family angry?" "Did they say 'yes' or 'no'?"

A *leading* closed question is dangerous, particularly with children. In earlier examples, you have seen that a long series of closed questions can bring out the story, but may provide only the client's limited responses to *your* questions rather than what the client really thought or felt. Worse, the client may end up adopting your way of thinking or may simply stop coming to see you.

Case Study: Questions Drawing Out Strengths With Difficult Clients
Group Practice Exercise: Group Practice With Open and Closed Questions

► Our Thoughts About Keneki (From p. 78)

We would likely begin by asking Keneki what he is thinking about his future after he completes school. We would start with informal conversation about current school events, or something personal we know about Keneki. The first question might be, "You'll soon be starting your senior year; what do you see ahead?" Likely he will talk about his feelings and thoughts about the coming year, but soon we predict that discussion about future plans would come up. We would encourage him to elaborate on his responses and explore possibilities.

If Keneki does not bring up future plans on his own, we'd likely use some of the following questions:

"Beyond meeting class requirements, what are some of your plans for after graduation?"

"How can I help you achieve your goals?"

"What have you done so far about reaching those goals?"

If he focuses on indecision about volunteering for the army, enrolling in a local community college, or attending the state university, we might ask him some of the following questions:

"What about each of these appeals to you?"

"Could you tell me about some of your strengths?"

"If you went to college, what might you like to study?"

"How do finances play a role in these decisions?"

"Are there any negatives about any of these possibilities?"

"How do you imagine your ideal life 10 years from now?"

On the other hand, Keneki just might look to you for guidance and say, "I don't know, but I guess I better start thinking about it." We might ask him to review his past likes and dislikes for possible clues to the future. Out of these questions, we might see patterns of ability and interest that suggest actions for the future.

"What courses have you liked best in high school?"

"What have been some of your activities?"

"Could you tell me about the jobs you've had in the past?"

"Tell me about your hobbies and what you do in your spare time."

"What gets you most excited and involved?"

"What did you do that made you feel most happy in the past year?"

If Keneki is uncomfortable in the counseling office, all of these questions might put him off. He might feel that we are grilling him and perhaps even see us as intruding in his world. Usually, getting this type of information and organizing it requires the use of questioning. But questions are only effective if you and the client have a good relationship and are working together.

Summarizing Key Points of "Questions: Opening Communication"

Defining Questions

▶ Open and closed questions are an essential part of helping theories and practices.

▶ Closed questions focus the interview, provide specific information, and are answered in few words; open questions allow more client talk time and exploration of client concerns.

Discerning the How of Using Questions

▶ Questions help begin the interview and open conversation.

▶ Initial questions may lead clients and determine their responses.

▶ Key sentence stems of certain open questions may predict client response. *What* tends to lead to facts, *how* to process and feelings, and *why* to reasons; *could/can/would* were described as maximally open.

▶ Potential difficulties with questions include client grilling and bombardment, the use of questions to make statements, and defensiveness in response to *why* questions.

▶ Many people have negative experiences with questions. We may have been "put on the spot" or grilled. It becomes key to determine when and how to use questions effectively.

▶ Open questions help clients elaborate and enrich their stories.

▶ Questions used to identify strengths can help clients face their issues with more confidence and ability.

▶ Questions' effectiveness can be enhanced when used respectfully and with consideration of multicultural dimensions of clients.

Observing Questions in Action: Interview Transcript, Conflict at Work

▶ Closed questions can bring out specific data, but if they are overused, the interviewer will run out of things to say and so will the client.

▶ Open questions give more control to the client and encourage more client talk time.

Practicing Questioning Skills

▶ Practice intentional prediction with classmates and volunteers to find out most likely outcome of questions.

▶ Practice open and closed questions. Focus on a particular issue, and create questions to search for positive strengths and resources.

▶ Record your role-play or practice and seek feedback afterward. Use group practice to get immediate comments from your classmates or observers. What kinds of comments do you get? What do they suggest about questions and your questioning style? How can you use this feedback to sharpen your skills?

Refining: Extensions of the Questioning Skill

▶ Questions help clients tell their story and bring out concrete details of the client's world.

▶ The "What else?" question brings out missing data. It is maximally open and allows the client considerable control.

▶ Questions may be seen as rude and intrusive and may be inappropriate. Sufficient trust needs to be present for questions to work effectively, particularly when multicultural differences are present.

▶ With less verbal clients, be particularly careful not to lead the client into your own frame of reference.

Assess your current level of knowledge and competence as you complete the chapter:

1. Flashcards: Use the flashcards to check your understanding of key concepts and facilitate memorization of key information.

2. Self-Assessment Quiz: The quiz will help you assess your current knowledge and prepare for course examinations.

3. Portfolio of Competencies: Evaluate your present level of competence on the ideas and concepts presented in this chapter using the Self-Evaluation Checklist. Self-assessment of your competencies demonstrates what you can do in the real world.

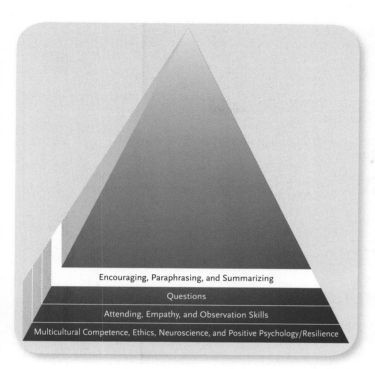

Encouraging, Paraphrasing, and Summarizing

Questions

Attending, Empathy, and Observation Skills

Multicultural Competence, Ethics, Neuroscience, and Positive Psychology/Resilience

Chapter 5
Encouraging, Paraphrasing, and Summarizing
Active Listening and Cognition

The possibility is only one sentence away. . . .
Our goal is to make the eyes shine!

—Andrew Zander

Emotion is the system that tells us how important something is. Attention focuses us on the important and away from the unimportant things. Cognition tells us what to do about it. Cognitive skills are whatever it takes to do those things.

—Alvaro Fernandez

Encouraging, paraphrasing, and summarizing are active listening skills that are the cognitive center of the basic listening sequence and are key in building the empathic relationship. When clients sense that their story is heard, they open up and become more ready for change. This leads to more effective **executive brain functioning** and thus to improved cognitive understanding, organization of issues, and decision making.

Emotional regulation is the other critical aspect of executive brain functioning. The next chapter on reflection of feelings discusses drawing out client emotions and balancing these feelings with cognitive reality. This chapter focuses primarily on cognition. Needless to say, we need a cognitive/emotional balance.

Chapter Goals

Awareness, knowledge, skills, and actions developed through the concepts of this chapter and this book will enable you to:

▶ Help clients talk in more detail about their issues of concern and help prevent the overly talkative client from repeating the same facts. Clarify for the client and you, the interviewer, what is really being said during the session.

▶ Check on the accuracy of what you hear by saying back to clients the essence of their comments and providing periodic summaries.

► Facilitate cognitive understanding for clearer decision making and more effective action.

► Promote development of the brain's executive functions, central for organizing thoughts, regulating emotions, planning action, and implementing planned actions.

Assess your current level of knowledge and competence as you begin the chapter:

1. **Self-Assessment Quiz:** The chapter quiz will help you determine your current level of knowledge. You can take it before and after reading the chapter.

2. **Portfolio of Competencies:** Before you read the chapter, please fill out the Self-Evaluation Checklist to assess your existing knowledge and competence on the ideas and concepts presented in this chapter. Then, at the end of the chapter, complete the checklist again to summarize your competencies after study and practice.

► Defining Active Listening

Active listening is a communication process that requires intentional participation, decision making, and responding. What we listen to (selective attention) and respond to has a profound influence on how clients talk to us about their concerns. When a client shares with us a lot of information all at once and talks rapidly, we can find ourselves confused and even overwhelmed by the complexity of the story. We need to hear this client accurately and often slow the story down a bit. Our accurate listening, in turn, leads to client understanding and synthesis, providing clients with a clearer picture of their own stories.

Active listening is central in facilitating our brain's executive functioning—the cognitive understanding and the emotional underpinnings. Without organization of our own stories and client stories, we will continue to live in indecision, confusion, and even chaos. Attending and active listening to those around us is fundamental. This chapter focuses on the brain's cognitive executive part of active listening; the following chapter on reflection of feelings examines the role of emotion.

Encouraging, paraphrasing, and summarizing are basic to empathic understanding and enable you to communicate to clients that they have been heard. You say back to clients what you have heard, using their key words. You help clients by distilling, shortening, and clarifying what has been said. When using empathic listening skills, be sure not to mix in your own ideas with what the client has been saying. Accurate empathic listening is not as common, nor as easy, as it may sound, but its impact is often profound.

Following are the responses you can anticipate from your client when you use these active listening skills. Remember to use the checkout frequently to obtain feedback on the accuracy of your listening skills.

ENCOURAGING	ANTICIPATED CLIENT RESPONSE
Encourage with short responses that help the client keep talking. These responses may be verbal (repeating key words and short statements) or nonverbal (head nods and smiling).	Clients elaborate on the topic, particularly when encouragers and restatements are used in a questioning tone of voice.

PARAPHRASING	ANTICIPATED CLIENT RESPONSE
Shorten or clarify the essence of what has just been said, but be sure to use the client's main words when you paraphrase. Paraphrases are often fed back to the client in a questioning tone of voice.	Clients will feel heard. They tend to give more detail without repeating the exact same story. They also become clearer and more organized in their thinking. If a paraphrase is inaccurate, the client has an opportunity to correct the interviewer.

SUMMARIZING	ANTICIPATED CLIENT RESPONSE
Summarize client comments and integrate thoughts, emotions, and behaviors. Summarizing is similar to paraphrasing but used over a longer time span.	Clients will feel heard and discover how their complex and even fragmented stories are integrated. The summary helps clients make sense of their lives and will facilitate a more centered and focused discussion. Secondarily, the summary also provides a more coherent transition from one topic to the next or a way to begin and end a full session. As a client organizes the story more effectively, we are seeing growth in brain executive functioning and better decision making.

CHECKOUT/PERCEPTION CHECK	ANTICIPATED CLIENT RESPONSE
Periodically, check with your client to discover how your interviewing lead or skill was received. "Is that right?" "Did I hear you correctly?" "What might I have missed?"	Interviewing leads such as these give the client a chance to pause and reflect on what they have said. If you indeed have missed something important or have distorted their story and meaning, they have the opportunity to correct you. Without an occasional checkout, it is possible to lead clients away from what they really want to talk about.

A client, Jennifer, enters the room and starts talking immediately:

I really need to talk to you. I don't know where to start. I just got my last exam back and it was a disaster, maybe because I haven't studied much lately. I was up late drinking at a party last night and I almost passed out. I've been sort of going out with a guy for the last month, but that's over as of last night. . . . [pause] But what really bothers me is that my mom and dad called last Monday and they are going to separate. I know that they have fought a lot, but I never thought it would come to this. I'm thinking of going home, but I'm afraid to. . . .

Jennifer continues for several more minutes in much the same manner, repeating herself, and she seems close to tears. Information is coming so fast that it is hard to follow her. Finally, she stops and looks at you expectantly.

Imagine you are listening to Jennifer's detailed and emotional story. What are you thinking about her at this moment? Write down what you could say and do to help her feel that you empathize with her and understand her concerns, but perhaps also to help her focus. Compare your ideas with the discussion that follows.

Jennifer is a client who does not have cognitive control of her thoughts and feelings. She needs help in organizing and making sense of her world and then deciding how to act appropriately.

When working with Jennifer, a useful first step is to summarize the essence of Jennifer's several points and say them back to her. As part of this initial response, use a checkout (e.g., "Have I heard you correctly?") to see how accurate your listening was. The checkout (sometimes called a perception check) offers her a chance to think about what she has said and the accuracy and completeness of your summary. You could follow this by asking, "You've talked about many things. Where would you like to start today?"

Choosing to focus first on the precipitating crisis is another possible strategy. We could start with Jennifer's parents' separation, as that seems to be the immediate precipitating crisis, and restate and paraphrase some of her key ideas. Doing this is likely to help her focus on one key issue before turning to the others. The other concerns clearly relate to the parental separation, and they will have to be dealt with as well.

EXERCISE 5.1 Reflection
How do our thoughts for counseling Jennifer compare with what you would do? What might you see as the most useful way to respond?

Interactive Exercise: Identifying Skills. Work with a client's statements to determine what skills are used by the interviewer.

Interactive Exercise: Identification and Classification. Work with a condensed portion of the Mary Bradford Ivey and Damaris interview to categorize each of Mary's leads to practice your ability to classify microskills.

Interactive Exercise: Basic Competence. How would you respond to the five client comments?

▶ Discerning the Basic Techniques and Strategies of Encouraging, Paraphrasing, and Summarizing

ENCOURAGERS

Encouragers are verbal and nonverbal expressions the counselor or therapist can use to prompt clients to continue talking. Encouragers include minimal verbal utterances ("ummm" and "uh-huh"), head nods, open-handed gestures, and positive facial expressions that encourage the client to keep talking. Silence, accompanied by appropriate nonverbal communication, can be another type of encourager. These encouragers are not meant to direct client talk; rather, they simply encourage clients to keep talking.

Repetition of key words can encourage a client and has more influence on the direction of client talk. Consider the following client statement:

> "And then it happened again. The grocery store clerk gave me a dirty look and I got angry. It reminded me of my last job, where I had so much trouble getting along. Why are they always after me?"

The counselor could use a variety of short encouragers in a questioning tone of voice ("Angry?" "Last job?" "Trouble getting along?" "After you?" "Tell me more."), and in each case the client would likely talk about a different topic. Note your selection of single-word encouraging responses as they may direct clients more than you think.

RESTATEMENT

A **restatement** is a type of extended encourager in which the counselor or interviewer repeats short statements of two or more words exactly as used by the client. "The clerk gave you a dirty look." "You got angry." "You had trouble getting along in your last job." "You wonder why they are always after you." Restatements can be used with a questioning tone of voice; they then function much like single-word encouragers. As with short encouragers, different types of restatements lead the client in different directions.

Well-timed encouragers maintain flow and continually communicate to the client that you are listening. All types of encouragers facilitate client talk unless they are overused or used badly. Picking out a key word or short phrase to use as an encourager often leads clients to provide you with their underlying thoughts, feelings, or behaviors related to that word or phrase. Just one well-observed word or restatement can open important new avenues in the session. On the other hand, the use of too many encouragers can seem wooden and unexpressive, whereas too few encouragers may suggest to clients that you are not interested.

Always remember smiling as an encourager. The warmth and caring you demonstrate may be the most important part of the relationship, even more important than what you say. Facial expression and vocal tone are the nonverbal components of encouraging.

PARAPHRASING

At first glance, **paraphrasing** appears to be a simple skill, only slightly more complex than encouraging. In encouraging and restating, exact words and phrases are fed back to the client. Paraphrasing covers more of what the client has just said, usually several sentences. Paraphrasing continues to feed back key words and phrases, but catches and distills the essence of what the client has said. Paraphrasing clarifies a confusing client story.

When you paraphrase, the tone of your voice and your body language indicate to the client whether you are interested in listening in more depth or would prefer that the client move on to another topic.

If your paraphrase is accurate, the client is likely to reward you with a "That's right" or "Yes . . . " and then go on to explore the issue in more depth. Once clients know they have been heard, they are often able to move on to new topics. The goal of paraphrasing is to facilitate client exploration and clarify issues.

Accurate paraphrasing will help the client stop repeating a story unnecessarily. Some clients have complex problems that no one has ever bothered to hear accurately, and they literally need to tell their story over and over until someone indicates they have been heard clearly.

How do you paraphrase? Observe the client, hear the important words, and use them in your paraphrase much as the client does. You may use your own words, but the main ideas and concepts must reflect the client's view of the world, not yours!

An accurate paraphrase usually consists of four dimensions:

1. A *sentence stem* sometimes using the client's name. Names help personalize the session. Examples: "Damaris, I hear you saying . . . ," "Luciano, sounds like . . . ," "Looks like the situation is. . . ."
2. The *key words* used by the client to describe the situation or person. Include main ideas and exact words that come from clients. This aspect of the paraphrase is sometimes confused with the encouraging restatement. A restatement, however, covers a very limited amount of client talk and is almost entirely in the client's own words.
3. The *essence of what the client has said* in briefer and clearer form. Identify, clarify, and feed back the client's sometimes confused or lengthy talk into succinct and meaningful statements. The counselor has the difficult task of staying true to the client's ideas but not repeating them exactly.

4. A *checkout* for accuracy. Here you ask the client for feedback on whether the paraphrase (or other skill) was correct and useful.

In the following example, a brief client statement is followed by key word encouragers, restatements, and a paraphrase that an interviewer might use to encourage client talk.

> "I'm really concerned about my wife. She has this feeling that she has to get out of the house, see the world, and get a job. I'm the breadwinner and I think I have a good income. The children view Yolanda as a perfect mother, and I do too. But last night, we really saw the problem differently and had a terrible argument."

- ▶ *Key word encouragers*: "Breadwinner?" "Terrible argument?" "Perfect mother?"
- ▶ *Restatement encouragers*: "You're really concerned about your wife." "You see yourself as the breadwinner." "You had a terrible argument."
- ▶ *Paraphrase*: "You're concerned about your picture-perfect wife who wants to work even though you have a good income, and you've had a terrible argument. Is that how you see it?"

As always, personalize and make your active listening real. A stem is not always necessary and, if overused, can make your comments seem like parroting. Clients have been known to say in frustration, "That's what I just said; why do you ask?" Again, smiling and warmth make a difference.

SUMMARIZING

This skill falls along the same continuum as the key word encourager, restatement, and paraphrase. **Summarizing**, however, encompasses a longer period of conversation than paraphrasing; at times it may cover an entire interview or even issues discussed by the client over several interviews. The summary essentially puts together and organizes client conversation, thus supporting the brain's executive functioning.

Many summaries also include client emotional and feeling tone, as well as do many mainly cognitive paraphrases. As emotions are often first reactions and typically occur before cognition regulates emotion, this area needs special attention. A major role of executive functioning is emotional regulation, outlined in the next chapter.

In summarizing, the interviewer attends to verbal and nonverbal comments from the client over a period of time and selectively attends to key concepts and dimensions, restating them for the client as accurately as possible. A checkout at the end for accuracy is a key part of summarizing. Following are some examples. The emotional words are in italics.

To begin a session: "Let's see, last time we talked about your *feelings* toward your mother-in-law, and we discussed the argument you had with her when the new baby arrived. You saw yourself as *guilty and anxious*. Since then you haven't gotten along too well. We also discussed a plan of action for the week. How did that go?"

Midway in the interview: "So far, I've seen that you felt *guilty* again when you saw the action plan as manipulative. Yet one idea did work. You were able to talk with your mother-in-law about her garden, and it was the first time you had been able to talk about anything without an argument and you *felt more comfortable*. You visualize the possibility of following up on the plan next week. Is that about it?"

At the end of the session: "In this interview we've reviewed more detail about your *feelings* toward your mother-in-law. Some of the following things seem to stand out: First, our plan didn't work completely, but you were able to talk about one thing without yelling. As we talked, we identified some behaviors on your part that could be changed. They include better eye contact, relaxing more, and changing the topic when you start to see yourself getting *angry.* Does that sum it up?"

Interactive Exercise: Writing Listening Responses. Your turn to practice what you learned!

Interactive Exercise: Encouraging, Paraphrasing, and Summarizing: Marshall and Jesse Interview. Here you will find five client comments for you to explore possible interviewer responses and determine which one may be the most helpful.

▶ Observing Listening Skills and Children

The observation case presented here is to remind us that the listening skills used with adults are also useful with children. Children too often go through life being told what to do. If we listen to them and their singular constructions of the world, we can reinforce their unique qualities and help them develop a belief in themselves and their own value. In this way, we are increasing their executive functioning, a necessary part of growth and development.

Children generally respond best if you seek to understand the world as they do. Smiling, warmth, and the active listening skills are essential. Frequently paraphrase or re-state what they have said, using their important words. Under stress, children may be confused, so be careful that the story you bring out is theirs, not yours.

Talk to children at their eye level whenever possible; avoid looking down at them. This may mean sitting on the floor or in small chairs. Be prepared for more topic jumps with children; use attending skills to bring them back to critical issues. They may need to expend excess energy by doing something with their hands; allow them to draw or play with sand or clay as they talk to you, or engage them in a game like Chutes and Ladders or checkers.

Questions can put off some children, although they remain one of the best ways to obtain information. Seek to get the child's perspective, not yours, while remaining carefully aware not to use leading questions that might bring out inaccurate stories. Children may have difficulty with a general open question such as "Could you tell me what happened?" Use short sentences, simple words, and a concrete language style. Break down abstract questions into concrete and situational language, using a mix of closed and open questions: "Where were you when the fight occurred?" "What was going on just before the fight?" "Then what happened?" "How did he feel?" "Was she angry?" "What happened next?" "What happened afterward?"

In questioning children on touchy issues, be especially careful of leading questions, which can easily lead to inaccurate understanding of situations. Furthermore, leading questions have been known to encourage development of false memories in both children and adults.

The following example interview is an edited version of a videotaped interview conducted by Mary Bradford Ivey with Damaris, a child actor, role-playing a problem based on a composite of real cases. Damaris is an 11-year-old sixth grader. The session presents a child's problem, but all of us, regardless of age, have experienced nasty teasing and put-downs, often in our closest relationships. If carried on too long, what you read here could easily become bullying.

Mary first draws out the child's story about teasing and then her thoughts and feelings about the teasing. Mary follows with a focus on the child's strengths, an example of the positive wellness approach.

Portions of this session are available in a video in the web resources. There you will see a very involved style with constant encouragement and smiles. This is a visual illustration of what we call "warmth," and it is basic to virtually all counseling relationships. Think about it—can you communicate warmth and caring to all your clients?

INTERVIEWER AND CLIENT CONVERSATION	PROCESS COMMENTS
1. Mary: (Smiling) Damaris, how're you doing?	The relationship between Mary and Damaris is already established; they know each other through school activities.
2. Damaris: Good.	She smiles and sits down.
3. Mary: I'm glad you could come down. You can use these markers if you want to doodle or draw something while we're talking. I know—you sort of indicated that you wanted to talk to me a little bit.	Mary welcomes the child and offers her something to do with her hands. Many children get restless just talking. Damaris starts to draw almost immediately. You may do better with an active male teen by taking him to the basketball court while you discuss issues. It can also help to have things available for adults to do with their hands.
4. Damaris: In school, in my class, there's this group of girls that keep making fun of my shoes, just 'cause I don't have Nikes.	Damaris looks down and appears a bit sad. She stops drawing. Children, particularly the "have-nots," are well aware of their economic circumstances. Some children have used sneakers; Damaris, at least, has newer sneakers.
5. Mary: They "keep making fun of your shoes"?	Encourage in the form of a restatement using Damaris's *exact key words*.
6. Damaris: Well, they're not the best; I mean—they're not Nikes, like everyone else has.	Damaris has a slight angry tone mixed with her sadness. She starts to draw again.
7. Mary: Yeah, they're nice shoes, though. You know?	It is sometimes tempting to comfort clients rather than just listen. We already know that Damaris is not satisfied with them. A simple "uh-huh" could have been more effective. However, positive comments and reassurance used later and more appropriately may be very effective. Too early use turns out to be subtractive empathy.
8. Damaris: Yeah. But my family's not that rich, you know. Those girls are rich.	Clients, especially children, hesitate to contradict the counselor. Notice that Damaris uses the word "*But . . .*" When clients say, "Yes, *but . . . ,*" interviewers are off track and need to change their style. Here we see implied that Damaris is less "OK" that her peers. Her self-cognitions leave her with a sense of inadequacy.
9. Mary: I see. And the others can afford Nike shoes, and you have nice shoes, but your shoes are just not like the shoes the others have, and they tease you about it?	Mary backs off her reassurances and paraphrases the essence of what Damaris has been saying using her key words. (Level 2 interchangeable empathy)
10. Damaris: Yeah. . . . Well, sometimes they make fun of me and call me names, and I feel sad. I try to ignore them, but still, the feeling inside me just hurts.	If you paraphrase or summarize accurately, a client will usually respond with *yeah* or *yes* and continue to elaborate the story.
11. Mary: It makes you feel hurt inside that they should tease you about shoes.	Mary reflects Damaris's feelings. The reflection of feeling is close to a paraphrase and is elaborated in the following chapter. (Interchangeable empathy)
12. Damaris: Mmm-hmm. [pause] It's not fair.	Damaris thinks about Mary's statement and looks up expectantly as if to see what happens next. She thinks back on the basic unfairness of the whole situation.

(continued)

INTERVIEWER AND CLIENT CONVERSATION	PROCESS COMMENTS
13. Mary: So far, Damaris, I've heard how the kids tease you about not having Nikes and that it really hurts. It's not fair. You know, I think of you, though, and I think of all the things that you do well. I get . . . you know . . . it makes me sad to hear this part because I think of all the talents you have, and all the things that you like to do and—and the strengths that you have.	Mary's brief summary covers most of what Damaris has said so far. Mary also discloses some of her own feelings. Sparingly used self-disclosure can be helpful. Mary begins the strength-based positive asset search by reminding Damaris that she has strengths to draw from. These strengths support the building of both cognitive and emotional resilience. (Additive empathy)
14. Damaris: Right. Yeah.	Damaris smiles slightly and relaxes a bit.
15. Mary: What comes to mind when you think about all the positive things you are and have to offer?	Here Mary is empowering executive functioning. An open question encourages Damaris to think about her strengths and positives. (Potentially additive empathy)
16. Damaris: Well, in school, the teacher says I'm a good writer, and I want to be a journalist when I grow up. The teacher wants me to put the last story I wrote in the school paper.	Damaris talks a bit more rapidly and smiles. (It was additive.)
17. Mary: You want to be a journalist, 'cause you can write well? Wow!	Mary enthusiastically paraphrases positive comments using Damaris's own key words. (Interchangeably empathy, but the "wow" and enthusiasm are also additive.)
18. Damaris: Mmm-hmm. And I play soccer on our team. I'm one of the people that plays a lot, so I'm like the leader, almost, but . . . [Damaris stops in midsentence.]	Damaris has many things to feel good about; she is smiling for the first time in the session. Seeing personal strengths facilitates positive cognition and executive functioning.
19. Mary: So, you are a scholar, a leader, and an athlete. Other people look up to you. Is that right? So how does it feel when you're a leader in soccer?	Mary is strengthening executive functioning by adding the names *scholar* and *athlete* for clarification and elaboration of the positive asset search. She knows from observation on the playground that other children do look up to Damaris. Counselors may add related words to expand the meaning. Mary wisely avoids leading Damaris and uses the checkout, "Is that right?" Mary also asks an open question about feelings. And we note that Damaris used that important word "but." Do you think that Mary should have followed up on that, or should she continue with her search for strengths?
20. Damaris: [small giggle, looking down briefly] Yeah. It feels good.	Looking down is not always sadness! The spontaneous movement of looking down briefly is termed the "recognition response." It most often happens when clients learn something new and true about themselves. Damaris has internalized the good feelings.
21. Mary: So you're a good student, and you are good at soccer and a leader, and it makes you feel good inside.	Mary summarizes the positive asset search using both facts and feelings. The summary of feeling *good inside* contrasts with the earlier feelings of *hurt inside*. (Interchangeable empathy)
22. Damaris: Yeah, it makes me feel good inside. I do my homework and everything [pause and the sad look returns], but then when I come to school, they just have to spoil it for me.	Again, Damaris agrees with the paraphrase. She feels support from Mary and is now prepared to deal from a stronger position with the teasing. Here we see what lies behind the "but" in 18 above. We believe Mary did the right thing in ignoring the "but" the first time. Now it is obvious that the negative feelings need to be addressed. When Damaris's wellness strengths are clear, Mary can better address those negative feelings.

(continued)

INTERVIEWER AND CLIENT CONVERSATION	PROCESS COMMENTS
23. Mary: They just spoil it. So you've got these good feelings inside, good that you're strong in academics, good that you're, you know, good at soccer and a leader. Now, I'm just wondering how we can use those good feelings that you feel as a student who's going to be a journalist someday and a soccer player who's a leader. Now the big question is how you can take the good, strong feelings and deal with the kids who are teasing. Let's look at ways to solve your problem now.	Mary restates Damaris's last words and again summarizes the many good things that Damaris does well. Mary changes pace and is ready to move to the problem-solving portion of the interview. This additive empathy is setting the stage for stronger executive functioning and resolution of the teasing before it becomes serious.

Positive stories and identified strengths when put next to the negative cognitions almost inevitably weaken the negative while simultaneously building executive functioning. This is the power of the positive asset search and positive psychology. Mary had an empathic warm relationship and was able to draw out Damaris's story fairly quickly. She focused on positive assets and wellness strengths to address Damaris's issues and challenges. Clients can solve problems best from their strengths. Be positive, but don't minimize why they came to see you.

Situations such as this one with Damaris may sound simple and basic, but what is happening here occurs in parallel form with teens and adults. Someone hassles or bullies us, and this enters our thoughts and cognitions—and over time can interfere with executive functioning. Adult bullying is beginning to be recognized as a significant personal issue for our clients. There is awareness of harassment, but bullying takes the issue even further. The approaches of positive psychology are a critical part of working with bullying and harassment, but they are not the whole solution.

School situations such as reported here often require the counselor to talk to teachers and make sure that the school bullying policy is followed. Mary often brought in those who were teased for small friendship groups. If the girls were "mean girls," Mary would set up a supportive group of friends for Damaris. Actually, one teacher commented briefly about this on the playground and the teasing ceased. But it is often not that easy. Be prepared to take supportive action outside the interview.

▶ Practicing: Becoming Competent in Encouraging, Paraphrasing, and Summarizing

EXERCISE 5.2 Self-Reflection: You as an Active Listener
The active listening skills can be seen as specific refinements of attending behavior. Before we asked you to think about yourself and your ability to attend. There we gave some focus to talk time. In a conversation, do you talk most of the time, or are you mainly a listener? Or perhaps you balance talk time. Think back and start observing yourself even more carefully on these dimensions.

Then, as a second part of self-reflection, ask yourself the following questions: (1) Am I an encouraging person, indicating to the other person that I am there through smiling and short minimal encourages (uh-huh, yes, tell me more)? (2) Do I paraphrase to make sure that I understand the cognitive frame of my friends and family, particularly when we see thing differently or even have conflict? (3) Less likely, have you ever listened so carefully to someone that you summarize what they have said so that they (and you!) truly understand?

EXERCISE 5.3 Practice Varying Styles of Vocal Tone

Practice different vocal tones.

"I really need to talk to you. I don't know where to start." (Pause, Jennifer looks at you expectantly).

Provide an encourager using different vocal tones, and test out different postures and gestures.

"I was up late drinking at a party last night and I almost passed out. I've been sort of going out with a guy for the last month, but that's over as of last night."

Paraphrase this statement using first a supportive tone of voice, then a judgmental tone. What would a neutral voice sound like?

"But what really bothers me is that my mom and dad called last Monday and they are going to separate. I know that they have fought a lot, but I never thought it would come to this. I'm thinking of going home, but I'm afraid to."

Summarize the three paragraphs using varying vocal tones.

EXERCISE 5.4 Record an Individual Practice Session and Obtain Feedback

Join with a classmate, or ask a friend or family member to work with you as a volunteer. Draw out the person's issues, concern, or story using attending, questioning, and the active listening skills of encouraging, paraphrasing, and summarizing. In addition, be sure to draw out positive strengths and stories and notice how that affects the client and possibly changes nature of the session. In this session, seek to be a little more active than usual so that you can test out the active listening skills.

Some possible topics include an issue in school or at work, future career plans, or a current local or national issue around some personally meaningful issue related to the RESPECTFUL model (e.g., race, spirituality, economic issues). After 15 minutes, conclude and seek verbal feedback from the volunteer. If possible, also obtain written feedback using a downloaded form from the web resources. Listen to or watch the recorded session, making note of both your own and client behavior.

EXERCISE 5.5 Record a Group Practice Session With Feedback

Especially effective is meeting with a group of three or four, taking turns as interviewer and client. One member can run the recording equipment and observe. The second observer takes notes with an emphasis on the skill of reflection of feeling. Both observers will be helpful in pointing out nonverbal behavior. In a group such as this, the downloaded forms from the web resources will be especially helpful. Use the Feedback Form: Encouraging, Paraphrasing, and Summarizing to help you achieve mastery of these skills.

Interactive Exercise: Toward Intentional Competence. This interactive exercise uses two activities to help you master the listening skills.

a. Generating Written Encouragers, Restatements, Paraphrases, and Summaries. This activity allows you to practice these listening skills.

b. Practice of Skills in Other Settings. This activity is designed to help you practice the skills beyond the classroom.

Group Practice Exercise: Group Practice With Active Listening Skills

Case Study: Meridith. This interview discusses the loss of power that women sometimes feel in the process of seeking medical assistance for life-threatening illness. Reflect on the case and discuss how would you use the microskills presented in this chapter to help Meridith.

▶ Refining: Executive Functioning, Microaggresions, and Diversity Issues

Active listening skills not only demonstrate empathy, they also facilitate organization of chaotic stories and troubling life experiences. In neuroscience terms, listening facilitates more effective brain executive functioning, critical for cognitive understanding, emotional regulation, and behavioral change.

—Carlos Zalaquett

The active listening skills of encouraging, paraphrasing, and summarizing are key to understanding client cognitions. It is the way that clients think about things and their beliefs that we seek to understand. Their stories are the key to their cognitions. To be successful in interviewing and counseling, we need to understand clients' cognitive styles—and, when appropriate, help them change their cognitions and thoughts so that they are more comfortable with themselves and with others.

When we attend to clients and use the active listening skills, we facilitate executive functioning and the development of new **neural networks** that become part of long-term memory in the hippocampus. Moreover, the very act of listening can lead to restorying—the generation of new meanings and new ways of thinking. That new story leads to more effective executive functioning in the brain and more effective action, as seen in overt behavior.

Executive functioning is also critical for *emotional regulation*. (See the following chapter for elaboration.) We use our cognitive capacity to regulate impulsive emotions and act appropriately in complex situations, especially when we feel challenged. This could range from eating less sugar despite being tempted by beautifully decorated cupcakes to not saying or doing something hurtful when we suddenly become angry with a loved one. It enables us to live in a complex social world. Ineffective executive functioning may lead to negative emotions, feelings of anxiety, or depression.

Cognitions may be defined as language-based thought processes underlying all thinking activities, such as analyzing, imaging, remembering, judging, and problem solving. Cognitive behavioral therapy (CBT), rational emotive behavioral therapy (REBT), and dialectical behavior therapy (DBT) are three examples of cognitive theories of counseling and psychotherapy that focus on changing cognitions to achieve human change.

Our interviewing and counseling skills affect the brain and the mind. We can improve cognitive functioning, emotional regulation, relationship with others, and intentional action. In this process, we access memories in the hippocampus and can potentially facilitate the development of new neural networks in the process of generating new stories.

As an example of the importance of how client concerns and issues arise, visit Box 5.1. Here we see how the behavior of others, starting with small hurts, can damage and limit one's self-concept and, ultimately, executive functioning as well.

| **BOX 5.1** | Cumulative Stress and Microaggressions: When Do "Small" Events Become Traumatic? |

At one level, being teased about the shoes one wears doesn't sound that serious—children will be children! However, some poor children are teased and laughed at throughout their lives for the clothes they wear. At a high school reunion, Allen talked with a classmate who recalled painful memories, still immediate, of teasing and bullying during school days. Child and adolescent trauma can affect one's whole life experience.

The microaggressions this classmate experienced in high school became part of her persona and left her with more limited executive functioning and fewer life possibilities. Small slights become big hurts if repeated again and again. Athletes and "popular" students may talk arrogantly and dismissively about the "nerds," "townies," "hicks," or other outgroups. Teachers, coaches, and even counselors sometimes join in the laughter. Over time, these slights mount inside the child or adolescent. Some people internalize their issues as psychological distress; others may act them out in a dramatic fashion—witness the recurring school, church, and workplace shootings throughout the United States.

Microaggressions is a term for these small hurts that accumulate and magnify over time (Sue, 2010). Discrimination and prejudice are other examples of cumulative stress and trauma. One of Mary's interns, a young African American woman, spoke of a recent racial insult.

At a restaurant she overheard two White people talking loudly about the "good old days" of segregation. Perhaps the remark was not directed at her, but still it hurt. She related how common racial insults and microaggressions were in her life, directly or indirectly. She could tell how bad things were racially by the number of calls she made to her family.. When an incident occurred that troubled her, she needed to talk to her sister or parents and seek support.

Out of continuing indignities can come feelings of underlying insecurity about one's place in the world (internalized oppression and self-blame) and/or tension and rage about unfairness (externalized awareness of oppression). Either way, the person who is ignored or insulted feels tension in the body, the pulse and heart rate increase, and—over time—hypertension (high blood pressure) may result. The psychological becomes physical, and cumulative stress becomes traumatic.

Soldiers, veterans, police officers, fire fighters, women who suffer sexual harassment, those who are short or overweight, the physically disfigured through birth or accident, gays and lesbians, and many others are at risk for having cumulative stress build to real trauma or posttraumatic stress.

Be alert for signs of cumulative stress in your clients. Are they internalizing the stressors by blaming themselves? Or are they externalizing and building a pattern of explosive rage and anger? All these people have important stories to tell, and at first these stories may sound routine. The occurrence of posttraumatic stress responses in later life may be alleviated or prevented by your careful listening and support.

Children, adolescents and adults may internalize harassment and bullying, somehow thinking it is "their fault." Our task is to help these clients name their very real concerns as externally caused, even to naming it as *oppression*. Next, we want to facilitate development of personal strategies and social support systems so that the individual facing these challenges is no longer weak and alone.

Finally, think of yourself as a potential social action agent in your community. What can you do to help groups of clients, such as those described above, deal with microaggressions and stressors more effectively? Understanding broad social stressors is part of being an effective helper—and taking action or organizing groups toward a healthier lifestyle and working with them to take action for betterment represents a challenge for the future.

EXERCISE 5.6 Overtly Racist/Oppressive Behavior

When a client shows clear racism, sexism, anti-Semitism, or other oppressive thinking in the interview, the interviewer is faced with a challenge. There is an increasing move in the field toward interviewer responsibility to instruct clients who may need to learn more tolerance and respect.

What are your thoughts?

What does the current code of ethics suggest?

What are you going to do when faced with these challenges?

DIVERSITY AND ACTIVE LISTENING

Periodic encouraging, paraphrasing, and summarizing are basic skills that seem to have wide cross-cultural acceptance. Virtually all your clients like to be listened to accurately. But it may take more time to establish a relationship with clients who are culturally different from you, and knowledge of their languages may help (see Box 5.2).

| BOX 5.2 | National and International Perspectives on Counseling Skills |

Developing Skills to Help the Bilingual Client
Azara Santiago-Rivera

It wasn't that long ago that counselors considered bilingualism a "disadvantage." We now know that a new perspective is needed. Let's start with two fundamental assumptions: *The person who speaks two languages is able to work and communicate in two cultures and, actually, is advantaged. The monolingual person is the one at a disadvantage!* Research actually shows that bilingual children have more fully developed capacities and a broader intelligence (Power & Lopez, 1985).

(continued)

BOX 5.2 (continued)

If your client was raised in a Spanish-speaking home, for example, he or she is likely to think in Spanish at times, even though having considerable English skills. We tend to experience the world nonverbally before we add words to describe what we see, feel, or hear. For example, Salvadorans who experienced war or other forms of oppression *felt* that situation in their own language.

You are very likely to work with clients in your community who come from one or more different language backgrounds. Your first task is to understand some of the history and experience of these immigrant groups. Then we suggest that you learn some key words and phrases in their original language. Why? Experiences that occur in a particular language are typically encoded in memory in that language. So, certain memories containing powerful emotions may not be accessible in a person's second language (English) because they were originally encoded in the first language (for example, Spanish). And if the client is talking about something that was experienced in Spanish, Khmer, or Russian, the *key words* are not in English; they are in the original language.

Here is an example of how you might use these ideas in the session:

Social worker: Could you tell me what happened for you when you lost your job?

Maria (Spanish-speaking client): It was hard; I really don't know what to say.

Social worker: It might help us if you would say what happened in Spanish and then you could translate it for me.

Maria: ¡Es tan injusto! Yo pensé que perdí el trabajo porque no hablo el inglés muy bien. Me da mucho coraje cuando me hacen esto. Me siento herido.

Social worker: Thanks; I can see that it really affected you. Could you tell me what you said now in English?

Maria (more emotionally): I said, "It all seemed so unfair. I thought I lost my job because I couldn't speak English well enough for them. It makes me really angry when they do that to me. It hurts."

Social worker: I understand better now. Thanks for sharing that in your own language. I hear you saying that *injusto* hurts and you are very angry. Let's continue to work on this and, from time to time, let's have you talk about the really important things in Spanish, OK?

This brief example provides a start. The next step is to develop a vocabulary of key words in the language of your client. This cannot happen all at once, but you can gradually increase your skills. Here are some Spanish key words that might be useful with many clients.

Respeto: Was the client treated with respect? For example, the social worker might say, "Your boss did not treat you with *respeto*."

Familismo: Family is very important to many Spanish-speaking people. You might say, "How are things with your *familia*?"

Emotions (see next chapter) are often experienced in the original language. When reflecting feeling, you could learn and use the following key words.

aguantar: endure	*miedo:* fear	*amor:* love
orgullo: proud	*cariño:* affection	*sentir:* feel
coraje: courage or anger		

We also recommend learning key sayings, metaphors, and proverbs in the language(s) of your clients. Dichos are Spanish proverbs, as in the following examples.

Al que mucho se le da, mucho se le demanda.
The more people give you, the more they expect of you.

Más vale tarde que nunca.
Better late than never.

No hay peor sordo que el no quiere oir.
There is no worse deaf person than someone who doesn't want to listen.

La unión hace la fuerza.
Union is strength.

Consider developing a list like this, learn to pronounce them correctly, and you will find them useful in counseling Spanish-speaking clients. Indeed, you are giving them respeto. You may wish to learn key words in several languages.

Carlos Zalaquett comments: The Spanish version of *Basic Attending Skills, Las Habilidades Atencionales Básicas: Pilares Fundamentales de la Comunicación Efectiva* (Zalaquett, Ivey, Gluckstern-Packard, & Ivey, 2008), can help both monolingual and bilingual helpers. The attending skills are illustrated with examples provided by Latina/o professionals from different Latin American countries. Using the information and exercises included in the book, you can sharpen your interviewing, counseling, and psychotherapeutic tools to provide effective services to clients who speak Spanish. Several programs in Guatemala use the textbook to teach basic attending skills (Ana Maria Jurado, personal communication, July 2014).

North American and European counseling theory and style generally expect the client to get at the problem immediately and may not place enough emphasis on relationship building. Some traditional Native American Indians, Dene, Pacific Islanders, Aboriginal Australians, and New Zealand Maori may want to spend a full interview getting to know and trust you before you begin. You may find yourself conducting interviews in homes or in other village settings. *Do not expect this to be true of every client who comes from an indigenous background.* If these clients have experienced the dominant culture, they are likely to be more comfortable with your usual style. Nonetheless, expect trust building and rapport to take more time.

Building trust requires learning about the other person's world. In general, if you are actively working as a counselor, involve yourself in positive community activities that will help you understand your clients better. If you are seen as a person who enjoys yourself in a natural way in the village, in the community, and at pow-wows or other cultural celebrations, this will help build general community trust. The same holds true if you are a person of color. It will be helpful for you to visit a synagogue or an all-White church, understand the political/power structure of a community, and view White people as a distinct cultural group with many variations. Each of these activities may help you to avoid stereotyping those who are culturally different from you. Keep in mind the multiple dimensions of the RESPECTFUL model—virtually all interviewing or counseling is cross-cultural in some fashion.

When you are culturally different from your client, self-disclosure and an explanation of your methods may be helpful. For example, if you use only questioning and listening skills, the client may view you as suspicious and untrustworthy. The client may want directions and suggestions for action.

A general recommendation for working cross-culturally is to discuss differences early in the interview. For example, "I'm a White European American and we may need to discuss whether this is an issue for you. And if I miss something, please let me know." "I know that some gay people may distrust heterosexuals. Please let me know if anything bothers you." "Some White people may have issues talking with an African American counselor. If that's a concern, let's talk about it up front." "You are 57 and I [the counselor] am 26. How comfortable are you working with someone my age?" There are no absolute rules here for what is right. A highly acculturated Jamaican, Native American Indian, or Asian American might be offended by the same statements. Also remember that you can, if necessary, refer the client to another colleague if you get the sense that your client is truly uncomfortable and does not trust you.

Some Asian (Cambodian, Chinese, Japanese, Indian) clients from traditional backgrounds may be seeking direction and advice. They are likely to be willing to share their stories, but you may need to tell them why you want to wait a bit before coming up with answers. To establish credibility, there may be times when you have to commit yourself and provide advice earlier than you wish. If this becomes necessary, be assured and confident; just let them know that you want to learn more and that the advice may change as you get to know them better.

Consider possible differences in gender. Even though there are many exceptions to this "rule," women tend to use more paraphrasing and related listening skills; men tend to use questions more frequently. You may notice in your own classes and workshops that men tend to raise their hands faster at the first question and interrupt more often.

 Video Activity: Toward Multicultural Competence. The video clip illustrates the importance of knowledge, awareness, and skills in multicultural areas. The goal in this exercise is to help you think a bit about diversity issues, your response to these issues, and what you might actually say to the client.

Summarizing Key Points of "Encouraging, Paraphrasing, and Summarizing: Active Listening and Cognition"

Defining Active Listening

▶ The purpose of the listening skills it to hear the client and feed back what has been said.

▶ Listening is active. How you selectively attend to clients affects how they tell their stories or discuss their concerns.

▶ Listening helps in exploring and clarifying issues during the session.

▶ Encouraging, paraphrasing, and summarizing represent different points on a continuum from single words and short phrases to summaries of a section of an interview, or even a series of sessions.

Discerning the Basic Techniques and Strategies of Encouraging, Paraphrasing, and Summarizing

▶ Encouraging, paraphrasing, and summarizing are key active listening skills that keep the client talking.

▶ Encouragers help clients elaborate their stories and include key words, uh-huh's, smiles, a warm style, and short comments.

▶ The restatement is a form of encourager that repeats short phrases back to the client.

▶ The paraphrase includes (1) a sentence stem, often with the client's name; (2) key words used by the client; (3) the essence of what a client has said in briefer and clarified form; and (4) a checkout for accuracy. The accurate paraphrase will catch the essence of and clarify what the client says.

▶ The summary covers a longer period of the interview. It is also often used to begin or end a session or repeat back to the client what was said in the previous session(s).

▶ Checking what you hear helps you better understand clients' comments and increases accuracy.

▶ All these skills communicate that you are listening and help the client feel heard, further explore his or her story, and gain a greater understanding of issues and concerns.

Observing Listening Skills and Children

▶ Children, adolescents, and some adults will be more comfortable if you provide something for them to do with their hands. Avoid towering over small children; sit at their level. Avoid abstractions, use short sentences and simple words, and focus on concrete, observable issues and behaviors.

▶ Note that the case of Damaris demonstrates effective verbal attending through encouraging, paraphrasing, and summarizing, which help the client explore the issues more effectively. Effective questions are used to bring in new data, organize the discussion, and point out positive strengths.

Practicing: Becoming Competent in Encouraging, Paraphrasing, and Summarizing

▶ Exercise the skills presented in this chapter to increase awareness, knowledge, and competence.

▶ To increase effective use of encouragers, paraphrases, and summaries, learn about your style and possible variations, and practice with others to receive feedback.

Refining: Executive Functioning, Microaggresions, and Diversity Issues

▶ Executive functions are the cognitive mental processes that regulate human behavior. The brain's frontal lobes provide the biological substrate of these functions.

▶ Cognitive skills help clients become aware of their emotions and regulate their emotional reactions.

▶ Cognitive-based counseling and therapy focus on changing clients' cognitions in order to achieve change.

▶ Microaggressions are small hurts that accumulate and magnify over time. They may cause stress and trauma.

▶ Discrimination and prejudice are other examples of cumulative stress and trauma.

▶ The listening skills are widely used cross-culturally and throughout the counseling and psychotherapy world, but more or less participation and self-disclosure on your part may be necessary depending on the characteristics of the interviewing context.

▶ Trust building occurs when you visit the client's community and learn about cultures different from your own. Best of all is having a varied multicultural group of friends.

Assess your current level of knowledge and competence as you complete the chapter:

1. **Flashcards:** Use the flashcards to check your understanding of key concepts and facilitate memorization of key information.

2. **Self-Assessment Quiz:** The quiz will help you assess your current knowledge and prepare for course examinations.

3. **Portfolio of Competencies:** Evaluate your present level of competence on the ideas and concepts presented in this chapter using the Self-Evaluation Checklist. Self-assessment of your competencies demonstrates what you can do in the real world.

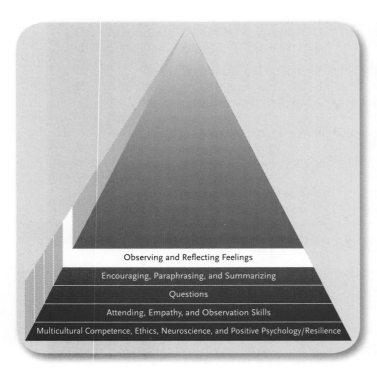

Observing and Reflecting Feelings

Encouraging, Paraphrasing, and Summarizing

Questions

Attending, Empathy, and Observation Skills

Multicultural Competence, Ethics, Neuroscience, and Positive Psychology/Resilience

Chapter 6
Observing and Reflecting Feelings
The Heart of Empathic Understanding

No cognition without emotion and no emotion without cognition.

—Jean Piaget

The artistic counselor catches the feelings of the client. Our emotional side often guides our thoughts and action, even without our conscious awareness.

—Allen Ivey

Emotions and feelings underlie our cognitions. Research shows clearly that emotions come first and may then be modified or elaborated later by cognitions. Emotions are also the engines of change, for without the support of our feelings, cognitive change is nearly impossible. Our words, thoughts, and behaviors are all intertwined with emotions—and the emotions often take the lead in what we say and do. Reflection of feelings is recognized by many as the most significant listening skill after attending behavior and is central to emotional regulation.

Chapter Goals

Awareness, knowledge, skills, and actions developed through the concepts of this chapter and this book will enable you to:

▶ Facilitate clients' awareness of their emotional world and its effect on their thoughts and behavior.

▶ Help clients sort out and organize their mixed feelings, thoughts, and behaviors toward themselves, significant others, or events.

▶ Clarify emotional strengths and use these to further client resilience.

▶ Center the counselor and client in fundamental emotional experience basic to resolving issues and achieving goals.

▶ Facilitate executive brain functioning through emotional regulation.

> **Assess your current level of knowledge and competence as you begin the chapter:**
>
> 1. **Self-Assessment Quiz:** The chapter quiz will help you determine your current level of knowledge. You can take it before and after reading the chapter.
>
> 2. **Portfolio of Competencies:** Before you read the chapter, please fill out the Self-Evaluation Checklist to assess your existing knowledge and competence on the ideas and concepts presented in this chapter. Then, at the end of the chapter, complete the checklist again to summarize your competencies after study and practice.

▶ Defining Reflection of Feelings

What is reflection of feeling, and what can the interviewer or counselor expect when using this skill? How might the client respond? Not all clients are comfortable with exploring emotions, so be ready for another response to keep the interview progressing—be intentional!

REFLECTION OF FEELINGS	ANTICIPATED CLIENT RESPONSE
Identify the client's key emotions and feed them back to clarify affective experience. With some clients, a brief acknowledgment of feelings may be more appropriate. Reflection of feelings is often combined with paraphrasing and summarizing. Include a search for positive feelings and strengths.	Clients will experience and understand their emotional state more fully and talk in more depth about feelings. They may correct the interviewer's reflection with a more accurate descriptor. In addition, client understanding of underlying feelings and emotions leads to emotional regulation with clearer cognitive understanding and behavioral action. Critical to lasting change is a more positive emotional outlook.

 Video Activity: Reflection of Feelings

COMPARING COGNITIVE PARAPHRASING/SUMMARIZING WITH EMOTIONAL REFLECTION OF FEELING

Reflection of feelings involves observing emotions, naming them, and repeating them back to the client. Paraphrasing and summarizing are more cognitively oriented, focusing primarily on words and concepts. The cognitive and emotional will often be integrated in counselor leads and statements; the critical distinction is how much one emphasizes cognitive content (paraphrase) or emotion (reflection of feelings). At the same time, you will often find yourself combining paraphrases with reflections of feeling. This is even more true in summarizing.

Note, for example, the confused cognitions and emotions expressed in Thomas's story.

My Dad drank a lot when I was growing up, but it didn't bother me so much until now. [pause] But I was just home and it really hurts to see what Dad's starting to do to my Mum—she's awful quiet, you know. [Looks down with brows furrowed and tense] Why she takes so much, I can't figure out. [Looks at you with a puzzled expression] But, like I was saying, Mum and I were sitting there one night drinking coffee, and he came in, stumbled over the doorstep, and then he got angry. He started to hit my mother and I stopped him. I almost hit him myself, I was so angry. [Anger flashes in his eyes.] I worry about Mum. [A slight tinge of fear seems to mix with the anger in the eyes, and you notice that his body is tensing.]

Paraphrase Thomas's main ideas—the cognitive content of his conversation; then focus on emotion and reflect his feelings. Use your attention and intuition to note the main feelings expressed, in words or nonverbally. Here are two possible sentence stems for your consideration:

Paraphrase: "Thomas, I hear you saying . . ."

Reflection of feelings: "Thomas, sounds like you feel . . ."

Thomas's cognitive content includes Dad's history of drinking, Mum's quietness and submission, and the difficult situation when Thomas was last home. The paraphrase focuses on the content, clarifies the essence of what the client said, indicates that you heard what was said, and encourages him to move the discussion further. The key cognitive concepts and words (besides mother and father) are *drinking*, *violence*, and *hit*.

> "Thomas, I hear you saying that your father has been *drinking* a long time, and your Mum takes a lot. But now he's started to be *violent*, and you've been tempted to *hit* him yourself. Have I heard you right?"

Emotion is expressed both verbally and nonverbally. Nonverbals are often the first and clearest sign of emotion. As the story is told, the interviewer sees Thomas look down with furrowed brow, followed by a puzzled look, all possibly indicating deep concern for his Mum. There is facial and body tension, and we see anger in his eyes. Underlying the puzzlement and confusion, we find key emotional words that include the relatively mild *bother* but also, more significantly, *hurt*, *fear*, and *anger*. We can assume that all of these exist in quiet Mum as well as in Thomas.

> "Thomas, sounds like you *fear* for your Mum. I see the *hurt*, *confusion*, and some *fear* in your eyes. A lot is happening inside you. I sense some real *tension*. Am I close to what you're feeling right now?"

Emotional regulation, a key goal or role of our executive functions, involves using the cognitive portions of the brain to *control* the more immediate and reactive emotional centers. This is one of the long-term goals of counseling. At the same time, many clients are unaware of their emotions. Then, our task is to help them discover the value of understanding feelings.

However, emotional regulation is equally about emotional freedom, the ability to experience feelings in the here and now rather than just controlling or suppressing them. In this example, the counselor would seek to help Thomas use his valid anger in useful ways, rather than letting the anger spill over in an emotional and cognitive disaster. In addition, Thomas likely needs to talk about his sadness. Unless we empathically acknowledge the difficult emotions, clients are less likely to respond to the positive approach.

Coupled with this, supported by positive psychology and therapeutic lifestyle changes, we need to look for strengths and resources from the past and present. These are important to build Thomas's resilience and emotional regulation as he finds new ways of thinking, feeling, and behaving. The positive language of hope and gladness breeds or promotes encouragement. As we work with client indecision and the many confusing things we face daily, understanding underlying emotions will enable better decisions and greater personal satisfaction.

 Interactive Exercise: Paraphrase vs. Reflect Feeling

▶ Discerning: The Basic Techniques and Strategies of Observing and Reflecting Feelings

Reason is the horse we ride after our emotions have decided where to go!
—Carlos Zalaquett

Emotional regulation allows us to manage the horse, its direction, and its speed.
—Mary Bradford Ivey

 Interactive Exercise: Developing a Basic Feeling Vocabulary

First, let us examine the observation of emotions. Then we will turn to specifics of the skill.

OBSERVING THE VERBAL AND NONVERBAL LANGUAGE OF EMOTIONS

As a first step toward naming and understanding emotions, seek to establish and increase your own **vocabulary of emotions** and your ability to observe and name them accurately. Counseling and therapy traditionally have focused on four primary emotions—*sad, mad, glad, scared*—as a mnemonic for memory, but two more need to be added: *disgust* and *surprise.* These six are the **primary emotions**, and their commonality, in terms of facial expression and language, has been validated throughout the world in all cultures (Ekman, 2007). Make special note that our field has historically listed four "negative" emotions and only one that is positive and life affirming. Surprise can be pleasant or very scary, depending on what is happening at the moment (Eckman & Cordaro, 2011; Liu, Zhang, & Luo, 2014).

For practical purposes, Ekman's six emotions are what you will be using in the context of the interview. However, important recent sociobiological research reveals that these emotions come from two basic dimensions of evolutionary and human survival—approach and avoidance (Jack, Garrod, & Schyns, 2014). In our language, these could be described as fear and happiness.

Why is the basic approach/avoidance paradigm important to counseling? Think of each as a basic approach to life. You will encounter clients who avoid contact with others in a variety of ways, including fear, anger, and disgust. Sadness may be described as a failure to meet approach needs. Those who are depressed may use their sadness as a way to avoid dealing with others and the world at large. Surprise often shows in quick body responses, which you will sometimes see when you confront clients on inconsistencies and discrepancies in their interpretation of the world. That surprise may show as a laugh as you help them clarify, or it may show as anger when they seek to deny what you just said.

Because we have a larger emotional vocabulary for the avoidance/fear dimensions, it means that we have to work all the harder with a positive psychology and wellness approach. It is far too easy to let clients talk about their avoidance issues; it may be more difficult, but all the more important, to facilitate approach/happy/love emotions.

Many of your clients will be expressing mixed emotions. In couple conflict, for example, there will be approach emotions as well as distancing avoidance feelings. In making a vocational decision or in dealing with a difficult supervisor, there will also be conflicting approach/avoidance issues. Decisional counseling (see Chapter 12) gives special attention to helping clients balance these highly emotions decisions. Thinking through the approach/avoidance issues may be helpful to your clients.

EXERCISE 6.1 Expanding Emotional Vocabulary

In the blanks below, brainstorm more emotional and feeling words. Think of related words with different intensities. As you do so, consider the approach/avoidance dimensions of each and how cognition and culture modify the meaning. For example, *mad* might lead you to think of *annoyed*, *angry*, and *furious*. Try this exercise, and then continue to brainstorm and build your vocabulary of emotions. A list of emotions developed for comparison and expansion of one's emotional vocabulary may be found at the end of this chapter.

Sad/ Unhappy/ Depressed	Mad/Angry/ Explosive	Glad/Happy/ Contented/ Love	Scared/Fear/ Anxious	Disgust/ Repugnance/ Loathing	Surprise/ Shock/ Wonder
_____	_____	_____	_____	_____	_____
_____	_____	_____	_____	_____	_____
_____	_____	_____	_____	_____	_____
_____	_____	_____	_____	_____	_____

Sad/unhappy, mad/angry, scared/fear are at the center of much of the work we do in counseling and therapy. Sadness can lead to depression, which leads to a cycle of inaction. Anger all too often leads to impulsive behavior that is destructive of self and others. It appears in spousal abuse, those who bully, oppositional defiant children and teens, and sociopathic behavior. Fear is related to anxiety, phobias, and an avoidant personality style. Clearly, if we are to work to improve executive functioning and cognitive competence, we must deal effectively with these challenging emotions, but this is best done from a base of positives and strengths (see discussion below).

When we have a significant loss of a loved one, the emotion of sadness needs to be worked through rather than denied. While anger management will be one of your important counseling and therapy skill sets, there are times when anger is appropriate and a motivator for action. Injustice, unfairness, bullying, and harassment are four examples where client anger is suitable.

The so-called negative emotions of sad, mad, and fear are primarily located in the limbic system. But calling them negative is not fully accurate. Fear in the face of danger is protective. For example, you see a car coming toward you, and you don't have time to think. Protective fear enables you to short-circuit the time it takes to think and enables you to swerve to save your life. When you duck at a baseball coming toward your head, protective fear is there again. When a woman fears abuse, it can serve as a motivator to find safety.

Disgust is an interesting addition to the original four primary emotions. It is thought to have evolved as a way to ensure avoidance of unhealthy objects, particularly when we recognize a "disgusting" odor. We say that "it stinks" or a certain person is "rotten" and a "rat." Disgust evokes dimensions of anger, but overlaps with fear as well. If you are counseling a couple thinking of separation or divorce, you likely will see anger. However, if "anger" turns to "disgust," saving the relationship is a larger challenge.

Two examples related to surprise are shock and wonder, one often negative, the other usually positive. When you counsel clients and hear them accurately, their surprise at being heard opens the way to change. Confrontation is a basic skill presented in Section III that can lead to change. The surprise behind confronting clients with the mixed messages and unseen conflicts in their lives can be an important moment. You will surprise some clients with the many good things in their lives, past and present, thus enabling to start breaking through disappointment with themselves and negative self-thoughts.

The importance of building solid positive emotions. You can enable clients to bring out stories of positive emotions and thoughts; this is particularly valuable when you work with clients who have a cognitive style of primarily negative thinking. Love and caring are complex positive social emotions. The left prefrontal cortex is the primary location of positive emotional experience (e.g., glad/happy) and is also where our executive decision-making functions lie. Both love and caring represent thoughts and feelings that are expressed *in relationship* with others, although self-love is real as well. Think of love as a positive emotion, but with a cognitive component relating to a loved one, animal, or even an object associated with joyful or caring experiences.

If you use a strength-based approach, based on positive psychology and therapeutic lifestyle changes, you can help clients gain mental health and effective problem solving much more quickly. There is evidence that positive emotions also lead to physical well-being, as well as life satisfaction (Carl , Soskin, Kerns, & Barlow, 2013). Historically, the counseling and therapy fields have focused on negative issues ("What's your problem?") with insufficient attention to client resources and personal assets. As stated in Chapter 2, your clients can best deal with the challenges of negative emotions and negative cognitions if we constantly build on their capabilities.

Clients often express emotions in unclear ways, demonstrating mixed and conflicting emotions. You can help them sort through these more complex feelings. Clients may experience many feelings all at once. A client going through a difficult separation or divorce may express feelings of love toward the partner at one moment and extreme anger the next. Words such as puzzled, sympathy, embarrassment, guilt, pride, jealousy, gratitude, admiration, indignation, and contempt are social emotions that derive from primary emotions. How we deal with emotional experience, of course, depends very much on our learning history. Basic emotions appear to be universal across all cultures, but the social emotions appear to be learned in the community and culture, from family members and peers. They are made more complex by the multifaceted and challenging world that surrounds us—and their meanings may change from day to day and even in one interview.

Confusion, frustration, and mixed feelings. While not recognized as part of the basic cross-cultural emotions, most of us and your clients are likely to recognize these words and the body feelings that often go with them. Stop for just a moment and recall when you might have been confused, frustrated, or undecided as to where to go. If you can take time to note the possible feelings in your stomach, the tension in your shoulders, and maybe even the clenching of your fists, it becomes apparent that these words do carry a heavy load of emotion.

When clients are torn and experience confusion and frustration, at another level they typically come with mixed feelings of sadness, anger, or fear. On the other hand, their feeling of being torn may come from an optimistic positive desire for resolution and deciding between good things. Relatively few clients will come to you with single, distinct emotions.

Even in the most discouraged of clients, you will want to search out pieces and past stories with feelings of success and pride.

Many clients will say, "I feel confused" or "I feel frustrated" early in the session. This is your opportunity to reflect that feeling, and as you hear the story and concerns around these words, you will discover a fuller understanding of the mixed emotions usually involved. In short, for practical purposes, confusion, frustration, and indecision need to be treated as emotional reactions to a literally confusing and often overwhelming world. As you listen to the cognitive content of the story—as you encourage, paraphrase, and summarize—the underlying basic emotions will appear.

Then our task is to understand the facts, issues, and stories around the emotions we discover. With emotional untanglement, cognitions and thoughts are likely to become clearer. With cognitive understanding, emotions also become better defined.

 Interactive Exercise: Increasing Your Feeling Vocabulary

THE SPECIFIC SKILLS OF REFLECTING FEELINGS

The first task, of course, is constant awareness and observation of emotions and feelings as clients show them explicitly through words in their conversation and nonverbally through their patterns of eye contact, facial expression, and body language. Vocal tone, speech hesitations, and variations in loudness are also useful clues to underlying feelings.

At the most basic level, reflection of feelings involves the following set of verbal responses:

1. *Sentence stem.* Choose a sentence stem such as "I hear you are feeling . . . ," "Sounds like you feel . . . ," "I sense you are feeling. . . ." Unfortunately, these sentence stems have been used so often they can sound like comical stereotypes. As you practice, you will want to vary sentence stems and sometimes omit them completely. Using the client's name and the pronoun *you* help soften and personalize the sentence stem.

2. *Feeling label.* Add an emotional word or feeling label to the stem ("Angelica, you seem to feel sad about . . . ," "Looks like you're happy," "Sounds like you're discouraged today; you look like you feel really down"). For mixed feelings, more than one emotional word may be used ("Miguel, you appear both glad and sad . . ."). Note that if you use the term "you feel . . . "too often that you may sound repetitious.

3. *Context or brief paraphrase.* You may add a brief paraphrase to broaden the reflection of feelings. The words *about*, *when*, and *because* are only three of many that add context to a reflection of feelings ("Angelica, you seem to feel angry about all the things that have happened in the past two weeks"; "Miguel, you seem to be excited and glad when you think about moving out and going to college, but also sad as you won't be with your friends and parents").

4. *Tense and immediacy.* Reflections in the present tense ("Right now, you look very angry") tend to be more useful than those in the past ("You felt angry when . . ."). Some clients have difficulty with the present tense and talking in the "here and now." "There and then" review of past feelings can be helpful and may feel safer for the client.

5. *Checkout.* Check to see whether your reflection of feelings is accurate. This is especially helpful if the feeling is unspoken ("You really feel angry and frustrated today—am I hearing you correctly?").

6. Bring out positive strengths and emotional stories to counter the negatives and difficulties. If you only focus on fearful, angry, sad emotions, you may find yourself reinforcing negative cognitions as well as negative emotions. If you employ positive psychology and bring out strengths close to the time you discuss serious concerns and issues, you will find that you have helped clients strengthen their abilities leading to resolution.

Your humanness and ability to be with the client through empathic understanding, attending behavior, and showing interest and compassion are basic when you deal with client emotional experience. Your vocal tone, your nonverbal behavior, and—particularly—your facial expressions together communicate appropriate levels of support and caring.

Reflection of feeling is a skill, but as we note at the beginning of the chapter, it is also an art. Please do not find yourself in a set routine that takes the authenticity out of the relationship. Being your authentic, natural self is equally important to the skill.

Acknowledgment of feelings. Sometimes a simple, brief recognition of feelings can be as helpful as a full reflection. In **acknowledging feelings**, you state the feeling briefly ("You seem to be sad about that," "It makes you happy") and move on with the rest of the conversation. With a harried and perhaps even rude busy clerk or restaurant server, try saying in a warm and supportive tone, "Being that rushed must make you tense." Often this is met with a surprised or relaxed look with an implicit thanks of appreciation. The same structure is used in an acknowledgment as in a full reflection, but with less emphasis and time given to the feeling.

Acknowledging and naming feelings may be especially helpful with children, particularly when they are unaware of what they are feeling. Children often respond well to the classic reflection of feelings, "You feel . . . [sad, mad, glad, scared] . . . because. . . ."

▶ Observing Reflection of Feelings in Action

The discovery of cancer, AIDS, or other major physical illness brings with it an immense emotional load. Busy physicians and nurses, operating primarily at the executive cognitive level, may fail to deal with patients' emotions or the worries of family members. In fact, the diagnosis of a life-threatening disease increases the risk of depression, general anxiety disorder, and suicide. Lorelei Mucci of Harvard University, best known for her wide-ranging research on suicide and cancer of various types, found, for example, that prostate cancer doubles the risk of suicide in affected patients (Fall et al., 2009).

The following transcript illustrates reflection of feelings in action. This is the second session, and Jennifer has just welcomed the client, Stephanie, into the room. After a brief personal exchange of greetings, it is clear, nonverbally, that the client is ready to start immediately.

INTERVIEW TRANSCRIPT: MY MOTHER HAS CANCER

INTERVIEWER AND CLIENT CONVERSATION	PROCESS COMMENTS
1. *Jennifer:* So, Stephanie, how are things going with your mother?	Jennifer knows what the main issue is likely to be, so she introduces it with her first open question.
2. *Stephanie:* Well, the tests came back and the last set looks pretty good. But I'm upset. With cancer, you never can tell. It's hard . . . [pause]	Stephanie speaks quietly and as she talks, her voice becomes even softer. At the word "cancer," she looks down. An issue with virtually all clients facing crisis is maintaining emotional balance and effective executive decision making.
3. *Jennifer:* Right now, you're really worried and upset about your mother.	Jennifer uses the emotional word ("upset") used by the client, but adds the unspoken emotion of worry. With "right now," she brings the feelings to here-and-now immediacy. She did not use a checkout. Was that wise? (Interchangeable empathy)

(continued)

INTERVIEWER AND CLIENT CONVERSATION	PROCESS COMMENTS
4. *Stephanie:* That's right. Since she had her first bout with cancer . . . [pause], I've been really concerned and worried. She just doesn't look as well as she used to, she needs a lot more rest. Colon cancer is so scary.	Often if you help clients name their unspoken feelings, they will verbally affirm or nod their head. Naming and acknowledging emotions helps clarify the total situation.
5. *Jennifer:* Scary?	Repeating the key emotional words used by clients often helps them elaborate in more depth. (Interchangeable empathy)
6. *Stephanie:* Yes, I'm scared for her and for me. They say it can be genetic. She had Stage 2 cancer and we really have to watch things carefully.	Stephanie confirms the intentional prediction and elaborates on the scary feelings. She looks frightened and physically exhausted. This is a clear example of how cognitions impact emotional experience.
7. *Jennifer:* So, we've got two things here. You've just gone through your mother's operation, and that was scary. You said they got the entire tumor, but your Mom really had trouble with the anesthesia, and that was frightening. You had to do all the caregiving, and you felt pretty lonely and unsupported. And the possibility of inheriting the genes is pretty terrifying. Putting it all together, you feel overwhelmed. Is that the right word to use, overwhelmed?	Jennifer summarizes what has been said. She repeats key feelings. She uses a new word, "overwhelmed," which comes from her observations of the total situation. Bringing in a new word to describe emotions needs to be done with care. In this example, Jennifer asked the client if that word made sense, and we see in the next exchange that Stephanie could use "overwhelmed" to discuss what was going on. Note how the counselor balances emotion and cognition. (Additive empathy due to the accuracy of the summary)
8. *Stephanie:* [immediately] Yes, I'm overwhelmed. I'm so tired, I'm scared, and I'm furious with myself. [pause] But I can't be angry; my mother needs me. It makes me feel guilty that I can't do more. [Starts to sob]	Stephanie is now talking about her issues at the here-and-now level. Stephanie has not cried in the interview before, and she likely needs to allow herself to cry and let the emotions out. Caregivers often burn out and need care themselves.
9. *Jennifer:* [sits silently for a moment] Stephanie, you've faced a lot and you've done it alone. Allow yourself to pay attention to you for a moment and experience the hurt. [As Stephanie cries, Jennifer comments.] Let it out . . . that's OK.	Stephanie has held it all in and needs to experience what she is feeling. If you are personally comfortable with emotional experience, this ventilation of feelings can be helpful. There will be a need to return later to a discussion of Stephanie's situation from a less emotional frame of reference. Jennifer hands Stephanie a box of tissues. A glass of water is available. (Additive empathy as Stephanie is there to support)
10. *Stephanie:* [continues to cry, but the sobbing lessens]	Sadness is a valid emotion, needs to be acknowledged and worked through. A later section of this chapter provides ideas for helping clients deal with deeper emotional experience.
11. *Jennifer:* Stephanie, I really sense your hurt and aloneness. I admire your ability to feel—it shows that you care. Could you sit up a little straighter now and take a deep breath? [Pause] How are you doing?	With the supportive positive reflection, the client sits up, the crying almost stops, and she looks cautiously at the interviewer. She wipes her nose and takes a deep breath. Jennifer did three things here: (1) She reflected Stephanie's here-and-now emotions; (2) she identified a positive asset and strength; and (3) she suggested that Stephanie take a breath. Conscious breathing often helps clients pull themselves together. (Additive empathy)

(continued)

INTERVIEWER AND CLIENT CONVERSATION	PROCESS COMMENTS
12. *Stephanie:* I'm OK. [pause]	She wipes her eyes and continues to breathe. She seems more relaxed now that she has let out some of her emotions. At this point, she can explore the situation more fully—both emotionally and content-wise—as she moves toward decisions. Emotional regulation and balance have returned.
13. *Jennifer:* You really feel deeply about what's happened with your mother, and it is so sad. At the same time, I see a lot of strength in you. We've talked earlier about how you supported her both before and after the operation. You did a lot for her. You care, and you also show her how you care by what you do.	Jennifer brings in the positive asset search. In the middle of a difficult time, Stephanie has shown she has the ability to handle the situation, even in the midst of worry and anxiety. This response from Jennifer reflects feelings, but also summarizes strengths that Stephanie can use to continue to deal with her concerns. (Additive empathy)

The major skill Jennifer used in this session was reflection of feelings, with a few questions to draw out emotions. Because human change and development are rooted in emotional experience, reflection of feelings is used in all theories of counseling and therapy. Humanistic counselors often consider eliciting and reflecting feelings the central skill and strategy of interviewing. The cognitive behavioral theories (e.g., CBT) use this skill, but their focus is on changing cognitions, which in turns often leads to emotional change.

You are most likely beginning your work and starting to discover the importance of reflecting feelings. It may take you some time before you are fully comfortable using this skill because it is seldom a part of daily communication. We suggest you start by first simply noting emotions and then acknowledging them through short reflections indicating that the emotions have been observed. As you gain confidence and skill, you will eventually decide the extent and place of emotional exploration in your helping repertoire.

 Interactive Exercise: Facilitating Clients' Exploration of Emotion at Varying Levels

▶ Practicing: Becoming Competent in Observing and Reflecting Feelings

As stressed throughout this book, reading about skills does not produce competence. It does not achieve mastery of skills when it comes to meeting clients in the session. Therefore, the exercises here may be the most important reading in this chapter.

EXERCISE 6.2 Self-Reflection: What Is Your Emotional Style?

If you are to observe clients' emotional experience and reflect their feelings, it is vital that you maintain access to your own emotions. Some of us are almost totally unaware our own emotional side; others may reach their emotions more easily, sometimes to the point of overemotionality. To be fully empathic means to be able to put yourself in the client's shoes, but with awareness that these are not your shoes. Some topics may be emotionally threatening to you. If so, the client might pick this up and not talk about them. Or a side topic that catches you at a feeling level may cause you to overemphasize something the client mentions.

What is your level of emotional awareness?

Can you identify how your body reacts emotionally to various topics and situations?

EXERCISE 6.3 Feeling Words Observed in Your First Interview

Go back to your recorded first interview (see Box 1.3 on page 16), transcribe the session if you have not done so already, and underline or highlight every emotionally oriented word. Then notice how the client responded when you used feelings. (You may use another interview if you prefer.)

What emotions did the client express nonverbally or verbally?

What emotions or feelings did you attend to?

How did you balance cognition and emotion?

What was your emphasis on positive strength emotions as compared to negative issues?

What might have you missed or overlooked?

What would you do differently as a consequence of your analysis of this interview?

EXERCISE 6.4 Acknowledgment of Feeling

Test out this brief skill the next time you are in a grocery store or restaurant. First observe behavior, then follow this with a smile and "Looks like a busy day" or "You look happy today" or "Must be frustrating with this many customers," communicating implicitly or explicitly that you understand the person's feelings. How did the person react?

EXERCISE 6.5 Record an Individual Practice Session and Obtain Feedback

Join a classmate, or ask a friend or family member to work with you as a volunteer. Request that the person tell you a story that has some emotional meaning to the storyteller. Video record the session (using a camera, computer, tablet, or smartphone), obtaining written or verbal permission as appropriate. Some possible topics include an enjoyable experience, ranging from a hike to travel to a date; interpersonal conflict; or a difficult decision. Draw out the story using attending, questioning, and active listening. Observe both nonverbal and verbal feelings. After 15 minutes, conclude and seek verbal feedback from the volunteer. If possible, obtain written feedback as well, using a downloaded form from the web resources. Listen to or watch the recorded session, making note of both your own and client behavior. Use the Feedback Form: Observing and Reflecting Feelings to increase mastery of this skill.

EXERCISE 6.6 Group Practice

Especially effective is meeting with a group of three or four, taking turns as interviewer and client. One member can run the recording equipment and observe. The second observer takes notes with an emphasis on the skill of reflection of feeling. Both observers will be helpful in pointing out nonverbal behavior. In a group such as this, the downloaded forms from the web resources will be especially helpful.

► Refining: Observation and Reflection of Feelings

Once you have a sense of the basics of the centrality of emotions in the helping session and the relationship of cognition and emotions to executive functioning and emotional regulation, the following issues may be useful elaborations of how to use the skill in multiple settings for maximal impact.

HELPING CLIENTS INCREASE OR DECREASE EMOTIONAL EXPRESSIVENESS*

You can assist your clients to get more accurately in touch with their emotions. Ultimately, this awareness will help both overall executive cognitive functioning and their own emotional regulation.

---

*Leslie Brain, graduate student at the University of Massachusetts, Amherst provided specific ideas for the recent updating provided in this section. We encourage all of you to share ideas for enriching this book with us.

Observe nonverbals. Breathing directly reflects emotional content. Rapid or frozen breath indicates contact with intense emotion, indicative of both limbic and reptilian brain involvement and potential loss of emotional regulation. Anticipate heart rate and blood pressure increase.

Also note facial flushing, eye movement, body tension, and changes in vocal tone. Especially, note hesitations. Attend to these clues and follow up at an appropriate point.

At times you may also find an apparent absence of emotion when discussing a difficult issue. This might be a clue that the client is avoiding dealing with feelings or that the expression of emotion is culturally inappropriate for this client. You can pace clients by listening and acknowledging feelings and then gradually leading them to fuller expression and awareness of affect. Many people get right to the edge of a feeling and then back away with a joke, change of subject, or intellectual analysis.

Pace and encourage clients to express more emotion by listening and mirroring body language at times.

▶ Say to the client that she looked as though she was close to something important. "Would you like to go back and try again?"
▶ Discuss some positive resource that the client has. This base can free the client to face the negative. You as counselor also represent a positive asset.
▶ Consider asking questions. Used carefully, questions may help some clients explore emotions.
▶ Use here-and-now strategies, especially in the present tense: "What are you feeling right now—at this moment?" "What's occurring in your body as you talk about this?" Use the word *do* if you find yourself uncomfortable with emotion: "What do you feel?" or "What did you feel then?" starts to move the client away from here-and-now experiencing.

When tears, rage, despair, joy, or exhilaration occur. Your comfort level with your own emotional expression will affect how clients face their feelings. If you aren't comfortable with a particular emotion or topic, your clients will likely avoid it, resulting in their handling the issue less effectively. Keep a balance between being very present with your own breathing and showing culturally appropriate and supportive eye contact, but still allowing room for the client to sob, yell, or shake. You can also use phrases such as these:

I'm here.

I've been there, too.

Let it out . . . that's OK.

These feelings are just right.

I hear you.

I see you.

Breathe with it.

Seek to keep deep emotional expression within a reasonable timeframe; 2 minutes is a long time when you are crying. Afterward, help the person reorient to the here-and-now present moment before reflecting and discussing the strong emotions. Tools for reorienting the interview that lead to emotional regulation include these:

▶ Help the client use slowed, rhythmic breathing. This may be accompanied by a brief acknowledgment of feeling.
▶ Discuss the client's positive strengths.

- ▶ Discuss direct, empowering, self-protective steps that the client can take in response to the feelings expressed.
- ▶ Suggest that the client stand and walk or center the pelvis and torso in a seated position.
- ▶ Explore the emotional outburst as appropriate to the situation. Cognitively reframe the emotional experience in a positive way.
- ▶ Comment that it helps to tell the story many times.

Caution. As you work with emotion, there is the possibility of reawakening issues in a client who has a history of painful trauma. When you sense this possibility, decide with the client in advance and obtain permission for the desired depth of emotional experiencing. The beginning interviewer needs to seek supervision and/or refer the client to a more experienced professional.

DIVERSITY ISSUES

We need to respect individual and cultural diversity in the way people express feelings. The student from China discussed in Box 6.1 is an example of cultural emotional control. Emotions are obviously still there, but they are expressed differently. Do not expect all

---

BOX 6.1	National and International Perspectives on Counseling Skills

Does He Have Any Feelings?

Weijun Zhang

Videotaped session: A student from China comes in for counseling, referred by his American roommate. According to the roommate, the client quite often calls his wife's name out loud while dreaming, which usually wakes the others in the apartment, and he was seen several times doing nothing but gazing at his wife's picture. Throughout the session the client is quite cooperative in letting the counselor know all the facts concerning his marriage and why his wife is not able to join him. But each time the counselor tries to identify or elicit his feelings toward his wife, the client diverts these efforts by talking about something else. He remains perfectly polite and expressionless until the end of the session.

No sooner had the practicing counselor in my practicum class stopped the videotape than I heard comments such as "inscrutable" and "He has no feelings!" escape from the mouths of my European American classmates. I do not blame them, for the Chinese student did behave strangely, judged from their frame of reference.

"How do you feel about this?" "What feelings are you experiencing when you think of this?" How many times have we heard questions such as these? The problem with these questions is that they stem from a European American counseling tradition, which is not always appropriate.

For example, in much of Asia, the cultural rationale is that the social order doesn't need extensive consideration of personal, inner feelings. We make sense of ourselves in terms of our society and the roles we are given within the society. In this light, in China, individual feelings are ordinarily seen as lacking social significance. For thousands

of years, our ancestors have stressed how one behaves in public, not how one feels inside. We do not believe that feelings have to be consistent with actions. Against such a cultural background, one might understand why the Chinese student was resistant when the counselor showed interest in his feelings and addressed that issue directly.

I am not suggesting here that Asians are devoid of feelings or strong emotions. We are just not supposed to telegraph them as do people from the West. Indeed, if feelings are seen as an insignificant part of an individual and regarded as irrelevant in terms of social importance, why should one send out emotional messages to casual acquaintances or outsiders (the counselor being one of them)?

What is more, most Asian men still have traditional beliefs that showing affection toward one's wife while others are around, even verbally, is a sign of being a sissy, being unmanly or weak. I can still vividly remember when my child was 4 years old, my wife and I once received some serious lecturing on parental influence and social morality from both our parents and grandparents, simply because our son reported to them that he saw "Dad give Mom a kiss." You can imagine how shocking it must be for most Chinese husbands, who do not dare even touch their wives' hands in public, to see on television that American presidential candidates display such intimacy with their spouses on the stage! But the other side of the coin is that not many Chinese husbands watch television sports programs while their wives are busy with household chores after a full day's work. They show their affection by sharing the housework!

Chinese or other Asians to be emotionally reserved, however. Their style of emotional expression will depend on their individual upbringing, their acculturation, and other factors. Many New England Yankees may be fully as reserved as or even more cautious in emotional expression than the Chinese student described by Weijun Zhang. But again, it would be unwise to stereotype all New Englanders in this fashion.

Openness to emotional roots varies wide from culture to culture and group to group. Generally speaking, citizens of the United States express opinions and emotions more freely than do people from other countries. As a result, Americans can be seen as rude and intrusive, while they may see other groups as cold and indifferent.

Regardless of multicultural background and experience, all clients have limbic systems and emotions. They breathe and experience varying heart rates. Their standards of what is appropriate emotional regulation, however, will vary. Expect emotional openness to be maximized with family and friends, but even here cultures vary.

THE PLACE OF POSITIVE EMOTIONS IN REFLECTING FEELINGS AND RESILIENCE

Contrasting the negative emotions such as sadness with joy can lead to inner peace and stoic equanimity.

—Antonio Damasio

When we learn how to become resilient, we learn how to embrace the beautifully broad spectrum of the human experience.

— Jaeda DeWalt

We grow best and become resilient with what we "can do" rather than what we "can't do." Positive emotions, whether joyful or merely contented, are likely to color the ways people respond to others and their environments. Research shows that positive emotions broaden the scope of people's visual attention, expand their repertoires for action, and increase their capacities to cope in a crisis. Research also suggests that positive emotions produce patterns of thought that are flexible, creative, integrative, and open to information (Gergen & Gergen, 2005). *Sad, mad, glad, scared* is one way to organize the language of emotion. But perhaps we need more attention to glad words, such as *pleased, happy, love, contented, together, excited, delighted, pleasured,* and the like.

> ### EXERCISE 6.7 Reflection: Here-and-Now Contrast of Positive and Negative Emotions
> Take just a moment now and think of specific situations in which you experienced each of the positive emotions listed immediately above. What changes did you notice in your body and mind? Compare these with the bodily experience of negative feelings and emotions.
>
> Very likely, if you smiled, your body tension would be reduced, and even your blood pressure could lower. Perhaps you even found a little bit of inner peace. Positive psychology shows us that attention to strengths lessens the load, takes our mind off our concerns, and empowers us when we recall our strengths and good experiences.
>
> How was your experience?
>
> What do you think of the changes you noticed in your body?
>
> In your mind?

When you experience emotion, your brain signals bodily changes. Positive, reinforcing neurotransmitters such as dopamine and serotonin increase. When you feel sad or angry, potentially destructive cortisol and hormones are released in the limbic system.

These chemical changes will show nonverbally. Emotions change the way your body functions and thus are a foundation for our thinking experience. As you help your clients experience more positive emotions, you are also facilitating wellness and a healthier body. The route toward health, of course, often entails confronting negative emotions.

Research examining the life of the long-living Mankato nuns found that women who had expressed the most positive emotions in early life lived longer than those who expressed a difficult past (Danner, Snowdon, & Friesen, 2001). Stress reactions to the 9/11 bombing disaster found that those who had access to the most positive emotions were more resilient and showed fewer signs of depression (Fredrickson, Tugade, Waugh, & Larkin, 2003). A resilient affective lifestyle results in a faster recovery and lower damaging cortisol levels.

These examples of well-being and wellness are located predominantly in the executive prefrontal cortex, with lower levels of activation in the amygdala (Davidson, 2004, p. 1395). Drawing on long-term memory for positive experiences is one route toward well-being and stress reduction. These strengths enable us to deal more effectively with life challenges.

Figure 6.1 shows the level of brain activity in a depressed and a nondepressed individual. In this picture, we see what happens in the absence of positive feelings and emotions. Executive functioning and emotional regulation have broken down. The lower brain limbic system has taken over. Working with depression can be challenging, but our goal is to increase positive functioning. If we focus only on negative, problematic stories and reflect these negative feelings, we are reinforcing depression. While it is necessary to hear these

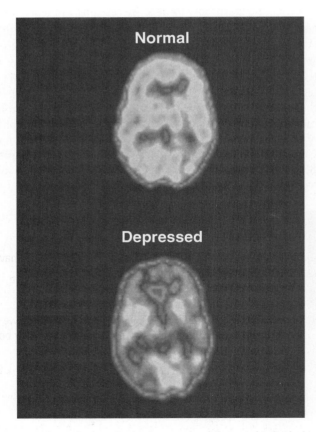

FIGURE 6.1 Brain activation in normal and depressed individuals.

Source: Science Source

stories, ultimately our goal with any client expressing sadness or depression is to find positives. (Needless to say, major depression as seen in Figure 6.1 will often require medication before counseling and therapy can be fully effective.)

STRATEGIES FOR POSITIVE REFLECTION

Several techniques can be used to help clients reflect on positive emotions and feelings. The first, from positive psychology, is searching for wellness and positive assets, as emphasized in Chapter 2. As part of a positive wellness assessment, be sure that you reflect the positive feelings associated with therapeutic lifestyle changes. For example, your client may feel safety and strength in the spiritual self, pride in gender and/or cultural identity, caring and warmth from past and/or present friendships, and the intimacy and caring of a love relationship. It would be possible to anchor these emotions early in the session and draw on these positive emotions during more stressful moments. Out of this can come a "backpack" of positive emotions and experiences that are always there and can be drawn on as needed.

Couples with relationship difficulties can be helped if they focus more on the areas where things are going well—what remains good about the relationship. Many couples focus on the 5% where they disagree and fail to note the 95% where they have been successful or enjoyed each other. Some couples respond well when asked to focus on the reasons they got together in the first place. These positive strengths can help them deal with very difficult issues.

When providing your clients with homework assignments, ask them to engage daily in activities associated with positive emotions. For example, it is difficult to be sad and depressed when running or walking at a brisk pace. Meditation and yoga are often useful in generating more positive emotions and calmness. Seeing a good movie when one is down can be useful, as can going out with friends for a meal. In short, help clients remember that they have access to joy, even when things are at their most difficult.

Finally, service to others often helps people feel good about themselves. When one is discouraged and feeling inadequate, volunteering for a church work group, working on a Habitat for Humanity home, or giving time to work with animal shelters can all be helpful in developing a more positive sense of self.

Caution. Please do not use the preceding paragraphs as a way to tell your clients that "everything will be OK." Some interviewers and counselors are so afraid of negative emotions that they never allow their clients to express what they really feel. *Do not minimize difficult emotions by too quickly focusing on the positive.* Nonetheless, our goal is resilience and growth!

Exercise 6.8 Positive Emotions
Take a moment to think of specific situations in which you have experienced positive emotions. Very likely when you think of these situations, you will smile, which will help reduce your overall body tension. Encouraging positive memories is a central route toward emotional regulation and executive functioning. It is even likely that your blood pressure will change in a more positive direction. Tension produces damaging cortisol in the brain, and we all need to learn to control our bodies more effectively. Positive imaging is a useful strategy for both you and your clients.

SOME LIMITATIONS OF REFLECTING FEELINGS

Reflection has been described as a basic skill of the counseling process, yet it can be overdone. With friends, family, and fellow employees, a quick acknowledgment of feelings ("If I were you, I'd feel angry about that . . ." or "You must be tired today") followed by continued normal conversational flow may be most helpful in developing better relationships. Similarly, with many clients, a brief acknowledgment of feelings may be more useful.

However, with complex issues, identifying unspoken feelings may be necessary. Sorting out mixed feelings is key to successful counseling, be it vocational interviewing, personal decision making, or in-depth individual counseling and therapy.

Be aware that not all clients will appreciate your comments on their feelings. Exploring the emotional world can be uncomfortable for those who have avoided looking at feelings in the past. An empathic reflection can have a confrontational quality that causes clients to look at themselves from a different perspective; therefore, it may seem intrusive to some. Timing is particularly important with reflection of feelings. Clients tend to disclose feelings only after rapport and trust have been developed. Less verbal clients may find reflection puzzling or may say, "Of course I'm angry; why did you say that?" Some men may believe that expression of feelings is "unmanly." Brief acknowledgment of feelings may be received with appreciation early on and can lead to deeper exploration in later interviews.

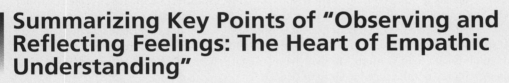

Case Study: Becoming Aware of and Skilled With Emotional Experience
Interactive Exercise: Identifying Skills. My Mother Has Cancer
Interactive Exercise: Becoming Aware of and Skilled With Emotional Experience: Naomi, Abusive Relationship
Group Practice Exercise: Group Practice With Reflection of Feeling

Summarizing Key Points of "Observing and Reflecting Feelings: The Heart of Empathic Understanding"

Defining Reflection of Feelings

▶ Both paraphrasing and reflection of feelings feed back to clients what they have been experiencing.

▶ A paraphrase focuses on the verbal content of what the client says; reflection of feelings centers on both verbal and nonverbal emotional underpinnings.

▶ Unspoken feelings may be seen in the client's nonverbal expression, may be heard in the client's vocal tone, or may be inferred from the client's language.

▶ Feelings are layered like an onion. Words like *frustration*, *mixed up*, and *confused* represent conflicting emotions underneath surface words.

▶ Checkouts can help confirm the accuracy of paraphrases and reflections of feelings.

Discerning: The Basic Techniques and Strategies of Observing and Reflecting Feelings

▶ People constantly feel and express emotion in verbal and nonverbal ways.

▶ The proficient interviewer will hear explicit emotions, observe implicit emotions, and feed these emotions back to the client.

▶ Sad, mad, glad, and scared are primary emotions used as root words for building a vocabulary of emotion. They appear to be universal across cultures.

▶ Social emotions (embarrassment, guilt, pride) are modified and built on primary emotions. They are learned in a family and cultural context.

► Everyone has complex emotions associated with people and events in their lives. Helping clients sort out these feelings is an important part of interviewing, counseling, and psychotherapy.

► Proficient interviewers are able to observe and reflect emotional dimensions accurately.

► The components of reflection of feelings are (1) a sentence stem, (2) the feeling word, (3) some context, (4) present tense, (5) a checkout, and (6) positive strengths. Beginning with the client's name is helpful, and present tense, here-and-now reflections are often more powerful than a review of past emotions.

► Research and clinical experience reveal that special attention to positive emotions can provide clients with strengths to better address their emotional challenges.

Observing Reflection of Feelings in Action

► Because human change and development are rooted in emotional experience, reflection of feelings is critical in all theories of counseling and therapy.

► Reflection of feelings clarifies the client's emotional state, leads clients in new directions, and results in new discoveries.

► Identify positive qualities and emotions to help clients deal more effectively with negative emotions.

► In the example interview, Jennifer reflects the main emotional words actually used by the client. She also points out unspoken feelings and checks out with the client whether the identified feeling is accurate. For example, "Is that close to what you feel?"

Practicing: Becoming Competent in Observing and Reflecting Feelings

► Practicing what you have read is essential for skill mastery.

► Start with self-reflection. How comfortable are you with emotions? Are there areas where clients may cause you to feel uncomfortable?

► Go back and review your first interview. Focus on feeling words observed in the interview. Notice what you did and what you learned.

► Practice acknowledgment of feelings, as a brief recognition may be more helpful sometimes. Use acknowledging and naming feelings.

► Work with another person and record an individual practice session reflecting feelings. Seek feedback. Review the session, noting the client's emotions and your responses to them.

► Practice with a group of three or four classmates. Use the group guidelines to facilitate practice in reflecting feelings.

Refining: Observation and Reflection of Feelings

► You can assist your clients to get more accurately in touch with their emotions. By focusing on breathing patterns and other nonverbal behaviors, you can help clients increase or decrease emotional expressiveness.

► Mirror body language to encourage clients to express more emotion if needed. Use here-and-now strategies such as "What are you feeling now?" to increase emotional awareness or "What did you feel then?" to move the client away from here-and-now experiencing.

► Check your own comfort level with your own emotional expression, as this will affect how clients face their feelings. If you are uncomfortable with a particular emotion or topic, your clients will likely avoid it.

▶ Show culturally appropriate and supportive behavior, as clients from diverse groups will present varying degrees of comfort working with emotions.

▶ Positive emotions broaden clients' behavioral responses. To facilitate growth, focus on what clients "can do" rather than what they "can't do."

▶ Help clients reflect on positive emotions and feelings. Use the positive wellness assessment and the therapeutic lifestyle changes. Reflect the feelings associated with these.

▶ Remember to use positive focus in a way that is helpful to the client. Avoid minimizing or being overly optimistic.

▶ Building the relationship is central, as some clients may be reticent to reveal their deepest feelings to others.

▶ Reflection of feelings can be overdone, particularly if the client comes from a family or culture that believes emotional expression is inappropriate.

▶ Not all clients are comfortable discussing emotion. Exploring emotions can be uncomfortable for some, especially if they have avoided this in the past. Also, less verbal clients and those who are not used to dealing with emotions may respond negatively to your reflections. Initially, brief acknowledgment of feelings may be a better option with these clients.

Assess your current level of knowledge and competence as you complete the chapter:

1. Flashcards: Use the flashcards to check your understanding of key concepts and facilitate memorization of key information.

2. Self-Assessment Quiz: The quiz will help you assess your current knowledge and prepare for course examinations.

3. Portfolio of Competencies: Evaluate your present level of competence on the ideas and concepts presented in this chapter using the Self-Evaluation Checklist. Self-assessment of your competencies demonstrates what you can do in the real world.

Emotional Words for Exercise 6.1

Sad/unhappy/depressed. Miserable, down, blue, lonely, pained, devastated, disillusioned, bitter, sorry, hopeless, grief-stricken, guilty, ashamed, dejected, despair, glum, gloomy, dismal, heartbroken, hurting, joyless, cheerless, fragile.

Mad/angry/explosive. Annoyed, bad, furious, seething, hostile, violent, jealous, vicious, irate, critical, competitive, cross, irritated, indignant, irked, enraged, hostile, pissed off, ticked off, fed up, stormy, volatile, bombed, kill, hit, rub, thoughtless.

Glad/happy/contented/love. Comfortable, content, pleased, open, safe, peaceful, aware, relaxed, proud, easy, pleased, supportive, kind, good, up, satisfied, grateful, hope, joy, sunny, cheer, cheery, strong, overjoyed, gleeful, thrilled, walking on air, fortunate, lucky, warmth, intimacy, attraction, tranquil, untroubled, soothe, soothing, harmony, friendly, nonviolent, sincere, genuine, humorous, empathic, sympathetic, thoughtful.

Scared/fear/anxious. Nervous, panic, terror, suffocated, panicked, on edge, edgy, horror, alarm, agitation, caged, dread, distress, uptight, surrounded, uneasy, trapped, troubled, tense, uncomfortable, fragile, threatened, vulnerable, unloved, timid, small, insecure, irrelevant, unimportant.

Disgust, revulsion, distaste. Revolted, revolting, repugnant, aversive, nausea, vomit, sicken, nauseating, loathing, contempt, outrage, shocking, stinks, stinky, tasteless, smells, horrifying, appalling, offending, foul, gross, incredibly bad, terrible, dreadful, grim, monstrous, yucky, out of it.

Surprise, shock, wonder. Admiration, amaze, astonish, awe, astound, bewilder, electrify, fascination, unexpectedness, strangeness, unusualness, stun, stupefy.

Confused, frustrated, mixed up

General words related to confusion ("I'm undecided"). Undecided, uncertain, vague, indistinct, blurred, imperfect, sketchy, unsure, foggy.

Words that suggest negative emotions underlying the confusion ("I feel powerless"). Frustrated, torn, lost, drained, empty, numb, exhausted, bewildered, overwhelmed, muddled, addled, befuddled, disoriented, unbalanced, unhinged, helpless, stupid, disappointed, dissatisfied.

Clients may express confusion and indecision in more positive ways. "I see the move offering a better job, but here I'm pretty comfortable." "I'm undecided between psychology and social work; they both appeal to me." "I want to make my life meaningful, but I'm not sure how." "Should I marry this person?" "Which school is best for me?" "How can I improve my relationship/lifestyle/study habits/nutrition?"

Chapter 7
How to Conduct a Five-Stage Counseling Session Using Only Listening Skills

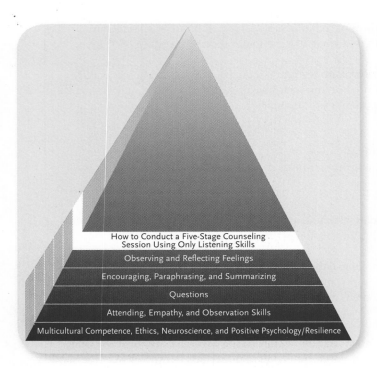

How to Conduct a Five-Stage Counseling Session Using Only Listening Skills

Observing and Reflecting Feelings

Encouraging, Paraphrasing, and Summarizing

Questions

Attending, Empathy, and Observation Skills

Multicultural Competence, Ethics, Neuroscience, and Positive Psychology/Resilience

A leader is best when people barely know that he [or she] exists,

Of a good leader who talks little,

When the work is done,

The aim is fulfilled

They will say, "We did this ourselves."

—Lao Tse

This quote from Lao Tse summarizes a major goal of this book. Our goal as competent interviewers and counselors is to facilitate clients' making their own decisions, finding their own direction, resolving their concerns, and discovering their true selves. Listening is at the heart of interviewing and counseling. If we truly listen, clients will often find their own resolution for many of their concerns and issues, thus finding new possibilities as they meet life challenges.

Chapter Goals

Awareness, knowledge, skills, and actions developed through the concepts of this chapter and this book will enable you to:

▶ Develop further competence with the basic listening sequence (BLS), the foundation of effective interviewing, counseling, and psychotherapy.

▶ Understand and become competent in the five stages of the well-formed session: *empathic relationship—story and strengths—goals—restory—action.*

▶ Learn the basic system of decisional counseling and how it relates to other theories of helping.

▶ Conduct a complete decisional session using only listening skills and the five stages.

▶ Have a good start on Rogerian person-centered counseling, as well as most other theories of counseling and therapy, through your awareness, knowledge, and skills with the basic listening sequence and the five stages.

Assess your current level of knowledge and competence as you begin the chapter:

1. Self-Assessment Quiz: The chapter quiz will help you determine your current level of knowledge. You can take it before and after reading the chapter.

2. Portfolio of Competencies: Before you read the chapter, please fill out the Self-Evaluation Checklist to assess your existing knowledge and competence on the ideas and concepts presented in this chapter. Then, at the end of the chapter, complete the checklist again to summarize your competencies after study and practice.

▶ Defining the Basic Listening Sequence: Foundation for Empathic Listening in Many Settings

To review, the listening skills of the first section of this book contain the building blocks for establishing empathic relationships for effective interviewing, counseling, and psychotherapy. In addition, empathic understanding and careful listening are valuable in all areas of human communication. When you go to see your physician, you want the best diagnostic skills, but you also want the doctor to listen to your story with understanding and empathy. A competent teacher or manager knows the importance of listening to the student or employee.

The microskills represent the specifics of effective interpersonal communication and are used in many situations, ranging from helping a couple communicate more effectively to enabling a severely depressed client to make contact with others. They also are used to train AIDS workers in Africa and in interviewing refugees around the world. Furthermore, teaching the social skills of listening has become a standard and common part of individual counseling and psychotherapy.

When you use the BLS, you can anticipate how others are likely to respond.

BASIC LISTENING SEQUENCE (BLS)	ANTICIPATED CLIENT RESPONSE
The basic listening sequence (BLS), based on attending and observing, consists of these microskills: using open and closed questions, encouraging, paraphrasing, reflecting feelings, and summarizing.	Clients will discuss their stories, issues, or concerns, including the key facts, thoughts, feelings, and behaviors. Clients will feel that their stories have been heard. In addition, these same skills will help friends, family members, and others to be clearer with you and facilitate better interpersonal relationships.

To review, resting on a foundation of attending and observation, these are the skills of the BLS and the central function of each:

▶ Questioning—open questions followed by closed questions to bring out client stories and concerns.

▶ Encouraging—used throughout the session to support clients and provide specifics around their thoughts, feelings, and behaviors.

▶ Paraphrasing—catches the cognitive essence of stories and facilitates executive functioning.

▶ Reflecting feelings—provides a foundation for emotional regulation, enabling examination of the complexity of emotions.

▶ Summarizing—brings order and makes sense of client conversation, thus facilitating executive functioning and emotional regulation.

When you add the checkout to the basic listening sequence, you have the opportunity to obtain feedback on the accuracy of your listening. Clients will let you know how accurately you have listened.

The skills of the BLS need not be used in any specific order, but it is wise to ensure that all are used in listening to client stories. Each person needs to adapt these skills to meet the requirements of the client and the situation. The competent counselor uses client observation skills to note client reactions and intentionally flexes to change style, thus providing the support the client needs.

Examples of how the BLS is used in counseling, management, medicine, and general interpersonal communication are shown in Table 7.1.

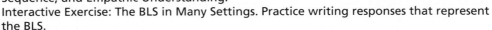

Interactive Exercise: I Don't Seem to Fit In: Predicting Results, the Basic Listening Sequence, and Empathic Understanding.

Interactive Exercise: The BLS in Many Settings. Practice writing responses that represent the BLS.

Video Activity: Reflection of Feelings. Mary and Sandra's interview demonstrates reflection of feelings. This exercise gives you the opportunity to analyze and comment on reflection of feelings, suggest other ways to accomplish this, and explore your own way of responding to emotional expressions.

TABLE 7.1   Four Examples of the Basic Listening Sequence

Skill Usage	Interviewing and Counseling	Management	Medicine	Interpersonal Communication (Listening to others, friends, or family)
Open questions	"Thomas, could you tell me what you'd like to talk to me about . . ."	"Assad, tell me what happened when the production line went down?"	"Ms. Santiago, Could we start with what you think is happening with your headache? Is this OK?"	"Kiara, how did the session with the college loan officer go?"
Closed questions	"Did you graduate from high school?" "What specific careers have you looked at?"	"Who was involved with the production line problem?" "Did you check the main belt?"	"Is the headache on the left side or on the right?" "How long have you had it?"	"Were you able to get a loan that covers what you need?" "What interest rate are they using?"
Encouraging/ Paraphrasing	"So you're considering returning to college."	"Sounds like you've consulted with almost everyone."	"I hear you saying that the headache may be worse with red wine or too much chocolate."	"Wow, I can understand what you are saying, Kiara, two loans are a lot."
Reflecting feelings	"You feel confident of your ability but worry about getting in."	"I sense you're upset and troubled by the supervisor's reaction."	"You say that you've been feeling very anxious and tense lately."	"I sense you feel somewhat anxious and worried when you think of paying it back."
Summarizing	In each case, the effective listener summarizes the cognitive/emotional story from the client's or other person's point of view *before* bringing in the listener's own point of view or perhaps an influencing skill.			

► Defining the Five-Stage Model for Structuring the Session

The five stages of the well-formed session (**empathic relationship—story and strengths—goals—restory—action**) provide an organizing framework for using the microskills with multiple theories of counseling and psychotherapy. All counselors and therapists need to establish an empathic relationship and draw out the client's story. In different ways, explicit or implicit goals are established, and all seek to develop new ways of thinking, feeling, and behaving.

At this point, please review Table 7.2, which summarizes the five stages of the session in detail. Note that listening skills are central at each stage. If you are sufficiently skilled in the microskills and the five stages, you are ready to complete a full session using only attending, observation, and the BLS. Many clients can resolve concerns and make decisions without your direct intervention. This was one of Carl Rogers's major goals in his person-centered approach. However, Rogers favored using as few questions as possible.

The second half of this book focuses on influencing skills, which are used primarily in stages 4 and 5. Thus, brief mention of some of the influencing skills is included in the table.

After you have mastered the five stages step by step, consider them a checklist to ensure that you have covered all the bases in any session. However, following the stages in a specific order is not essential. Many clients will discuss their issues moving from one stage to another and then back again, and you will frequently want to encourage this type of recycling. New information revealed in later counseling stages might result in the need for more data about the basic story, thus redefining client concerns and goals in a new way. Often you will want to draw out more strengths and wellness assets.

The circle of the five stages of a counseling session in Figure 7.1 reminds us that helping is a mutual endeavor between client and counselor. We need to be flexible in our use of skills and strategies. A circle has no beginning or end; rather, a circle is a symbol of an

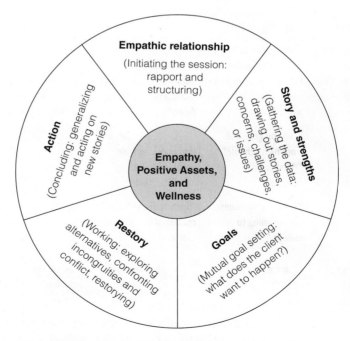

FIGURE 7.1 The circle of counseling stages.

TABLE 7.2 The Five Stages of the Microskills Session

Stage	Function and Purpose	Commonly Used Skills	Anticipated Client Response
1. *Empathic relationship.* Initiate the session. Develop rapport and structuring. "Hello, what would you like to talk about?" "What might you like to see as a result of our talking today?"	Build a working alliance and enable the client to feel comfortable with the counseling process. Explain what is likely to happen in the session or series of sessions, including informed consent and ethical issues. Discover client reasons for coming to you.	Attending, observation skills, BLS, information giving to help structure the session. If the client asks you questions, you may use self-disclosure.	The client feels at ease with an understanding of the key ethical issues and the purpose of the session. The client may also know you more completely as a person and a professional—and has a sense that you are interested in her or his concerns.
2. *Story and strengths.* Gather data. Use the BLS to draw out client stories, concerns, problems, or issues. "I'd like to hear your story." "What are your strengths and resources?"	Discover and clarify why the client has come to the session and listen to the client's stories and issues. Identify strengths and resources as part of a strength-based positive psychology approach.	Attending and observation skills, especially the basic listening sequence and the positive asset search.	The client shares thoughts, feelings, and behaviors; tells the story in detail; presents strengths and resources.
3. *Goals.* Set goals mutually. The BLS will help define goals. "What do you want to happen?" "How would you feel emotionally if you achieved this goal?" One possible goal is exploration of possibilities, rather than focusing immediately.	If you don't know where you are going, you may end up somewhere else. In brief counseling (later in this chapter), goal setting is fundamental, and this stage may be part of the first phase of the session. All the same, openness to change and exploration are good places to start.	Attending skills, especially the basic listening sequence; certain influencing skills, especially confrontation (Chapter 9), may be useful.	The client will discuss directions in which he or she might want to go, new ways of thinking, desired feeling states, and behaviors that might be changed. The client might also seek to learn how to live more effectively with stressful situations or events that cannot be changed at this point (rape, death, an accident, an illness). A more ideal story might be defined.
4. *Restory.* Explore alternatives via the BLS. Confront client incongruities and conflict. "What are we going to do about it?" "Can we generate new ways of thinking, feeling, and behaving?"	Generate at least *three* alternatives that might resolve the client's issues. Creativity is useful here. Seek to find at least three alternatives so that the client has a choice. One choice at times may be to do nothing and accept things as they are. The system of restorying will vary extensively with different theories and approaches.	Summary of major discrepancies with a supportive confrontation. More extensive use of influencing skills, depending on theoretical orientation (e.g., psychoeducation, interpretation, reflection of meaning, feedback). But this is also possible using only listening skills. Use creativity to solve problems.	The client may reexamine individual goals in new ways, solve problems from at least those alternatives, and start the move toward new stories and actions.
5. *Action.* Plan for generalizing interview learning to "real life." "Will you do it?"	Generalize new learning and facilitate client changes in thoughts, feelings, and behaviors in daily life. Commit the client to homework and action. As appropriate, plan for termination of sessions.	Influencing skills, such as directives and information/explanation, plus attending and observation skills and the basic listening sequence to check out client understanding.	The client demonstrates changes in behavior, thoughts, and feelings in daily life outside of the interview. Or the client explores new alternatives and reports back discoveries.

egalitarian relationship in which counselor and client work together. The hub of the circle is empathy, positive assets, and wellness, a central part of all stages.

▶ Discerning Specifics of the Well-Formed Five-Stage Interview

STAGE 1. EMPATHIC RELATIONSHIP—INITIATING THE SESSION: RAPPORT, TRUST BUILDING, STRUCTURING, PRELIMINARY GOALS ("HELLO")

Introducing the session and building rapport are most critical in the first session, but they will remain central in all subsequent sessions. Most sessions begin with some variation of "Could you tell me how I might be helpful?" or "What would you like to talk about today?" "Hello, Lynette." "It's good to meet you, Marcus." A prime rule for establishing rapport is to use the client's name and repeat it periodically throughout the session, thus personalizing the session. Once you have completed the necessary structuring of the session (informing the client about legal and ethical issues), many clients are immediately ready to launch into a discussion of their issues. These clients represent instant trust. Our duty is to honor the client and continue to work on relationship issues throughout the session.

Some situations require more extensive time and attention to the rapport stage than others. Rapport building can be quite lengthy and blend into treatment. For example, in reality therapy with a delinquent youth, playing Ping-Pong or basketball and getting to know the client on a personal basis may be part of the treatment. It may take several sessions before clients who are culturally different from you develop real trust.

Structure. Structuring the session includes informed consent and ethical issues, as outlined in Chapter 2. Clients need to know their rights and the limitations of the session. If this is part of an ongoing series of sessions, you can help maintain continuity by summarizing past sessions and integrating them with the current session. At times, even telling clients about the stages of the session may be useful, so that they know what they are about to encounter.

Listen for preliminary goals. Although more definitive setting of short- and long-term goals happens at stage 3, initial goals are helpful during the first session(s). These early goals provide an initial structure for you as you seek to understand and empathize with the client. The early goals are revised and clarified after a fuller story has been brought out. Goal setting at this first stage is particularly important in brief counseling and coaching. However, early goals are often helpful at the beginning of most sessions.

Share yourself as appropriate. Be open, authentic, and congruent. Encourage clients to ask you questions; this is also the time to explore your cultural and gender differences. What about cross-cultural counseling when your race and ethnicity differ significantly from your clients'? Authorities increasingly agree that cultural, gender, and ethnic differences need to be addressed in a straightforward manner relatively early in counseling, often in the first session (for example, see Sue & Sue, 2013).

Observe and listen. The first session tells you a lot about the client. Note the client's style and, when possible, seek to match her or his language. As the relationship becomes more comfortable, you may note that you and the client have a natural mirroring of body language. This clearly indicates that the client is ready to move on to telling the story and finding strengths (stage 2).

STAGE 2. STORY AND STRENGTHS—GATHERING DATA: DRAWING OUT STORIES, CONCERNS, AND STRENGTHS ("WHAT IS YOUR CONCERN?" "WHAT ARE YOUR STRENGTHS AND RESOURCES?")

Draw out the client's story. What are the client's thoughts, feelings, and behaviors related to her or his concern? We draw out stories and concerns using the skills of the basic listening sequence. Open and closed questions will help define the issue as the client views it. Encouragers, paraphrases, and checkouts will provide additional clarity and an opportunity for you to verify whether you have heard correctly. Reflection of feeling will provide understanding of the emotional underpinnings. Finally, summarizing provides a good way to put the client's conversation into an orderly format.

Elaborate the story. Next, explore related thoughts, feelings, and interactions with others. Gather information and data about clients and their perceptions. The basic journalistic outline of *who, what, where, when, how,* and *why* provides an often useful framework to make sure you have covered the most significant items. In your attempts to define the central client concerns, always ask yourself, what is the client's real world and current story?

Draw out strength and resource stories. Clients grow best when we identity what they can do rather than what they can't do. Don't focus just on the difficulties and challenges. The positive asset search should be part of this stage of the session. This might be the place for a comprehensive wellness search, as described in Chapter 2.

Failure to treat. Failure to treat can be the cause of a malpractice suit. This most often occurs when counselors fail to draw out stories and clients are unclear about what they really want, often resulting in no clear goals for the session. When the story is clear and strengths are established, you can review and clarify goals. Clients who participate in goal setting and understand the reasons for your helping interventions may be more likely to participate in the process and be more open to change.

STAGE 3. GOALS—MUTUAL GOAL SETTING ("WHAT DO YOU WANT TO HAPPEN?")

Mutuality and an egalitarian approach. Your active involvement in client goal setting is essential. If you and the client don't know where the session is going, the session may wander with no particular direction. Too often the client and counselor assume they are working toward the same outcome when actually each of them wants something different. A client may be satisfied with sleeping better at night, but the counselor wants complete personality reconstruction. The client may want brief advice about how to find a new job, whereas the counselor wants to give extensive vocational testing and suggest a new career.

Refining goals and making them more precise. If you searched for broad goals early in the relationship, they can provide a focus and general direction. In stage 3, it helps to review early goals, divide them into subgoals if necessary, and make them truly clear and doable. It has been suggested that if you don't have a goal, you're just complaining. Authorities on brief counseling and coaching favor setting goals in the first part of the session, together with relationship building. With high school discipline problems, less verbal clients, and members of some cultural groups, setting a clear joint goal may be the key factor in relationship building. If you adapt your counseling style to each client, you will have a better chance of succeeding.

Incremental goals. It is often best to start with specific needed steps to reach a larger goal. Thus, help your clients understand and break down issues into manageable units and measure progress in a step-by-step fashion. Listen and learn their strengths and what has worked with them in the past. The client's positive assets and resources are often the best route toward resolution of issues.

Approach and avoidant goals. Approach goals are what client seeks to improve or achieve (intimacy, career), while avoidant goals are those that the client wants to prevent or decrease (anxiety, depression). In truth, these are closely related—for example, "I want to learn how to get closer to my spouse with our old warm relationship" versus "I want to stop fighting" or worse "I want to stop her/him from behaving that way." With an approach goal, we can establish specific behaviors that may help clients reach what they want to have happen and use their strengths (e.g., identify positive past experiences, learn to listen, develop impulse control). An avoidant goal is more difficult (Conoley, 2014).

Depression and anxiety can be reframed with approach goals of managing stress and client-defined specifics such as sleeping better, have more friends, dealing with a difficult relationship or employer—specific behaviors and actions that are part of alleviating the larger issues. Depression and anxiety might be avoided via medication. But a positive approach could include breaking down change into behavioral specifics, including therapeutic lifestyle changes such as exercise, meditation, returning to church, and learning new social skills. In each of these areas, focusing on strengths (past and present) through positive psychology and positive reframing can make a difference.

Exploration. With many clients, the first step toward clear goals is exploration of possibilities rather than tying things down too quickly. For example, a client may be undecided about a college major or career. While the ultimate goal is that decision, first comes exploration of alternative majors and possible careers. The action goal is really starting to look seriously.

Summarizing the differences between the present story and the preferred outcome. Once the goal has been established, a brief summary of the original presenting concern as contrasted with the defined goal can be very helpful. Consider the model below as a basic beginning to working through client issues. This is a possible opening to stage 4, restorying, using a supportive confrontation.

> Kaan, on the one hand, your concern/issue/challenge is [summarize the situation briefly], but on the other hand, your goal is [summarize the goal]. What occurs to you as possibilities for resolution?

Needless to say, this will be expressed in more words and differently than in these model sentences. Nonetheless, as part of goal setting, the present situation and the desired situation need to be contrasted. Interesting, even Carl Rogers has been known to ask clients what their session goals might be.

Define both the concern and the desired outcome in the client's language. The summary confrontation should list several alternatives that the client has considered. The client ideally should generate more than one possibility before moving on to stage 4. You may want to use hand movements, as if balancing the scales, to present the real and the ideal. Using such physical movements can add clarity to the summary confrontation of key issues.

Define a goal, make the goal explicit, search for assets to help facilitate goal attainment, and only then return to examine the nature of the concern.

STAGE 4. RESTORY—WORKING: EXPLORING ALTERNATIVES, CONFRONTING CLIENT INCONGRUITIES AND CONFLICT, RESTORYING ("WHAT ARE WE GOING TO DO ABOUT IT?")

Starting the exploration process. How does the counselor help the client work through new solutions? Summarize the client conflict as described in the preceding section, "On the one hand. . . ." Be sure that your summary of the issue is complete, including both the facts of the situation and the client's thoughts and feelings.

Use the BLS to facilitate the client's resolution of the issue(s). Imagine a school counselor talking with a teen who has just had a major showdown with the principal. Establish rapport, but expect the teen to challenge you; he or she likely expects you to support the principal. Do not judge, but gather data from the teen's point of view. If you have developed rapport (stage 1) and listened during data gathering (stage 2), the teen will likely search for solutions in a more positive fashion. Follow by asking what he or she would like to have happen in terms of a positive change. Work with the teen to find a way to "save face" and move on.

Encourage client creativity. Your first goal in restorying is to encourage your clients to discover their own solutions. To explore and create with the teen above, listen well and summarize: "You see the situation as . . . and your goal is. . . . The principal tells a different story and his goal is likely to be. . . ." If you have developed rapport and listened well, many teens will be able to generate ideas to help resolve the situation.

Skilled questioning can be useful in helping clients find positive solutions. Needless to say, these questions are best used after you have heard the client's original story. They are also useful in the fourth stage of the interview as we work toward restorying. These questions are derived from brief counseling's solution-oriented approach.

- ▶ "Can you brainstorm ideas—just anything that occurs to you?"
- ▶ "What other alternatives can you think of?"
- ▶ "Tell me about a success that you have had."
- ▶ "What has worked for you before?"
- ▶ "What part of the problem is workable if you can't solve it all right now?"
- ▶ "Which of the ideas that we have generated appeals to you most?"
- ▶ "What are the consequences of taking that alternative?"

Interviewers and counselors all try to resolve issues in clients' lives in a similar fashion. The counselor needs to establish rapport, define the issue, and help the client identify desired outcomes.

Relate client issues and concerns to desired outcomes. The distinction between the problem and the desired outcome is the major incongruity that may be resolved in three basic ways. First, the counselor uses attending skills to clarify the client's frame of reference and then feeds back a summary of client concerns and the goal. Often clients generate their own synthesis and resolve their challenges. Second, counselors can use information, directives, and psychoeducational interventions to help clients generate new answers. Third, if clients do not generate their own answers, the counselor can use interpretation, self-disclosure, and other influencing skills to resolve the conflict. Finally, in systematic problem solving and decision making, counselor and client generate and brainstorm alternatives for action and set priorities among the most promising possibilities.

Aim for a decision and a new story. This process of exploring, brainstorming, and testing theoretical strategies facilitates client decision making and the generation of a new story. Once a decision has been made or a new workable story developed, see that plans are made to put these ideas into action in the real world. You need to help clients generalize feelings, thoughts, behaviors, and a plan for action beyond the session itself.

The preceding outline is specific to decisional counseling, but you will find it virtually identical to motivational interviewing and crisis counseling. Different theories will vary in their use of skills and emphasis at this stage. However, the earlier parts of the session tend to be relatively similar among varying theoretical viewpoints.

STAGE 5. ACTION—CONCLUDING: GENERALIZING AND ACTING ON NEW STORIES ("WILL YOU DO IT?")

The complexities of life are such that taking a new behavior back to the home setting may be difficult. How do we generalize thoughts, feelings, and behaviors to daily life? Some counseling theories work on the assumption that behavior and attitude change will come out of new unconscious learning; they "trust" that clients will change spontaneously. This indeed can happen, but there is increasing evidence that planning for change greatly increases the likelihood that it will actually occur in the real world.

Consider the situation of the teen in conflict with the principal. Some good ideas may have been generated, but unless the teen follows up on them, nothing is likely to change in the conflict situation. Find something that "works" and leads to changes in the repeating behavioral problems. As you read through the list of generalization suggestions below, consider what you would do to help this teen and other clients restory and change their thoughts, feelings, and behaviors.

Change does not always come easily, and many clients revert to earlier, less intentional behaviors. Work to help the client plan for change to ensure real-world relevance. Following are three techniques that you can use to facilitate the transfer of learning from the session, increasing the likelihood that the client will take new learning home and try new behaviors.

Contracting. The most basic way to help clients maintain and use new learning is the establishment of an informal or written contract to do something new and different. Ideally, this should be clear and specific enough that the client can actually do it easily. With more complex issues, contract for things that represent part of the solution. If we ask clients to change their behavior totally, they likely will fail and may not return for another session.

Homework and journaling. Assigning homework so that the effect of the session continues after the session ends has become increasingly standard. Some counselors use "personal experiments" with clients who do not like the idea of doing "homework." Negotiate specific tasks for the client to try during the week following the session. Use very specific and concrete behavioral assignments, such as "To help your shyness, you agree to approach one person after church/synagogue/mosque and introduce yourself." Ask the client to keep a journal of key thoughts and feelings during the week; this can become the basis of the follow-up session. Another possibility is paradoxical intention: "Next week, I want you to deliberately do the same self-defeating behavior that we have talked about. But take special notice of how others react and how you feel." This helps the client become much more aware of what he or she is doing and its results.

Follow-up and support. Ask the client to return for further sessions, each with a specific goal. The counselor can provide social and emotional support through difficult periods. Follow-up is a sign that you care. Use the telephone for behavior maintenance checks.

Using email is possible, but may result in loss of privacy. Many counselors and therapists call clients but do not give them their email. If you work in an agency and give a phone number to a client, use an office number that is always attended.

Behaviors and attitudes learned in the session do not necessarily transfer to daily life without careful planning. Consider asking your client at the close of the interview, "Will you do it?"

 Interactive Exercise: The Impossible Boss. Review and analyze Machiko and Robert's interview, and score your responses.

TAKING NOTES IN THE SESSION

Beginning helpers typically question whether they should take notes during the session, and it's not hard to find opinions for and against note taking. We are going to share our opinions based on our experience, recognizing that individual views vary on this issue. Most important, follow the directions of your agency.

Intentionality in counseling and psychotherapy requires accurate information. Therefore, we recommend that you listen intentionally and take notes. This is our opinion, but some people will disagree with us. You and your client can usually work out an arrangement suitable for both of you. If you personally are relaxed about note taking, it will seldom become an issue. If you are worried about taking notes, it likely will be a problem. When working with a new client, obtain permission early about taking notes. We suggest that any case notes be made available to clients and in practice sessions. Volunteer client feedback on the session can be most helpful in thinking about your own style of helping. Using your own natural style, you might begin:

> I'd like to take a few notes while you talk. Would that be OK? I'll also write down your exact key words so both of us can refer to them. I'll make a copy of the notes before you leave, if you wish. As you know, all notes in your file are open to you at any time.

In-session note taking is often most helpful in the initial portions of counseling and psychotherapy and less important as you get to know the client better. Audio or video recording the session follows the same guidelines. If you are relaxed and provide a rationale to your client, making this type of record of the session generally goes smoothly. Some clients find it helpful to take audio recordings of the session home and listen to them, thus enhancing their learning from the session. There is nothing wrong with *not* taking notes in the session, but records typically need to be kept, and we recommend writing session summaries shortly after the session finishes. In these days of performance accountability for your actions, a clear record can be helpful to you, the client, and the agency with which you work.

Some expert interviewers and counselors object to note taking, arguing that when the counselor is closely in tune and has great listening skills, there is no need for notes. Indeed, too much attention to note taking detracts from the central issues of rapport building and active listening. The most famous listener of all, Carl Rogers, found that when one records an interview, one often finds that what actually happened is different from what is found in notes. Thus, a balance is clearly needed, particularly in a time of increased scrutiny and possible legal issues.

HIPAA (Health Insurance Portability and Accountability Act) legal requirements regarding note taking are not always clear. Some rules give certain aspects of counseling more detailed protection than general medical records, but these rules are sometimes written vaguely, and they mention the possibility of maintaining dual records of psychotherapy. The agency you work with in practicum or internship can guide you in this area. You will also find Zur's (2011) *The HIPAA Compliance Kit* helpful.

▶ Observing The BLS and Five Stages in Decisional Counseling

The five stages are also a structure for decision making. Eventually, all clients will be making decisions about behavior, thoughts, feelings, and meanings. Each theory gives different attention to these, and they use varying language, names, and techniques. But the client makes the decisions, not us.

How are the five stages specifically related to decisions? Many see Benjamin Franklin as the originator of the systematic decision-making model. He suggested three phases of **problem solving**: (1) identify the problem clearly (draw out the story and strengths, along with goal setting); (2) generate alternative answers (restory); and (3) decide what action to take (action). However, the ancient Franklin model misses the importance of empathic relationship (stage 1), the need for clearer goal setting, and ensuring that the client takes action (stage 5) after the session in the real world. Another term for decisional counseling is problem-solving counseling. The essential issue is the same regardless of the terms we use: How can we help clients work through issues and come up with new answers?

Virtually all interviewing and counseling sessions involve decisions. Whether you become oriented to person-centered, cognitive behavior, narrative, behavior therapy, neurocounseling, coaching, or others, decisions will always be part of the counseling and therapy process. Decisions are a central issue in crisis counseling. Decisions are involved in millions of interviews conducted daily throughout the world in counseling, therapy, medicine, business, sports, government, and many other fields.

Decisional counseling is straightforward; you already have the most important basic understanding through the BLS and the five stages. Decisional counseling will be discussed in more detail in Chapter 12. In addition, a full transcript of an interview that Allen conducted with Mary can be found in the web resources. Perhaps the reason you will not find decisional counseling in most books on interviewing and counseling is that it is so basic it is easily overlooked. As it is a foundation, it is all the more essential that you become competent in decisional counseling.

As decisional counseling (or problem-solving counseling) is the most widely practiced form of helping, we urge you to take some time to master the concepts, which will come easily with an understanding of the microskills and the five-stage interview. In addition, you will find that this foundation enables you to become competent more rapidly in the many other theories of interviewing and counseling.

EXAMPLE INTERVIEW: I CAN'T GET ALONG WITH MY BOSS

It requires a verbal, cooperative client to work through a complete interview using only listening skills. This interview has been edited to show portions that demonstrate skill usage and levels of empathy. Robert, the client, is 20 and a part-time student who is in conflict with his boss at work. Machiko, the counselor, finds him relatively verbal and willing to work on the problem with her assistance.

Machiko and Robert know each other from class, so little time is spent on relationship in her transcript, although they did talk while her computer was being set up and tested for picture and sound. The agreement here was that the camera would be focused on Machiko during the session. If a personal or college video cam were available, then the two of them should be on camera. If you use a cell phone with a good microphone, again the two of you could both be shown.

Stage 1: Empathic relationship

INTERVIEWER AND CLIENT CONVERSATION	PROCESS COMMENTS
1. *Machiko:* Thanks for volunteering to enable me to do a practice session for class. After we review this session, we can plan for you to interview me. Robert, do you mind if we tape this interview? It's for a class exercise in interviewing. I'll be making a transcript of the session, which the professor will read. Okay? We can turn the recorder off at any time. I'll show you the transcript if you are interested. I won't use the material if you decide later you don't want me to use it. Could you sign this consent form? (Robert signs and looks up.)	A small video cam has been set up. Machiko opens with a closed question followed by structuring information. It is critical to obtain client permission and offer client control over the material before recording. As a student you cannot legally control confidentiality, but it is your responsibility to protect your client.
2. *Robert:* Sounds fine; I do have something to talk about. Okay, I'll sign it. [Pause as he signs]	Robert seems at ease and relaxed. As the taping was presented casually, he is not concerned about the use of the recorder. Rapport was easily established.
3. *Machiko:* What would you like to share?	The open question, almost social in nature, is designed to give maximum personal space to the client.
4. *Robert:* My boss. He's pretty awful.	Robert indicates clearly through his nonverbal behavior that he is ready to go. Machiko observes that he is comfortable and decides to move immediately to gather data (stage 2). With some clients, several interviews may be required to reach this level of rapport.

Stage 2: Story and strengths

INTERVIEWER AND CLIENT CONVERSATION	PROCESS COMMENTS
5. *Machiko:* Could you tell me about it?	This open question is oriented toward obtaining a general outline of the problem the client brings to the session. (Potentially additive empathy)
6. *Robert:* Well, he's impossible . . . (Hesitates and looks at Machiko)	Instead of the expected general outline of the concern, Robert gives a brief answer. The predicted consequence didn't happen, but we see in the next two exchanges that 5 and 7 together are indeed additive.

(continued)

INTERVIEWER AND CLIENT CONVERSATION	PROCESS COMMENTS
7. *Machiko:* Impossible? . . . Go on . . .	Encourager with warm, supportive vocal tone. Intentional competence requires you to be ready with another follow-up response. Tone of voice is especially important here in communicating to the client.
8. *Robert:* Well, he's impossible. Yeah, really impossible. It seems that no matter what I do he is on me, always looking over my shoulder. I don't think he trusts me.	We are seeing the story develop. Clients often elaborate on the specific meaning of a concern if you use the encourager. In this case, the prediction holds true.
9. *Machiko:* Could you give me a more specific example of what he is doing to indicate he doesn't trust you?	Robert is a bit vague in his description. Machiko asks an open question eliciting concreteness in the story. (A search for concreteness is typically additive.)
10. *Robert:* Well, maybe it isn't trust. Like last week, I had this customer lip off to me. He had a complaint about a shirt he bought. I don't like customers yelling at me when it isn't my fault, so I started talking back. No one can do *that* to me! . . . (pause) And of course the boss didn't like it and chewed me out. It wasn't fair.	As events become more concrete through specific examples, we understand more fully what is going on in the client's life and mind. When Robert said, "No one can do *that* to me!" he briefly spoke angrily and clenched his teeth and tightened his fist briefly. After the pause, he spoke more softly and seemed puzzled.
11. *Machiko:* As I hear it, Robert, it sounds as though this guy gave you a bad time and it made you angry, and then the boss came in. I hear some real anger about the customer, but I'm not so clear about the feelings toward your boss.	Machiko's response is relatively similar to what Robert said. Her paraphrase and reflection of feeling represents basic interchangeable empathy. Picking up on the nonverbal mixed message was wise. (Interchangeable at first and then additive in the last sentence as she encourages client to explore emotions felt)
12. *Robert:* Exactly! It really made me angry. I have never liked anyone telling me what to do. But when the boss came after me, I almost lost it. I left my last job because the boss was doing the same thing.	Accurate listening often results in the client's saying "exactly" or something similar. Robert loosens up and starts talking about the boss as the real challenge.
13. *Machiko:* Difficult customers are hard to take, but your boss coming in like that is the real issue. He really bugged you. (brief pause) . . . And I hear that your last boss wasn't fair either?	Machiko's vocal tone and body language communicate nonjudgmental warmth and respect. She catches underlying emotions while bringing back Robert's key word *fair* by paraphrasing with a questioning tone of voice, which represents an implied checkout. This is an interchangeable empathic response, again with an additive "tone."

The interview continues to explore Robert's conflict with customers, his boss, and past supervisors. There appears to be a pattern of conflict with authority figures over the past several years. This is a common pattern among young males in their early careers. After a detailed discussion of the specific conflict situation and several other examples of the pattern, Machiko decides to conduct a positive asset search to discover strengths.

| --- | --- |
14. *Robert:* You got it.	Robert here is speaking to Machiko's understanding of his situation.
15. *Machiko:* Robert, we've been talking for a while about difficulties at work. I'd like to know some things that have gone well for you there. Could you tell me about something you feel good about at work?	Paraphrase, structuring, open question, and beginning positive asset search. (Additive positive empathy)
16. *Robert:* Yeah; I work hard. They always say I'm a good worker. I feel good about that.	Robert's increasingly tense body language starts to relax with the introduction of the positive asset search. He talks more slowly.
17. *Machiko:* Sounds like it makes you feel good about yourself to work hard.	Reflection of feeling, emphasis on positive regard (Interchangeable with positive additive dimensions)
18. *Robert:* Yeah. For example, . . .	

Robert continues to talk about his accomplishments. In this way Machiko learns some of the positives Robert has in his past and not just his difficulties. She has used the basic listening sequence to help Robert feel better about himself. Important here was emphasizing strengths and positive emotions. Machiko learns that Robert has positive assets, such as determination and willingness to work hard, to help him resolve his own problems. He takes pride in his ability to work hard.

Stage 3: Goals

INTERVIEWER AND CLIENT CONVERSATION	PROCESS COMMENTS
19. *Machiko:* Robert, given all the things you've talked about, could you describe an ideal solution? How would you like things to be?	Open question. The addition of a new possibility for the client represents additive empathy. It enables Robert to think of something new.
20. *Robert:* Gee, I guess I'd like things to be smoother, easier, with less conflict. I come home so tired and angry.	
21. *Machiko:* I hear that. It's taking a lot out of you. Tell me more specifically how things might be better.	Paraphrase, open question oriented toward concreteness. Restating "taking a lot out of you" keeps awareness of emotions present. (Interchangeable; the last sentence is potentially additive.)
22. *Robert:* I'd just like less hassle. I know what I'm doing, but somehow that isn't helping. I'd just like to be able to resolve these conflicts without always having to give in.	Robert is not as concrete and specific as anticipated. But he brings in a new aspect of the conflict—giving in.
23. *Machiko:* Give in?	Encourager. (Another potentially additive encourager)
24. *Robert:* Yeah. And, you know what . . .	

Machiko learns another dimension of Robert's conflict with others. Subsequent use of the basic listening sequence brings out this pattern with several customers and employees. As new data emerge in the goal-setting process, you may find it necessary to change the definition of the concern and perhaps even return to stage 2 for more data gathering.

INTERVIEWER AND CLIENT CONVERSATION	PROCESS COMMENTS
25. *Machiko:* So, Robert, I hear two things in terms of goals. One, that you'd like less hassle, but another, equally important, is that you don't like to give in. But all this makes you tired, irritable, and discouraged. Have I heard you correctly?	Machiko uses a summary of both cognitions and emotions to help Robert clarify his problem, even though no resolution is yet in sight. From a neuroscience frame of reference, she is seeking to help Robert's pattern of emotional regulation lead to better executive decisional functioning. She checks out the accuracy of her hearing. (Additive empathy)
26. *Robert:* You're right on, but what am I going to do about it?	At this point, it is clear that Robert feels heard and listened to. He leans forward with some anticipation of working toward resolution of his concerns.

Stage 4: Restory

INTERVIEWER AND CLIENT CONVERSATION	PROCESS COMMENTS
27. *Machiko:* So, Robert, on the one hand I heard you have a long-term pattern of conflict with supervisors and customers who give you a bad time. On the other hand, I also heard just as loud and clear your desire to have less hassle and not give in to others. We also know that you are a good worker and like to do a good job. Given all this, what do you think you can do about it?	Machiko remains nonjudgmental and appears to be very congruent with the client in terms of both words and body language. In this clear, positive, additive summary, she distills and clarifies what the client said. (Clear summaries typically include both interchangeable and additive empathy, hopefully including positive dimensions.)
28. *Robert:* Well, I'm a good worker, but I've been fighting too much. I let the boss and the customers control me too much. I think the next time a customer complains, I'll keep quiet and fill out the refund certificate. Why should I take on the world?	Robert talks more rapidly. He, too, leans forward. However, his brow is furrowed indicating some tension. He is "working hard." At the same time, Robert starts with a positive self-statement that has earlier been reinforced by Machiko.
29. *Machiko:* So one thing you can do is keep quiet. You could maintain control in your own way, and you would not be giving in.	Paraphrase, interchangeable empathy. Machiko is using Robert's key words and feelings from earlier in the interview to reinforce his present thinking. But she waits for Robert's response.
30. *Robert:* Yeah, that's what I'll do, keep quiet.	He sits back, his arms folded. This suggests that the "good" response above was in some way actually subtractive. There is more work to do.
31. *Machiko:* Sounds like a good beginning, but I'm sure you can think of other things as well, especially when you simply can't be quiet. Can you brainstorm more ideas?	Machiko gives Robert brief feedback. Her open question is an additive response. She is aware that his closed nonverbals suggest more is needed.

Clients are often too willing to seize the first idea as a way to agree and avoid looking fully at issues. It is helpful to use a variety of questions and listening skills to further draw out the client. Later in the interview, Robert was able to generate two other useful suggestions: (1) to talk frankly with his boss and seek his advice; and (2) to plan an exercise program to blow off steam and energy. In addition, Robert began to realize that his problem with his boss was only one example of a continuing problem with anger.

He and Machiko discussed the possibilities of continuing their discussions so that he could seek to keep his temper in control. Robert decided he'd like to talk with Machiko a bit more. A contract was made: If the situation did not improve within 2 weeks, Robert could consider anger management and seek someone with more experience at the campus counseling center.

Stage 5: Action—generalization and transfer of learning

INTERVIEWER AND CLIENT CONVERSATION	PROCESS COMMENTS
32. *Machiko:* So you've decided that the most useful step is to talk with your boss. But the big question is "Will you do it?"	Paraphrase, open question. (Potentially additive) Note that many questions are potentially additive, but questions can also be subtractive—we need to wait and see how useful they were.
33. *Robert:* Sure, I'll do it. The first time the boss seems relaxed.	He appears more confident and sits up.
34. *Machiko:* As you've described him, Robert, that may be a long wait. Could you set up a specific plan so we can talk about it the next time we meet?	Paraphrase, open question. To generalize from the interview, it is important to encourage specific and concrete action in your client so that something actually does happen. Too many of us are far too willing to accept that early acknowledgment of willingness to act. It typically is not enough.
35. *Robert:* I suppose you're right. Okay, occasionally he and I drink coffee in the late afternoon at Rooster's. I'll bring it up with him tomorrow.	A specific plan to take action on the new story is developing.
36. *Machiko:* I hear you ready to act. You've got the confidence and ability. What, specifically, are you going to say?	Paraphrase, open question, again eliciting concreteness. The statement on confidence is an interpretation of his overall verbal and nonverbal style. (Interchangeable followed by a potentially additive question)
37. *Robert:* I could tell him that I like working there, but I'm concerned about how to handle difficult customers. I'll ask his advice and how he does it. In some ways, it worries me a little; I don't want to give in to the boss . . . but maybe he will have a useful idea.	Robert is able to plan something that might work. With other clients, you may role-play, give advice, actually assign homework. You will also note that Robert is still concerned about "giving in."
38. *Machiko:* Would you like to talk more about this the next time we meet? Maybe through your talk with your boss we can figure out how to deal with this in a way that makes you feel more comfortable. Sounds like a good contract. Robert, you'll talk with your boss, and we'll meet later this week or next week.	Open question, structuring. If Robert does talk to his boss and listens to his advice—and actually changes his behavior—then this interview could be rated holistically at Level 3, additive empathy. If not, then a lower rating is obvious.

PERSON-CENTERED COUNSELING, THE BLS, AND THE FIVE STAGES

Theoretically and philosophically, this decisional interview using only listening skills is closely related to Carl Rogers's **person-centered counseling** (Rogers, 1957). Rogers developed empathic guidelines for the "necessary and sufficient conditions of therapeutic personality change." In the Machiko-Robert transcript, you saw a decisional counseling model combined with a modified person-centered approach. In Chapter 12 and Allen's web interview with Mary, you will see more use of influencing skills than you do here.

Rogers originally was opposed to the use of questions but in later life modified his position so that in some interviews a very few questions might be asked. These would be quite open and as "nondirective" as possible. "What is your goal?" and "What meaning does that have for you?" are two examples of very open questions that tend not to box the client into the interviewer's perspective or theory.

It was Carl Rogers who truly brought the ideas of empathic understanding to the interviewing process. Many would say that he humanized counseling and psychotherapy by stressing the importance of relationship, respect, authenticity, and positive regard, also called the working alliance. *Working alliance* is a useful term as it stresses the way we need to *work with*, rather than *work on*, the client. A good relationship and working alliance may be in itself sufficient to produce positive change.

You will encounter more about the centrality of the working alliance, the relationship, and the underlying value of person-centered theory in future study. However, at this time you can gain a beginning understanding of some aspects of the approach by focusing on listening skills and using as few questions as possible. In the process, think constantly of "being in the client's shoes" and experiencing the world as he or she does.

If you practice with this in mind, you will gain a better understanding what Carl Rogers is seeking—to discover the client's outer and inner world as the client experiences it.

▶ Practicing The Five Stages and Decisional Counseling

Here, there is a real need for audio or video recorded role-playing practice and feedback. This should come fairly easily, as the basic skills of brief counseling are similar to those you have learned through earlier chapters.

▶ Work with a partner, switching the roles of client and counselor. Plan for a minimum interview of 15 minutes.

▶ Select a concern for the role-play. This time the issues need to be very specific—for example, dealing with a specific conflict on the job, in the family, or with a partner. Academic stress is obviously a good topic for practice. Aim for concreteness and clarity throughout the storytelling.

▶ Record the session on audio or video using a computer, cell phone, or personal camera to provide some instant feedback.

The Microskills Interview Analysis Form (MIAF) in Box 7.1 can be a valuable aid for detailed analysis of what you have done. It will enable you to examine your skills and unique personal style in this and other sessions. The form can be completed by an observer watching the here and now of the session for immediate feedback. The form will also be valuable for you as you listen to and/or watch an audio or video recording.

The MIAF includes skills that have not yet been presented. We suggest that you focus on the listening skills presented in the first half of this book. These skills are foundational for any approach or theoretical orientation you select. Even the most directive and instructive interviewer, counselor, or therapist needs to be able to return to basic listening skills through the session.

Consider using the MIAF consistently as you practice the skills in later chapters of this book.

Microskills Interview Analysis Form (MIAF)

Lori Russell-Chapin and Allen E. Ivey*

INTERVIEWER/COUNSELOR _____ DATE _____

OBSERVER(S) _____ SUPERVISOR _____

Note: This form can be completed by an observer watching the here and now of the session for immediate feedback. The form will also be valuable for reflection and more attention to key details as you listen to and/or watch an audio or video recording.

Make a frequency count of every time a skill is used. With focusing, periodically note change of topic area emphasized.

Competency ratings are to be completed at the completion of the session.

1. *Basic competency.* The skill is used in the session and no empathy rating is possible.

2. *Interchangeable empathic competency.* The skill is used with empathic understanding, hearing the client accurately, and appears to encourage further client conversation.

3. *Additive empathic competency.* The skill has a positive impact on client verbal and nonverbal behavior. Questions are additive only if the client uses the question for additional exploration or understanding.

Interview stages need not always follow a specific order, but all issues within the stages usually need to be addressed at some point.

Stage 1: Empathic Relationship	Frequency	Competence Level	Comments on Skill and Client Reactions
Getting started			
Greeting			
Role definition/expectations			
Administrative tasks			
Skills			
Overall attending rating (eye contact, verbal following, vocal tone)			
Overall nonverbal rating (body language, note specifics)			
Open questions			
Closed questions			
Checkout/perception check			
Silence			
Encouraging/restating			
Paraphrasing			
Reflecting feelings			
Summarizing			
Other skills			
Early client goal established			
Overall rating of Stage 1			

*Special thanks to Dr. Lori Russell-Chapin, Bradley University, who suggested this form. It is adapted from *Your Supervised Practicum and Internship,* coauthored with Allen Ivey. Brooks/Cole.

Stage 2: Story and Strengths	Frequency	Competence Level	Comments on Skill/Client Reactions
Skills			
Overall attending rating			
Overall nonverbal rating			
Open questions			
Closed questions			
Checkout/perception check			
Silence			
Encouraging/restating			
Paraphrasing			
Reflecting feelings			
Summarizing			
Other skills			
Client's issues/concerns brought out clearly			
Client's strengths and assets brought out clearly			
Overall rating of Stage 2			

Stage 3: Goals	Frequency	Competence Level	Comments on Skills and Client Reactions
Skills			
Overall attending rating			
Overall nonverbal rating			
Open questions			
Closed questions			
Checkout/perception check			
Silence			
Encouraging/restating			
Paraphrasing			
Reflecting feelings			
Summarizing			
Focusing: client, client concerns, significant others, key individuals and groups, cultural-env.-context, client/ counselor relationship; specify critical areas of conversation.			
Confrontation			
Interpretation/reframing			
Reflection of meaning			
Feedback			
Self-disclosure			
Directives			
Providing information, psychoeducation			

(continued)

BOX 7.1 (continued)

Stage 3: Goals	Frequency	Competence Level	Comments on Skills and Client Reactions
Natural and logical consequences			
Client's goals defined clearly			
Flexibility and exploration of client goals encouraged			
Overall rating of Stage 3			

Stage 4: Restory	Frequency	Competence Level	Comments on Skills and Client Reactions
Skills			
Overall attending rating			
Overall nonverbal rating			
Open questions			
Closed questions			
Checkout/perception check			
Silence			
Encouraging/restating			
Paraphrasing			
Reflecting feelings			
Summarizing			
Focusing (client, client concern, significant others, key individuals and groups, CEC, immediate interpersonal relationship in the session; note primary areas emphasized)			
Confrontation			
Interpretation/reframing			
Reflection of meaning			
Feedback			
Self-disclosure			
Directives			
Providing information, psychoeducation			
Natural and logical consequences			
Was homework included?			
Follow-up agreement			
Overall rating of Stage 4			

Stage 5: Action	Frequency	Competence Level	Comments on Skills and Client Reactions
Skills			
Overall attending rating			
Overall nonverbal rating			
Open questions			
Closed questions			
Checkout/perception check			

Stage 5: Action	Frequency	Competence Level	Comments on Skills and Client Reactions
Silence			
Encouraging/restating			
Paraphrasing			
Reflecting feelings			
Summarizing			
Focusing (client, client concern, significant others, key individuals and groups, CEC, immediate interpersonal relationship in the session; note primary areas emphasized)			
Confrontation			
Interpretation/reframing			
Reflection of meaning			
Feedback			
Self-disclosure			
Directives			
Providing information, psychoeducation			
Natural and logical consequences			
Client's goals defined clearly			
Flexibility and exploration of client goals encouraged			
Conclusion/rescheduling			
Overall rating of Stage 5			

This form is adapted from Lori Rusell-Chapin and Allen Ivey's text: *Your Supervised Practicum and Internship: Field Resources for Turning Theory Into Action.*

 Case Study: Couple Conflict

▶ Refining: Demystifying the Helping Process

Demystify: make less mysterious or remove the mystery from
—*Webster's Online Dictionary*

Demystify: to make something easier to understand
—*Cambridge Advanced Learner's Dictionary & Thesaurus*

What makes counseling and psychotherapy successful? For years counseling and psychotherapy remained a mysterious, puzzling, and difficult-to-decipher process. Carl Rogers was the first to demystify the counseling process. Through the then new technology of the wire recorder, he showed us what is actually done in the session. Before that, therapists only reported what they remembered. Early research after Rogers's contribution found that what counselors said they did in the session was not what they actually did.

Rogers brought us the first conceptions for understanding the therapeutic process and empathy. However, the *behaviors* of a quality session remained vague. What are the observable counselor actions that make a session successful? What are the key therapist behaviors facilitating client growth? With the number of new therapies increasing over time, the uniqueness of the therapist and client, and the diversity of perspectives, issues, and goals, identifying the central components of the counseling and therapy relationship remained elusive.

Demystifying the counseling process. Until the microskills approach came along, the counseling and psychotherapy field had not identified the behaviors of the effective helper. With colleagues at Colorado State University, Allen obtained the first video grant to examine more precisely what are the *observable behavioral* skills of the competent counselor and therapist. For the first time, videotaping equipment enabled the recording of sessions for nonverbal as well as verbal behavior.

Attending, the first behavioral skill. The importance of attending behavior came to our attention almost by chance. We videotaped our secretary in a demonstration session. She totally failed to attend to the student she was counseling—looking down, nervous movements, and topic jumping. On reviewing the video, we identified the concrete specifics of attending behavior for the first time: appropriate eye contact, comfortable body language, a pleasant and smooth vocal tone, and verbal following—staying with the client's topic. When she went back for another video practice session, she listened effectively, and even looked like a counselor. All that happened in a half-hour!

Taking microskill learning home via action. The next level of demystification came when she returned after the weekend. "I went home, I attended to my husband, and we had a beautiful weekend!" We had not anticipated that session behavior would generalize to real life. I became aware of the importance of teaching communication microskills to clients and patients. Children, couples, families, management trainees, psychiatric patients, refugees, and many others have now been taught specifics of communication using the skills taught in this book.

Microtraining goes viral and multicultural. Soon after the microskills and microtraining became available, reverberations began. More than 450 data-based studies followed, and Allen's work was translated into multiple languages, becoming a regular part of the university curriculum in counseling, social work, and psychology. The adaptability to many different situations and cultures continues even today—ranging from Aboriginal social workers in Australia to AIDS peer counselors in Africa to managers in Japan, Sweden, and Germany. I (Carlos) was trained in the microskills as part of my graduate program and had the privilege of teaching them as part the first course on effective psychotherapy I taught in Chile. My students liked the framework and demonstrated the competencies.

Transtheoretical character of the microskills. Microskills bridge the gap between what is said to be done in the session and what actually happens. Following the tradition of Carl Rogers, the session is demystified. At the same time, the microskills provide a foundation for understanding widely different theories of therapy. Virtually all use some variation of the BLS and some form of the five-stage session, but you will find widely varying beliefs and actions when it comes to the influencing and action skills of the second half of this book. The five stages are also a basic decisional system. The discussion of decisional counseling in Chapter 12 emphasizes this point.

Video recording your sessions. This is why we hope that you will video your own sessions as frequently as possible, both during this course and when you move on as a professional. Video is now easily available through inexpensive video cams, miniature cameras, and even cell phones. Demystify your own behavior in the session. Watching the video together with your clients is invaluable. Use the microtraining model to become aware of your natural competencies and learn additional skills to work effectively with more people.

Becoming more specific about multicultural issues. About a year after the identification of the attending skills, paraphrasing, reflecting feelings, and summarizing, Allen was enthusiastically teaching a workshop on microskills. After he provided information on attending behavior and the importance of eye contact, a quiet student came up to him and described her experience in Alaska with Native Inuits. She commented that direct eye contact could be seen as an uncaring or even hostile act by traditional people. We can still attend, but we need to consider the natural communication style of each culture. This led Allen to a personal commitment to discover more and more about how culture affects the person and how it must be considered in all sessions.

And now, neuroscience. As a group, our most recent venture has been into this newly relevant field, and we now present workshops and teaching on "brain-based counseling and therapy." Research in neuroscience is further demystifying the helping process. Not too surprisingly, almost all of what has been done in our field is validated by neuroscience. But we can now be more precise in understanding what is going on inside the client, thus leading to the most appropriate interventions. Neuroscience research has led us to an increased belief in the need for a positive approach to counseling, with minimal attention to the negative. For practical purposes, expect your clients to come to you with a fair amount of brain knowledge, just from the popular media and TV. Neuroscience is leading toward a paradigm shift in our field, and we need to be ready to use this research in our practice. You will find that including instruction on how the brain can change in the process of counseling helps clients become more committed and hopeful as we work together in the session.

To be continued. The learning process of demystification constantly brings something new and exciting. Professional research and practice, along with student feedback, change what we say in this book. Join us in the learning process and share ideas with us. You can make a difference in our exciting field.

Neuroscience and the Centrality of Active Listening. The first three sections of *Essentials of Intentional Interviewing* have continually emphasized the importance of listening and developing a solid relationship and working alliance. Recent fMRI research reveals that your active listening skills literally activate positive regions in the brain. Specifically, it has been found that active listening impacts the reward system—a "warm glow" for receiving positives (ventral striatum). In addition, active listening increases emotional appraisal, mentalizing, and understanding the emotional communication of others (right anterior insula and medial prefrontal cortex). (Kawamichi, 2014) Once again we see that the theologian, Paul Tillich, is right—"Love is listening."

The National Institute of Mental Health brain-based assessment and treatment plan will be influencing the future of counseling and therapy practice. You can expect major impact of neuroscience and neurobiology on counseling and therapy practice. Expect to see major changes in the Diagnostic and Treatment Manual (DSM) and the International Classification of Disease (ICD). What is wonderful about the new area of neuroscience is that it shows us that most of what we are already doing is already correct—and it will inform our practice so that our work will be more precise and effective.

Summarizing Key Points of How to Conduct a Five-Stage Counseling Session Using Only Listening Skills

Defining the Basic Listening Sequence: Foundation for Empathic Listening in Many Settings

▶ The basic listening sequence (BLS) is built on attending and observing the client, but the key skills are using open and closed questions, encouraging, paraphrasing, reflecting feelings, and summarizing.

▶ When we listen to clients using the BLS, we want to obtain the overall background of the client's story and learn about the facts, thoughts, feelings, and behaviors that go with that story.

▶ Use checkouts to obtain feedback on the accuracy of your listening. Clients will let you know how accurately you have listened.

Defining the Five-Stage Model for Structuring the Session

▶ The five stages of the session are *empathic relationship—story and strengths—goals—restory—action*. These stages provide an organizing framework for using the microskills with multiple theories of counseling and psychotherapy.

Discerning Specifics of the Well-Formed Five-Stage Interview

▶ Stage 1: Empathic Relationship includes initiating the session, establishing rapport, building trust, structuring the session, and establishing early goals.

▶ Stage 2: Story and Strengths focuses on gathering data. Draw out stories, concerns, and strengths.

▶ Stage 3: Goals establishes goals in a collaborative way.

▶ Stage 4: Restory includes working with the client to explore alternatives, confronting client incongruities and conflict, and rewriting the client's narrative.

▶ Stage 5: Action involves collaborating with the client to take steps toward achieving desired outcomes and achieving change.

▶ Consider the five stages a checklist to be covered in each session.

▶ If you personally are relaxed about note taking, it will seldom become an issue. If you are worried about taking notes, it likely will be a challenging session.

Observing The BLS and Five Stages in Decisional Counseling

▶ Regardless of varying counseling and therapy theories, most interviews involve making some sort of decision, including defining the problem, defining the goal, and selecting from alternatives.

▶ Decisions are involved in millions of interviews conducted daily throughout the world in counseling, therapy, medicine, business, sports, government, and many other fields.

▶ Decisional counseling follows the five-stage structure of the interview.

▶ In the example interview, Machiko uses the basic listening sequence to help her client resolve his issue.

▶ Decisional counseling and the microskills provide a foundation that you can use to become competent more easily in other theories of helping.

Practicing The Five Stages and Decisional Counseling

▶ Practicing what you have read is essential for skill mastery. Use a role-played practice to apply the listening skills you have learned through earlier chapters.

▶ Work with a partner, switching the roles of client and counselor. Plan for a minimum interview of 15 minutes. Provide feedback and use the suggested form, if at all possible.

Refining: Demystifying the Helping Process

▶ Demystification is oriented to clarifying the "secrets" of successful interviewing and counseling.

Assess your current level of knowledge and competence as you complete the chapter:

1. Flashcards: Use the flashcards to check your understanding of key concepts and facilitate memorization of key information.

2. Self-Assessment Quiz: The quiz will help you assess your current knowledge and prepare for course examinations.

3. Portfolio of Competencies: Evaluate your present level of competence on the ideas and concepts presented in this chapter using the Self-Evaluation Checklist. Self-assessment of your competencies demonstrates what you can do in the real world.

SECTION III
Helping Clients Generate New Stories That Lead to Action: Influencing Skills and Strategies

Bravo, you have completed the first two sections of this book and are ready to conduct an interview using only listening skills! Never forget the watchwords "When in doubt, listen!" and "Love is listening!" Using the *empathic relationship—story and strengths—goals—restory—action* model you can help clients find new stories and use them for growth and change. The five-stage interview provides a structure that, for many clients, will be complete in itself—simply telling the story in a positive, supportive atmosphere is often sufficient for positive change to occur. Because clients have changed the way they look at things, the things they look at have changed as well.

Now we turn to becoming more actively involved in the change process by focusing on influencing skills. Look carefully at the word **influencing**. A web definition is "a power affecting a person, thing, or course of events, especially one that operates without any direct or apparent effort" (Free Dictionary, 2011). This definition is helpful because it (1) reminds us that interviewers and counselors are in a position of power relative to clients, even when we seek to be egalitarian, and (2) emphasizes that we can exert influence without "any apparent effort."

The influencing skills and strategies of Section III present many specifics for helping clients restory their lives, develop new ways of thinking and feeling, and take action to achieve desired goals. Both students and professionals in counseling, psychology, social work, and other mental health disciplines should listen and start using their power to help clients meet their needs, achieve wellness, and build new stories.

Throughout this section, think of counseling and therapy as a creative process. Creativity is the root of change. The famed theologian/philosopher Paul Tillich spoke of the "creation of the *New*." To help clients change and create the *New*, we need to establish an empathic relationship, listen to the client's story, and summarize it fully, recognizing the emotional base of client concerns. Then, if necessary, we can bring influencing skills into the session to encourage client intentionality, for intentionality is another way to think about the creative individual.

Chapter 8, Focusing the Interview: Exploring the Story From Multiple Perspectives, illustrates how multiple frameworks for examining stories and personal issues can result in breakthroughs in client understanding. Too many counselors and interviewers assume that their clients' concerns are self-created and miss contextual issues such as the impact of family, significant others, and multicultural and social factors.

Chapter 9, Empathic Confrontation: Supporting While Addressing Client Conflict, recognizes that most clients come to an interview experiencing ambivalence, feeling conflicted, having difficulties working with others, and seeking some form of resolution. Along with listening, empathic confrontation helps them move beyond their issues and minimize resistance. The Client Change Scale, included in this chapter, enables you to assess your client's responses and evaluate larger developmental change over a series of interviews.

Chapter 10, Reflection of Meaning and Interpretation/Reframing: Restorying Client Lives Through Meaning Making, discusses the use of these closely related skills to provide a deeper understanding of each client's issues and history. Both skills are concerned with finding new perspectives on life and its meaning.

Chapter 11, Empathic Self-Disclosure and Feedback: Relationship, Immediacy, and Genuineness, explores specific strategies for facilitating client change. Use of appropriate self-disclosure and feedback in a genuine way makes the interview more immediate and personal, and helps clients' creation of new stories.

Chapter 12, Influencing Client Actions and Decisions: Directives, Psychoeducation, Logical Consequences, and Decisional Counseling, provides you with the basics of four additional influencing skills. Each of these skills has specific strategies that can help clients create new stories and generalize and act on what is discovered in the session.

Take your time and work intentionally to achieve competence in the concepts, skills, and strategies of this section. Empathic confrontation and decisional counseling, for example, are rich and complex skills that even the most experienced counselor or therapist can always improve. The learning in interviewing, counseling, and therapy never ends. *Be patient, listen, learn new things constantly, practice new skills and ideas until you can use them effectively in the interview, and do not forget to create your New in this process.*

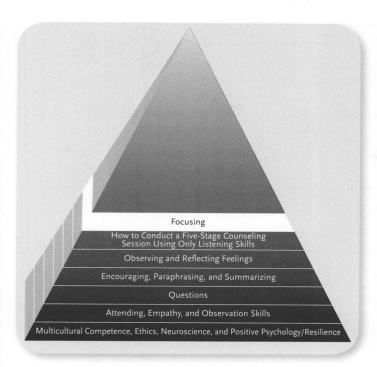

Focusing

How to Conduct a Five-Stage Counseling Session Using Only Listening Skills

Observing and Reflecting Feelings

Encouraging, Paraphrasing, and Summarizing

Questions

Attending, Empathy, and Observation Skills

Multicultural Competence, Ethics, Neuroscience, and Positive Psychology/Resilience

Chapter 8
Focusing the Interview
Exploring the Story From Multiple Perspectives

Every client has a unique individual self, but that self has been generated in web of relationships. We had best think of our clients as persons-in-relationship, persons-in-community.

—Mary Bradford Ivey

Alternative descriptions of the same reality evoke different emotions and different associations.

—Daniel Kahneman

Focusing is a skill that enables multiple tellings of the story and will help you and clients think of new possibilities for change. Client issues are often more complex than they seem, but clients often look at them from only one frame of reference. Their past experiences, cultural biases, fears, or denial further contribute to this narrowing view. Focusing can help clients clarify their issues, engage in more holistic restorying, and find more satisfying decisions and actions. Focusing also encourages you, the interviewer or counselor, to think about alternative frames of reference, broader systems, and cultural issues that might affect the client.

Chapter Goals

Awareness, knowledge, skills, and actions developed through the concepts of this chapter and this book will enable you to:

▶ Increase clients' awareness that the issues they present are not just about themselves and their internal thoughts and feelings, but also interconnected in a web of interpersonal and cultural/environmental relationships.

▶ Help clients expand their stories and describe their issues from multiple frames of reference.

▶ Operate more effectively in our diverse society by enabling clients to see themselves as selves-in-relation and persons-in-community through community and family genograms.

▶ Include advocacy, community awareness, and social change as part of your interviewing practice.

> **Assess your current level of knowledge and competence as you begin the chapter:**
>
> 1. Self-Assessment Quiz: The chapter quiz will help you determine your current level of knowledge. You can take it before and after reading the chapter.
>
> 2. Portfolio of Competencies: Before you read the chapter, please fill out the Self-Evaluation Checklist to assess your existing knowledge and competence on the ideas and concepts presented in this chapter. Then, at the end of the chapter, complete the checklist again to summarize your competencies after study and practice.

▶ Defining Focusing

Focusing is an important skill that enables clients to gain a more complete understanding of issues affecting their lives. It is the job of the counselor to help clients steer through these difficulties by helping them to focus on central issues that affect their decisions. Focusing is a skill that will make a difference in your practice and facilitate cognitive/emotional development. Focusing enables clients to understand the web of relationships in which they live and the cultural/environmental context surrounding their behavior.

Selective attention is closely related to the focusing skill. In Chapter 3, we paid attention to (focused on) various topics that the client brought up. Generally speaking, we want to attend to what the client is saying in the here and now. Traditionally, this has meant focusing on the individual clients before us and their immediate issues. By way of contrast, the focusing skill discussed here is more broadly based and encourages client attention to the multiple factors that relate to their decisions, concerns, and issue. Thus, when you use the focusing skill, you deliberately attend to broader dimensions to ensure that the client is able to resolve issues in the context in which they live.

FOCUSING	ANTICIPATED CLIENT RESPONSE
We seek to help clients become aware of where they place responsibility for their issues and concerns. Are they blaming themselves or others? The skill of focusing will enable them to take a more comprehensive view. Traditionally, interviewing focuses on the client and her or his central issue or concern. The skill of focusing helps enlarge the picture by focusing also on varying views of significant others (e.g., the troubling individual, the family, the partner). Often missed in traditional counseling theory and practice are factors related to the cultural/environmental context.	In response to your focus on selected dimensions, client understanding of the total situation will become enlarged. An issue that was once seen as a simple decision now has multiple dimensions. This expanded knowledge is likely to lead to more comprehensive and effective resolution of client concerns.

Empathically drawing out the story as the client sees it is obviously the way to begin any session. We need to understand the individual's cognitions and emotions. We do this with a focus on the individual client. However, that individual comes to us with a social and cultural history. The family, a close relationship, friends, and important "others" include many significant people ranging from teachers to employers, sport coaches to scoutmasters, biking groups to book clubs. We exist in a *web of relationship* with multiple dimensions of personal identity.

The web of relationship, in turn, is deeply affected by the fact that all exist in a *cultural/environmental context* (CEC). Each person in the web is affected in some way by the RESPECTFUL model—for example, differences in gender, sexual orientation, socioeconomic class, and ethnicity/race make our life web full of difference. The region or the country in which we live, political preferences, military experience, and the type of education we receive affects all these.

Moreover, what is the *historical moment* that surrounds the web of relationship? The great economic recession is a recent example. People living 25 years ago could count on finding satisfactory employment and a living wage. Automation has eliminated millions of jobs. Increasing income disparity, political conflict, war, and decisions of state and supreme courts massively affect the individual and the web of relationship.

 Interactive Exercise: Identify Focus Dimensions. Work through the Samantha and Janet session to identify the responses using focus analysis and to score your responses.

▶ Discerning the Skill of Focusing

Our clients do not come to us alone. Rather, they bring with them multiple voices from their web of relationships.

—Mary Bradford Ivey

Focusing will enable you to develop a more comprehensive view of your clients' web of relationships, their background, their present situation, and the many influences on their current issues. We do this first by focusing on the individual client, and also focusing on the concerns and issues the client brings. But focusing on significant others and key individuals and their place in the web brings context to the client's situation. Furthermore, if we encourage clients to discuss aspects of their RESPECTFUL background, we learn much about the cultural/environmental context—thus developing an understanding of the multiple voices that the client brings to us.

> Vanessa walks swiftly into the office and starts talking even before she sits down: "I'm really glad to see you. I need help. My sister and I just had an argument. She won't come home for the holidays and help me with Mom's illness. My last set of exams was a mess and I can't study. I just broke up with the guy I was going with for three years. And now I'm not even sure where I'm going to live next term. And my car wouldn't start this morning. . . ." (She continues with her list of issues and begins repeating stories almost randomly, but always with energy and considerable emotion.)

There are many clients like Vanessa, who have several issues in their lives. Interviewers frequently feel overwhelmed when a client spends 5 or more minutes telling problematic stories, jumping rapidly from topic to topic. Sometimes there is an insistence that we do "something" immediately and start resolving concerns. We may even fall into the trap of trying to solve client concerns for them, but then they may refuse to listen to us, thus adding resistance to the general confusion.

Focusing on the client. One rule helps us settle down and start working with clients: Counseling is for each unique individual client. It is best when the interview empowers clients *to resolve their own issues.* Vanessa is talking "all over the place," but we can help her focus and clarify her issues and her thoughts and feelings by attending to her. We could gently interrupt and say something like the following. Note the frequent use of the word *you* and the client's name, *Vanessa*.

> *Vanessa*, could we stop for a moment? I really hear *you* loud and clear. There are a lot of things happening right now. One of the best ways to approach these issues is to focus on how you are doing and feeling. Once I understand *you* a bit more, we can work on the issues that *you* describe.
>
> I get the sense that *you, Vanessa,* are hurting a lot right now and are confused about what to do next. Could *you* take a deep breath and tell me what *you* are feeling and thinking right now? What are these things doing to *Vanessa*? What's happening for *you*?

In these two paragraphs, we have used the words *Vanessa* and *you* more than 10 times. The goal here is to help Vanessa focus on internal cognitions and feelings. Interviewing and counseling are first and foremost for the client who sits in front of you. Summarize the several issues, then ask her what she would like to start working on first. Explore other issues later. Once we have a better grasp of the person(s) before us, we can work more effectively toward resolution, but usually only one issue at a time.

Selective attention (Chapter 3) is basic to focusing—clients tend to talk about and focus on topics to which you give your primary attention. Through your attending skills (visuals, vocal tone, verbal following, and body language), you indicate to your client that you are listening. But we all tend to focus or listen in different ways. It is important to be aware of both your conscious and unconscious patterns of selective attention; clients sometimes follow your lead rather than talk about what they really want to say. If you focus solely on individual issues, clients will talk about themselves and their frame of reference. If you focus on sexual concerns, they likely will talk about sex. Fail to include sex as a focus, and the topic will soon drift away.

Focusing on client issues, concerns, and challenges. This is the next most critical focus dimension. It sometimes is tempting to first draw out the specifics of the reasons clients come to us. Then, sometimes the client's individuality is lost, unless we personalize the session by integrating that word *you* and the client's name. Some beginning interviewers, and even experienced professionals, miss the uniqueness of the client while they draw out stories and issues.

Nonetheless, focusing on client issues, concerns, and challenges is the second most important focus dimension. Here we draw out stories so that we can understand the concrete specifics. Sometimes the stories are so fascinating we forget the client before us. Some may fail to give sufficient attention to the client as a person and use the interview as a chance to hear interesting problematic stories and details—almost as a voyeur. Following are some examples of how you can help the client focus more clearly on the issue that he or she considers most important at the moment.

▶ Vanessa, I hear a lot of issues—which would you like to focus on first?
▶ Vanessa, listening to you, I sensed that a lot of things are happening, but you sounded like your worries about your mother and challenges with your sister may be the most important. Would you like to start here?
▶ Could you tell me a story about how you think all this developed in the first place?

Expanding focus to understanding the web of relationships. Individuals live in a broader context of multiple systems. Increasingly, theorists are challenging the idea of a totally autonomous self, separated from other people, friends, family, and the cultural context.

We need to encourage understanding of the web of relationships and the impact of cultural/environmental issues on daily living.

Several important concepts have surfaced to help expand the idea of the individual self. The term **self-in-relation** defines the "individual self" as really only existing in relationship to others. The term **being-in-relation**, suggested by the feminist authors Jordan, Hartling, and Walker (2004), is similar to self-in-relation, but points out that all of us are beings who grow in the here-and-now moment as we interact with others. The **person-as-community** was stressed from an Afrocentric frame by Obonnaya (1994), who points out that our family and community history live within each of us. Clients come to the interview with many "internalized voices" affecting cognition and emotion. Bringing these to awareness through focusing is a vital, but often missing, part of counseling and therapy.

You will have a chance later in this chapter to develop a **community genogram**. This is a visual map of the key voices that affect the client even today. In this picture of the client's community of origin, you will learn much about the person's family and cultural background. The community genogram will enable you to draw out stories of strengths, and will also provide a background of how certain stressors impact the client. You can develop community genograms of clients' current living situation, thus generating a much broader sense of who they are both in the present and past.

Much individual counseling focuses on issues of internal conflict around clients' mistaken beliefs about themselves and their decisions. However, there can also be incongruity and discrepancies between the individual and family and friends. In addition, many client issues and events are related to a broader social context (e.g., poor schools, floods, job issues). If you help clients to see themselves and their issues as *persons-in-community*, they can learn new ways of thinking about themselves and use existing support systems more effectively. It is not the individual versus the social context; rather, social context enriches our understanding and enhances the uniqueness of each person and client.

The following list offers sample comments and questions that allow the interviewer to focus the session and help the client see the web of relationships.

Significant others (partner, spouse, friends, family)
► Vanessa, tell me a bit more about your mother and how she is doing.
► How are your friends helpful to you?
► What's going on with your sister?
► You seem to be thinking a lot about the breakup after three years of living together. Let's explore that a bit more now.
► Your grandmother was very helpful to you in the past. What would she say to you now?

Other key figures impacting the client currently (teacher, employer, medical staff, even a rude mechanic)
► Has your instructor been at all helpful?
► How do you feel about the physicians and nurses working with your mom?
► What did the mechanic do to upset you?

Cultural/environmental context (the client's unique RESPECTFUL multicultural background, as well as broader systemic issues such as the economy)
► Finding a new place to stay is difficult, and you think that landlords won't rent to culturally diverse renters.
► What are some strengths that you gain from your spiritual orientation?
► So finances are becoming a problem, and you are worried about taking out another loan.

- ► Could you share a bit about your community and things there that might be available for help and support?
- ► I hear that your sister just lost her job when the company was downsized. I wonder how that relates to the difficulties you have with her.

Mutual focus. As an interviewer, counselor, or psychotherapist, you are now also part of Vanessa's web of relationships. It may be helpful to briefly and carefully share experiences similar to those of the client. And, at times, it is helpful to discuss the here-and-now relationship between you and the client. The mutual focus is elaborated in Chapter 11 on self-disclosure and feedback.

- ► Vanessa, sounds like you feel overwhelmed with all that. I, too, felt overwhelmed when I found my mom had cancer. Like you I had to handle it alone. Then in the middle of it all, I found that insurance would not pay for it. Is that close to what you feel? (Self-disclosure and note the check-out, ending the self-disclosure and returning the focus to the client.)
- ► (Halfway through the first session) Vanessa, may I offer some feedback? I see you as facing many issues, but at the same time, I hear quite a few resources. I'll remind you of some positives that will help. For example, despite all this, your boss likes you and the way you handle the part-time job. And you seem strong enough to support your mother, . . . (continues with a few more resources and strengths and then ends the feedback with the following) So, in summary, I see you already brightened up here in the interview and looking more hopeful. Let us both remember those strengths and build on them as we work on your issues one by one.
- ► Vanessa, we have met now for three sessions. I wonder how you feel about what we have done. And perhaps I have missed something important. (Here-and-now discussion of the relationship.)

As an interviewer, be aware of how you focus an interview and how you can broaden the session so that clients are aware of themselves and their web of relationships. You can help them encounter and understand how they are persons-in-relation, persons-in-community. In a sense, you are like an orchestra conductor, selecting which instruments (thoughts, emotions, and behaviors) to focus on, thus enabling a better understanding of the whole. Some of us focus exclusively on the client and their immediate concern. Thus, we may fail to recognize the impact of broad contextual issues. An in-depth multicultural perspective on focusing is presented in Box 8.1.

 Interactive Exercise: Community Genogram. An exercise you can do by yourself or with a classmate to practice drawing data from a genogram.

► Observing the Skill of Focusing

We live in a culture that focuses on internal issues, individuality, and individual responsibility; thus, we most often focus using the words *I, you,* or the client's name. Indeed, it is the individual client that needs to work through internal or external issues and conflict. At the same time, awareness of the web of relationship reveals the importance of external causes of individual issues. For example, clients can "blame" themselves or others as the cause of issues. If we broaden understanding, the actual causes of life challenges and conflict will be better understood. This makes more effective thoughts, feelings, and behavior beyond the interview more easily achieved.

The following interview with Janet focuses first on her individual issue of taking virtually all the responsibility for difficulties in her relationship with Sander. This is the second interview. During the first session with couples, families, or groups, multiple issues can occur, and selective attention is important. You can be a much more effective counselor if you are able to help them focus together on one thing at a time. Often the function of group therapy is to reduce the clutter of voices and focus in on one person at a time, making sure that his/her thoughts and feelings are being heard. One suggestion is to have the person who is talking hold an object—a cup or a ball; when that person has finished talking, he/she passes the object to the next person. The idea of holding something is similar to giving that person the "podium." This gives all participants a better chance to express their ideas.

Janet has completed a community genogram of the home where she grew up in Eugene, Oregon (see "Practicing: Mapping the Web of Relationships, Community and Family Genograms," page 170). Please take a moment to look at Janet's web of relationships as seen in the community genogram.

The community genogram is used here to help Janet understand her self-blame and resultant passive acceptance of a situation for which there are multiple causes in the web of relationships. Once she starts becoming aware of external issues, she attributes less blame to herself.

INTERVIEWER AND CLIENT CONVERSATION	PROCESS COMMENTS
1. *Samantha:* Nice to see you, Janet. How have things been going?	A solid relationship was established during the first session. When Samantha met Janet outside her office, Janet seemed anxious to start the session.
2. *Janet:* Well, I tried your suggestion. I think I understand things a bit better.	Last week, Samantha suggested homework that asked Janet to spend the week listening carefully to Sander and trying to identify what he was really saying. She also advised continuing her own behavioral patterns so that she could note patterns of behavior more easily.
3. *Samantha:* You understand better? Could you tell me more?	Encourage. Focus on the main concern. (Interchangeable empathy)
4. *Janet:* Well, I listened more carefully to Sander so I could understand what he really wants from me. He really wants a lot. He wants me to be a better cook, he doesn't like the way I keep house, and the more I listened, the more he wanted. I guess I haven't given him enough attention. I should be doing more and doing it better.	Self-disclosure on observations. Focus on self (the client), the significant other, and concerns around the relationship.
5. *Samantha:* Janet, it sounds like you feel sad and a little guilty for not doing more. Am I hearing you right?	Reflection of feeling, focus on Janet with secondary focus on her presenting issue. (Interchangeable empathy)
6. *Janet:* Yes, I do feel sad and guilty; it's my responsibility to keep things together at home. He works so hard.	The key word *responsibility* appears. Janet seems to believe that she is responsible for keeping the relationship together. Here we see self-blame.

(continued)

INTERVIEWER AND CLIENT CONVERSATION	PROCESS COMMENTS
7. *Samantha:* Sounds like you're punishing yourself, when you're trying so hard and being so responsible. Could you tell me more about how Sander reacts?	Samantha focuses first on the client and then on the significant other. The change of focus to Sander by means of the open question may lead to a broader understanding. (Potentially additive empathy)
8. *Janet:* Yes, I get angry with myself for not doing better. It's hard with work too. But Sander wants the house to run perfectly. I try and try, but I always miss something. Then Sander blows up.	Here we see a triple focus on self, the issues between the couple, and Sander.
9. *Samantha:* I hear you trying very hard. Could you give me a specific example of a time that Sander blew up?	Paraphrase and open question, seeking concreteness. The focus changes from Janet to Sander in a search for concreteness. (Potentially additive empathy)
10. *Janet:* Last night I made a steak dinner; I try to fix what he wants. He seemed so pleased, but I forgot to buy steak sauce and he blew up. It really shook me up . . . then, when we went to bed, he wanted to make love, but I was so tense that I couldn't. He got angry all over again.	Examples help us understand the specifics of situations. Some clients talk in vague generalities and asking for specifics can make a real difference.
11. *Samantha:* What happened then?	Open question/encourager, focus on concern. Search for more concreteness.
12. *Janet:* He went to sleep. I lay there and shook. As I calmed down, I realized that I need to do a better job and maybe he won't get so angry.	Janet continues to internally attribute the issues to herself through self-blame.
13. *Samantha:* Let's see if I understand what you are saying. You are doing everything you can to make the relationship work. Sander blows up at little things, and you try harder. And then you feel that it is your fault. Have I heard you correctly?	Samantha summarizes the situation. She recognizes it as a pattern in Janet's relationship with her husband, but that Janet's internal beliefs led her to self-blame. Focus on Janet, Sander, and the central issue. (Additive empathy)
14. *Janet:* It is my fault, isn't it?	Self-focus. Note the acceptance of individual responsibility for the difficulties.
15. *Samantha:* I'm not so sure. We've talked about your relationship with Sander. Now let's talk about something that went right for you in the past. As you look at your community genogram, what reminds you of good times, when things were going well?	The community genogram often helps clients see more of the web of relationships. Focus shifts to cultural/environmental context and a search for positive assets and wellness strengths. Strengths and resources are helpful in understanding present situations. (Additive empathy)
16. *Janet:* (pause) Well, I loved visiting Grandma and Grandpa. They were friendly and helped me with my problems. Mom was the same way—she kept pointing out that I could do things.	Focus on family. Janet perks up and smiles.
17. *Samantha:* Could you tell me a story about Grandma and Grandpa or your Mom when they made things better for you?	Open questions focused on positive stories in the family. (Potentially additive)

(continued)

INTERVIEWER AND CLIENT CONVERSATION	PROCESS COMMENTS
18. *Janet:* Well, Mom had to work and carry the family, and I had the most fun with Grandma and Grandpa. One time kids were teasing me at school about my braces. I thought I was funny looking because I was different. They listened to me; they told me that I was beautiful and smart. They taught me to ignore the others, just try harder, and it would all work out. I found that it does—if I try harder, it usually does get better.	We learn about a supportive family. We also hear about a lifestyle that focuses on ignoring underlying issues and trying harder—exactly what Janet is doing with Sander.
19. *Samantha:* So you learned in your family that ignoring things and trying harder usually helped work things out. Where did your desire to try so hard come from?	We see the emphasis on individual responsibility. Paraphrase with focus on family and client. Linking of past history with present situation. (Additive empathy)
20. *Janet:* (pauses and looks at community genogram) Dad made impossible demands of Mom. After she talked with the minister of our church, she started standing up for herself. I learned in church that it is important to care for others, but I guess I can't care for others unless I care for myself.	Focus on community, the family and the church. Janet is beginning to draw on resources from the past. She is becoming aware of being a person-in-community. This helps her to start thinking of herself in new ways. Janet, drawing on these resources, learns that her issues are not just "her issues." They developed in an external context.
21. *Samantha:* Connections and relationships are very important to you. There is also a need to care for yourself if you are to care for others. Too many women fall into the trap of always caring for others and not caring for themselves. That seems to represent the family lesson of caring. We also need to focus on your Mom starting to care for herself, how that made a difference in your family. I wonder if it would make the same difference for you now?	A brief summary followed by reframing what Janet has said with a new perspective. This is a slight extension of what Janet seemed to be already saying. Reframing and interpretation tend to be best received when they are not too far from the client's present thinking. (Additive empathy)
22. *Janet:* I think so. I so appreciate your listening to me. I wonder if I've become too much of a doormat for Sander. But I worry about what he would do if I stood up more for myself. That seems to be my family history. Thank heavens for the church.	Janet focuses on herself as a person in a family within the community—family and church have been added to the web of relationships. The interviewing session can now start moving to restorying.
23. *Samantha:* I really like what you are saying. You seem ready to stand up for yourself, that you are not to blame. Could you tell me more about how the church and spirituality might provide support?	Focus on the interviewer's impressions of what Janet is saying. Open question with focus on the client and the cultural/environmental context in terms of the church, which seems important in developing resources. (Additive empathy)
24. *Janet:* Connections with friends are important, but church was a place where I could quietly meditate and work through issues. I don't seem to give myself time for that anymore.	Janet is identifying herself as a self-in-relation and draws on community resources for additional strength. Meditation is a well-known and important strategy that helps build a more positive self.
25. *Samantha:* Meditation and spirituality often give us a foundation for deciding what is best. Let us explore that in more detail. We also need to explore the community genogram for more sources of strength.	The focus turns to external sources that may help Janet build on the past to work through current issues. (Additive empathy)

Samantha helped Janet examine her situation in its broader context. What started as attributed self-blame has now been externalized. The family and/or community genogram can be helpful in aiding a person who places too much responsibility on herself or himself to balance causation between internal and external issues. It can help both clients and helping professionals to see the client as a person-in-relation or a person-in-community. This is a more leading approach, although the professional must continue to listen carefully to what the person has to say. As focus shifts to various dimensions of the larger client story, the client and the counselor

| BOX 8.1 | National and International Perspectives on Counseling Skills |

Where to Focus: Individual, Family, or Culture?

Weijun Zhang

Case study: Carlos Reyes, a Latino student majoring in computer science, was referred to counseling by his adviser because of his recent academic difficulties and psychosomatic symptoms. The counselor was able to discern that Carlos's major concern was his increasing dislike of computer science and growing interest in literature. He was intrigued about changing his major but felt overwhelmed by the potential consequences for his family, in which he is the oldest of four siblings. He is also the first in his family ever to attend college. Carlos has received some limited financial support from his parents and one of his younger siblings, and the family income is barely above the poverty line. The counseling was at an impasse. Carlos was reluctant to take any action and instead kept saying, "I don't know how to tell this to my folks. I'm sure they'll be mad at me."

During class discussion of this case, almost everyone argued that Carlos's problem is that he does not give priority to his personal career interests, that he should learn to think about what is good for his own mental health, and that he needs assertiveness training. I did not quite agree with my fellow students, who are all European Americans. I thought they were failing to see a decisive factor in the case: Carlos is Latino!

In traditional Hispanic culture, the extended family, rather than the individual, is the psychosocial unit of cooperation. The family is valued over the individual, and subordination of individual wants to the family needs is assumed. Also, traditional Hispanic families are hierarchical in form; parents are authority figures and children are supposed to be obedient. Given this cultural background, to encourage Carlos to make a major career decision totally by himself was impossible. Any counseling effort that did not focus on the whole family was doomed to fail.

Because it is the financial support from the family that made his college education possible, Carlos may be expected to contribute to the family when he graduates. This reciprocal relationship is a lifelong expectation in Hispanic culture, and the oldest son is especially responsible in this regard. Changing his major in his junior year

does not only mean he will be postponing the date when he will be able to help his family financially, but it also means he may not be able to do so at all, for we all understand how hard it is to find a good-paying job in the field of literature. When interdependence is the norm among Hispanic Americans, how can we expect Carlos to focus entirely on his personal interests without giving more weight to his family's pressing economic needs?

If I were Carlos's counselor, rather than focusing immediately on his needs, I would first support him with his family loyalty and then help him understand that there are not just two solutions: either . . . or. . . . Together, we might brainstorm to generate some alternatives, such as having literature as his minor now and as his pastime after he graduates, changing his career when his younger siblings are off on their own, or exploring possibilities that may combine the two. He could, for example, design computer programs to help schoolchildren learn literature. Each of these takes into account family needs as well as those of Carlos.

The professor praised me highly for my "different and sensitive perspective," but I shrugged it off; this is just common sense to most Third World minority people and, probably, many Italian and Jewish Americans as well. (I remember years ago, when I was trying to make major career decisions with my parents, at least ten of my relatives were involved. And these days, I am still obligated to help anyone in my extended family who is in financial need.) It took me almost a year to realize that when I am asked, here in the United States, "How's your family?" I usually need tell only how my wife and child are faring, not my parents, grandparents, and siblings.

If the meaning of family in Hispanic culture is confusing to many counselors, the traditional extended family clan system of Native American Indians, Canadian Dene, or New Zealand Maori can be even more difficult for them to grasp. This family extension can include at times several households and even a whole village. Unless majority group counselors are aware of these differences in family structure, they may cause serious harm through their own ignorance.

begin to understand the multidimensionality of the issue more fully. Janet's concerns are not just "Janet's concerns." Rather, her issues interact with many aspects within her web of relationships.

 Interactive Exercise: Focusing on Resources. This exercise gives you an opportunity to select the most helpful focusing response. Can you find a balance among the different possibilities?

► Practicing Mapping the Web of Relationships, Community and Family Genograms

Making your own community genogram is a good way to understand the importance of the web of relationships that clients bring to us. We go so far as to suggest placing the genogram prominently in some convenient place so that you and your client can refer to it as needed. Furthermore, the community genogram is an especially good format for practicing the skill of focusing. Thus, here we are asking you first to develop a genogram and then to analyze the genogram in terms of your own web of relationships. Finally, we hope that you and a partner will share the genograms and practice bringing out the multiple focus dimensions.

DEVELOPING A COMMUNITY GENOGRAM

At this point, we suggest that you develop a community genogram as presented in Box 8.2. Develop a community genogram for yourself, using your own style of presentation. This will help you, and later on your clients, to see how we are connected to many influences.

BOX 8.2 Developing a Community Genogram: Identifying Personal and Multicultural Strengths

The community genogram is a "free-form" activity in which clients are encouraged to present their community of origin or current community, using their own unique style. Two visual examples of community genograms are presented here. Through the community genogram, we can better grasp the developmental history of our clients and identify client strengths for later problem solving. Clients may construct a genogram by themselves or be assisted by you through questioning and listening to the things that they include.

Step 1: Develop a Visual Representation of the Community

► Select the community in which you were primarily raised. The community of origin is where you tend to learn the most about culture, but any other community, past or present, may be used.
► Represent yourself or the client with a significant symbol. Use a large poster board or flipchart paper.

Place yourself or the client in that community, either at the center or at another appropriate place. Encourage clients to be innovative and represent their communities in a format that appeals to them. Possibilities include maps, constructions, or star diagrams (used by Janet, below).
► Place family or families, nuclear or extended, on the paper, represented by the symbol that is most relevant for you or the client.
► Place important, most influential groups in the community genogram, represented by distinctive visual symbols. School, family, neighborhood, and spiritual groups are most often selected. For teens, the peer group is often particularly important. For adults, work groups and other special groups tend to become more central.

You may wish to suggest relevant aspects of the RESPECTFUL model discussed in Chapter 2. In this way, diversity issues can be included in the genogram.

(continued)

BOX 8.2 (continued)

Step 2: Search for Images and Narratives of Strengths

▶ Post the community genogram on the wall during counseling sessions.

▶ Focus on one single dimension of the community or the family. Emphasize positive stories even if the client wants to start with a negative story. Do not work with the negatives until positive strengths are solidly in mind, unless the client clearly needs to tell you the difficult story.

▶ Help the client share one or more positive stories relating to the community dimension selected. If you are doing your own genogram, you may want to write it down in journal form.

▶ Develop at least two more positive images and stories from different groups within the community. It is often useful to have one positive family image, one spiritual image, and one cultural image so that several areas of wellness and support are included.

We encourage clients to generate their own visual representations of their community of origin and/or their current community support network. The examples presented here are only two of many possibilities.

1. The star: Janet's world during elementary school tells us a good bit about a difficult time in her life. Nonetheless, note the important support systems.

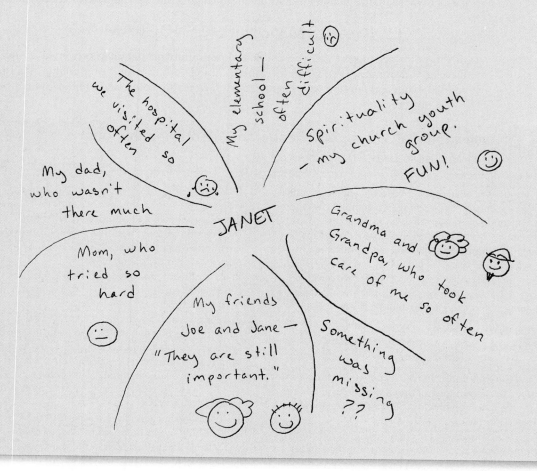

(continued)

BOX 8.2 (continued)

2. The map: The client draws a literal or metaphoric map of the community, in this case a rural setting. Note how this view of the client's background reveals a close extended family and a relatively small experiential world. The absence of friends in the map is interesting. Church is the only outside factor noted.

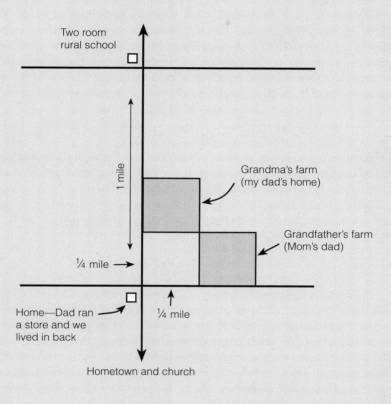

We are persons-in-relation to others and the social context around us. Think about your own life story and how it has been affected by the many relationships in your community of origin. Usually, the family is a critical part of the community genogram. However, the **family genogram** can be a useful supplement to the community genogram, bringing out additional details of family history (see Appendix B).

DEBRIEFING YOUR COMMUNITY GENOGRAM

First, search for stories of strength and support in your community genogram. There indeed may be concerns and issues from the past, but we resolve these with our strengths. Many of these strengths come from our cultural/environmental context, often best represented by our communities of origin.

At a cognitive level, note the array of community experiences that helped shape you as the multicultural being you are today. What forces have been most influential in your growth and development? While it may be necessary to think of hindrances, difficulties, and hassles, keep these to a minimum. Focus on the several strengths that your community brought to you. How does this background relate to what you are doing now?

Review the community once again. What focus dimensions are there? What might be missing? What could you add at this point?

Now look at the emotional roots of community stories. Take just one of those experiences and focus on one story of strength. Close your eyes and visualize what happened, with a focus on the positives. What are you thinking? Feeling? Can you feel that strength in some specific place in your body? That last question may seem strange, but most often we find that individuals who commit to this process can find places in their bodies that represent the strength of survival, the joy of being, or feelings of calm and safety. These are often associated with key figures or places in the past. We can go to those feelings and those places when we have need.

DEBRIEFING WITH A PARTNER

The community genogram provides you with a chance to learn about the developmental history and cultural background of your client. It will provide you with considerable data so that you can spotlight and focus on key issues. Where possible, post the community genogram on the wall.

Start the debriefing by asking your volunteer client to describe the community and things that he or she considers most important in development. Obtain an overview of the client's community. As part of this, remind the client that we are looking for strengths and positives. Later, we can review difficult events and those that relate to present concerns. You will want to seek stories from several areas of the community genogram, but you often will not have time for them all.

Search for and encourage images of strength and awareness of past and present resources. Help the client share one or more positive stories relating to the community dimension selected. Follow this by asking for a story about each element of the genogram. Seek to obtain positive stories of fun and support, strength, courage, and survival. Bring out the facts, feelings, and thoughts within the client's story. Many of us have been raised in communities that have been challenging and sometimes even oppressive, so a positive orientation can focus on the positives and strengths that have helped the client. That platform of positives makes it possible to explore problematic issues with a greater sense of hope. For more ideas on the community genogram, see Rigazio-DiGilio et al. (2005).

Develop at least two more positive images and stories from different groups within the community. For example, it may be useful to have one positive family image, one spiritual image, and one cultural image so that several areas of wellness and support are included.

Families and family stories are typical features of every community genogram. Again, search for positive family stories. Often you will obtain as much information as needed from the family as described in the community genogram. However, the family genogram in Appendix B offers a more detailed picture. The family genogram often focuses on intergenerational issues, and historically it has tended to focus on "problems." Information such as this can be useful, but we should also remember to look for survival strengths in family members. Help the client notice stories of strengths and survival in the face of difficulty.

 Interactive Exercise: Writing Alternative Focusing Statements. A good way to practice your focusing skills.
Group Practice Exercise: Group Practice With Focusing

► Refining Focusing on Challenging Issues

Two major issues where focusing can be helpful are the difficult matter of abortion and the totally different area of social justice advocacy. As a counseling issue, the first is very much a personal issue, but is affected deeply by the cultural/environmental context. The second is an area that some in the helping fields believe is not their responsibility.

THE CHALLENGE OF PRO-LIFE AND PRO-CHOICE

Too many women are forced to abort by poverty, by their menfolk, by their parents. . . . A choice is only possible if there are genuine alternatives.

—Germaine Greer

Even the smallest person can change the course of the future.

—J. R. R. Tolkien

Take some time to think through your own thoughts on a difficult and challenging issue—abortion, the termination of a fetus. This remains a central issue in what is now called the "culture wars." We have chosen this issue as it is about as challenging a moral and practical life decision issue as any you will face. The question you need to ask yourself is "How can I help clients, even when I disagree with them on what should happen when a woman becomes pregnant, either through accident or (even more challenging) rape?"

As an interviewer, counselor, or psychotherapist, you will encounter controversial cases and work with clients who have made different decisions than you would. Even the language of "pro-choice" and "pro-life" beliefs can be upsetting to some. It is best to begin with awareness of your standing on this issue: What is your personal position around termination of a fetus?

Review the multiple dimensions of focus. What do your family and those closest to you think and feel emotionally about abortion? Your friends? What do your community and church, both past and present, say and think? And how does your understanding of state laws and the extensive national media coverage affect your thinking? From a more complex, contextual point of view, spend a little time thinking about what has influenced your thinking on this issue; record what you discover. Our individual thoughts and decisions on critical issues are deeply intertwined in our social context.

As a counselor, it is vital that you understand the situations, thoughts, and feelings of those who take varying positions around controversial issues, whether you agree with them or not. Can you identify and understand some of the thoughts and feelings of those who have a different position from your own?

Counseling is not teaching clients how to live or what to believe. It is helping clients make their own decisions. Regardless of your personal position, you may find yourself unconsciously using the interview to further your beliefs. Most would agree that counselors should avoid bias in counseling. You may need to help your client understand more than one position on controversial issues. You will work with clients who are intolerant of differences or may demonstrate unconscious sexism, racism, anti-Semitism, anti-Islamism, or other forms of intolerance. The art and mastery of competent interviewing and counseling merge awareness and respect in the interest of client self-discovery, autonomy, and growth.

School systems and agencies may have written policies forbidding any discussion of abortion. Due to complex legal issues, abortion is nearly impossible in some areas. Further, if you are working within certain agencies (e.g., a faith-based agency or a pro- or antiabortion counseling clinic), the agency may have specific policies regarding counseling around abortion. Ethically, clients should be made aware of specific agency positions before counseling begins. It is vital to share these policies immediately with the client at the beginning of counseling so that client intentionality and choice are fully preserved.

Choosing an appropriate focus can be most challenging. Too many beginners focus only on the immediate problematic issue. The prompt "Tell me more about the abortion" may draw out details of the procedure but little about the client's distinctive personal experience. An extremely important task is drawing out the client's story: "I'd like to hear *your* story" or "What do *you* want to tell me?" There are no final rules on where to focus, but generally, we want to hear the client's unique experience. Focusing on the individual is usually the place to start; review pages 161 and 162 where the word *you* was used in the example interviewer comments.

Other key figures (the male, family, friends, or the threatening crowd outside the clinic) are part of the larger picture. What are their stories? How do they relate to the client as a person-in-relation? It is invaluable to keep the full context in mind in the process of examination and resolution of issues. For a full understanding of the client's experience, all pertinent relationships eventually need to be explored.

ADVOCACY AND SOCIAL JUSTICE

Injustice anywhere is a threat to justice everywhere.

—Martin Luther King, Jr.

Yes I am [a Hindu]. I am also a Muslim, a Christian, a Buddhist, and a Jew.

—Mahatma Gandhi

Social justice is a moral orientation toward a just and fair world. It is a value stance that emphasizes human rights and equality. *Tikkum olim* comes from the Jewish tradition and means "repairing the world." The Roman Catholic Church stresses social justice—the life and dignity of the human person. The United Methodist Church traditionally has studied and taken positions on abortion, alcohol abuse, gambling, capital punishment, gay and lesbian rights, and inclusivity of all races, and seriously examines the meaning of military service and war. Islam forbids discrimination, cruelty, and exploitation; in addition, it pays special attention to physical health—"We are all the offspring of the same parents." Selflessness and caring for others is one of the foundations of the Buddhist tradition.

Given these powerful statements, we can see that interviewing and counseling constantly involve moral issues, values, and social justice. The RESPECTFUL diversity model as applied to the web of relationships makes it clear that we cannot avoid these issues in most interviews. By becoming involved with helping others, you are already taking the route of selflessness and have made an excellent start on discerning meaning in your own life.

In the individual counseling session, you have the opportunity to help clients become aware of how social justice and cultural/environmental contextual issues affect their lives as persons-in-community. As a very concrete example, we suggest having a client's community genogram available for discussion and enlightenment. A depressed client learns that he or she is not totally at fault for the depression and discovers that family and contextual factors can be addressed, thus externalizing blame as part of preparing the client to return to life with new meanings and purposes. Oppressive life experience, such as being abused by societal poverty or joblessness, is not just an individual issue but is deeply entwined with social history. Discovering oneself as a person-in-community broadens perspective and presents new promise for the future.

You are going to face situations in which your best counseling efforts are insufficient to help clients resolve their issues and move on with their lives. The social context of homelessness, poverty, racism, sexism, and other societal issues may leave clients in an impossible situation. The challenge may be bullying on the playground, an unfair teacher, or an employer who refuses to follow fair employment practices. It is critical that we examine the societal stressors that our clients may face. Traditional approaches to interviewing and counseling may not be enough.

Advocacy is speaking out for your clients; working in the school, community, or larger setting to help clients; and also working for social change. What are you going to do on a daily basis to help improve the systems within which your clients live? Advocacy counseling competencies have been defined in detail by the American Counseling Association, and we highly recommend the recent book *ACA Advocacy Competencies: A Social Justice Framework for Counselors* (2010) by Ratts, Toporek, and Lewis for serious study and consideration.

Here are some examples of situations in which simply talking with clients about their issues may not be enough:

▶ As an elementary school counselor, you counsel a child who is being bullied on the playground.
▶ You are a high school counselor and work with a tenth grader who is teased and harassed about being gay while the classroom teacher quietly watches and says nothing.
▶ As a personnel officer, you discover systematic bias against promotion of women and minorities.
▶ Working in a community agency, you have a client who speaks of abuse in the home but fears leaving because she sees no future financial support.
▶ You are working with an African American client who has dangerous hypertension. You know that there is solid evidence that racism influences blood pressure.

The elementary school counselor can work with each grade or groups of students and educate them about bullying—what it means, how it feels, and ways they can empower themselves to avoid being bullied. The counselor can also work to invite parents and/or caregivers to classroom lessons on bullying so that skills learned at school are reinforced in the home. Another course of action might be for the counselor to work with school officials to set up policies around bullying and harassment, actively changing the environment that allowed bullying to occur. The high school counselor faces an especially challenging issue as interview confidentiality may preclude immediate classroom action. If this is not possible, then the counselor can initiate school policies and awareness programs against oppression in the classroom. The passive teacher may become more aware through training offered to all the teachers. You can help the African American client understand that hypertension is not just "his concern," but that his blood pressure is partially related to the constant stressors of racism in his environment. You can work to eliminate oppression in your community.

"Whistle-blowers" who name unethical or illegal situations that others prefer to avoid can face real difficulty. They will need your support and advice on how to proceed. For example, the company may not want to have its systematic bias exposed. On the other hand, through careful consultation and data gathering, the human relations staff may be able to help managers develop promotion programs that are fairer. Again, the issue of policy becomes important. Counselors can be advocates for policy changes in their work settings.

The counselor in the community agency knows that advocacy is the only possibility when abuse is apparent. For these clients, advocacy in terms of support in getting out of the home, finding new housing, and learning how to set up a restraining order may be far more important than self-examination and understanding.

Counselors who care about their clients also become their advocates when necessary. They are willing to move out of the counseling office, meet clients in the home, connect with the family's larger network, and create larger networks of support and social empowerment. You may work with others on a specific cause or issue to facilitate general human development and wellness (e.g., preterm pregnancy care, child care, fair housing, aid for the homeless, athletic fields for low-income areas). This requires you to speak out, to develop skills with the media, and to learn about legal issues. *Ethical witnessing* moves beyond working with victims of injustice to the deepest level of advocacy (Ishiyama, 2006). Counseling, social work, and human relations are inherently social justice professions, but speaking out for social concerns also needs our time and attention.

 Case Study: Career Counseling in Social Context: The "System" and the Client. A good exercise to think about the issues affecting Anne and how you would collaborate with her in facing these issues.

Summarizing Key Points of Focusing the Interview: Exploring the Story From Multiple Perspectives

Defining Focusing

▶ Selective attention will broaden the depth and breadth of a client's story. Through your attending skills (visuals, vocal tone, verbal following, and body language), you consciously or unconsciously indicate to your client the topics that you consider most important.

▶ If we help clients understand their web of relationships, they will be better prepared to understand themselves, have less blame, and see their issues in a more complete context.

Discerning the Skill of Focusing

▶ The interviewer and counselor may facilitate understanding of the client's web of relationships through focusing on the client, the client's issues and concerns, significant others, other key individuals and groups, the cultural/environmental context, and/or the immediate interpersonal relationship of the session.

Observing the Skill of Focusing

▶ Client issues are seldom one-person matters. Focusing on family background, community, and spirituality enriches the session and helps the client focus on positive strengths.

▶ The community genogram provides a clear visual picture of the client's background and helps the counselor understand the social context more quickly and fully.

Practicing Mapping the Web of Relationships, Community and Family Genograms

▶ The community genogram is a visual map of the client in relation to his or her environment, showing both stressors and assets. This type of genogram is useful for understanding a client's history and identifying strengths and resources.

▶ The family genogram can be a useful supplement to the community genogram, bringing out additional details of family history.

▶ You as an interviewer or counselor also come from a social context, and all the focus dimensions apply to you as well. You may consciously or unconsciously focus on certain issues and dimensions while ignoring others.

Refining: Focusing on Challenging Issues

▶ You are going to face situations in which your best counseling efforts are insufficient to help your clients resolve their issues and move on with their lives. The social context of poverty, racism, sexism, and many other forms of unfairness may leave your clients in nearly impossible situations.

▶ Those who adopt a social justice orientation with ethical witnessing move beyond understanding and take both remedial and preventive action for their clients.

Assess your current level of knowledge and competence as you complete the chapter:

1. **Flashcards:** Use the flashcards to check your understanding of key concepts and facilitate memorization of key information.

2. **Self-Assessment Quiz:** The quiz will help you assess your current knowledge and prepare for course examinations.

3. **Portfolio of Competencies:** Evaluate your present level of competence on the ideas and concepts presented in this chapter using the Self-Evaluation Checklist. Self-assessment of your competencies demonstrates what you can do in the real world.

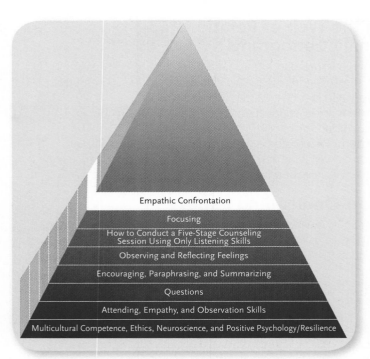

Chapter 9
Empathic Confrontation
Supporting While Addressing Client Conflict

Nowadays, if you are afraid of conflict, you are not going to do very well.

—Bill Parcells

I'm not interested in looking for who's responsible, but rather what's troubling you, and what you might be able to do about it.

—William Miller and Stephen Rollnick

Facing challenges is basic to client restorying. Empathic confrontation clarifies the conflict, bringing out internal and external discrepancies. It promotes change, as well as activating resilience and creativity. Most clients come to an interview experiencing ambivalence, feeling conflicted about issues, and seeking some form of resolution. Some may resist your efforts to bring about the very transformation they seek. Listening and empathic confrontation will help clients move beyond their issues and minimize resistance.

Chapter Goals

Awareness, knowledge, skills, and actions developed through the concepts of this chapter and this book will enable you to:

▶ Identify conflict, incongruity, ambivalences, and mixed messages in behavior, thought, feelings, or meanings.

▶ Identify client change processes occurring during the interview and throughout the treatment period, using the Client Change Scale.

▶ Define and practice the skills of empathic confrontation.

▶ Understand how to use empathic confrontation with sensitivity to diverse clients.

▶ Become aware of the perils of nonempathic confrontation and the importance of respecting resistance.

> **Assess your current level of knowledge and competence as you begin the chapter:**
>
> 1. Self-Assessment Quiz: The chapter quiz will help you determine your current level of knowledge. You can take it before and after reading the chapter.
>
> 2. Portfolio of Competencies: Before you read the chapter, please fill out the Self-Evaluation Checklist to assess your existing knowledge and competence on the ideas and concepts presented in this chapter. Then, at the end of the chapter, complete the checklist again to summarize your competencies after study and practice.

▶ Defining Empathic Confrontation

Effective empathic confrontation of conflict and incongruity leads clients to create new ways of thinking and increased intentionality. It may be described as follows.

Empathic confrontation is *not* a direct, harsh challenge but is a more gentle skill that involves listening to clients carefully and respectfully and helping them examine themselves or their situations more fully. Empathic confrontation is not "going against" the client; it is "going with" the client, seeking clarification and the possibility of a new resolution of difficulties. Think of empathic confrontation as a supportive challenge. Only provide a challenging confrontation in a solid supportive relationship.

Most clients come to an interview "stuck" or feeling "ambivalent"—having no alternatives for solving a problem or feeling torn between competing options. They may be in personal conflict with significant others and/or encounter bullying and harassment in their work setting. Others fail to see the internal contradictions of their own behaviors, thoughts, and feelings, engage in sheer denial, or refuse to take responsibility for their actions. Some engage in internal self-blame, while others never hesitate to attribute blame and responsibility to others.

Traditional treatment methods attempted to raise awareness of ambivalence and break down clients' denial or resistance as perceived by the interviewer. This confrontational style is ineffective (Norcross, 2012). Expressing empathy, using empathic confrontation, and building on the strengths and competencies of the client are more effective. If the conflict is defined clearly, many clients will resolve the contradictions themselves and move on to action. Motivational interviewing's "rolling with resistance" is another way to describe empathic confrontation. "What are you going to do about it?" (Miller & Rollnick, 2013)

Empathic confrontation as described below may lead clients to new ways of thinking and increased intentionality.

EMPATHIC CONFRONTATION	ANTICIPATED CLIENT RESPONSE
Supportively challenge the client: 1. Listen, observe, and note client conflict, mixed messages, and discrepancies in verbal and nonverbal behavior. 2. Point out and clarify internal and external discrepancies by feeding them back to the client, usually through the listening skills. 3. Evaluate how the client responds and whether the confrontation leads to client movement or change. If the client does not change, the interviewer flexes intentionally and tries another skill.	Clients will respond to the empathic confrontation of discrepancies and conflict with new ideas, thoughts, feelings, and behaviors, and these will be measurable on the five-point Client Change Scale. Again, if no change occurs, *listen*. Then try an alternative style of confrontation.

Our clients often become stuck because of internal and/or external conflict. **Internal conflict** or **incongruity** occurs when clients have difficulty making important decisions or feel confusion or sadness. They may have mixed feelings and thoughts about themselves or about their identity. **External conflict** may be with others (friends, family, coworkers, or employers) or with a difficult situation, such as being prevented from achieving an important social or career goal.

External conflict may lead to internal feelings of self-blame. And internal conflict can lead to external conflict with others—for example, the inability to make a decision may result in conflict with a significant other who needs to know what is going to happen. Clients may be struggling to navigate their cultural or ethnic identities, a mixture of both internal and external conflict.

Your listening skills are vital in helping the client identify conflicts, incongruities, and discrepancies. Once you have established sufficient rapport, developed a working relationship, and heard the client's story, you will identify instances of internal and external conflict. Questions, coupled with paraphrases, reflections of feeling, and summaries, will help clarify the conflict issues, but eventually you will want to provide a supportive challenge/empathic confrontation to reach the next step of personal growth. Positive growth and development occur with the resolution of conflict and incongruity. Through this process of transformation and change, clients learn new ways to manage their lives.

EMPATHIC CONFRONTATION IS A COMBINATION OF SKILLS

Empathic confrontation differs in a major way from all skills previously discussed, and from the influencing strategies as well. Confrontation does not stand alone as a skill but rather appears as a dimension of other skills. For example, imagine that a client talks to you about possibly leaving a job because of interpersonal conflict but also speaks of real anxiety about what this action might mean for the future. After hearing the full story, you might provide a brief summary confrontation as follows:

> Shavon, sounds as if you really have mixed feelings about this choice. On one hand, I hear some anger with your boss and a real desire to get out and move on. But on the other hand, I sense that you feel a bit confused and anxious about where you'd go next. Have I got that right?

In this summary, the interviewer confronts Shavon with the essence of her internal conflict and implicitly challenges her to think through what she really wants. A more forceful challenge would be "We've gone through this for some time. What do you, Shavon, really want?" If Shavon understands the challenge, she will generate her own ideas for possible resolution. The words "on one hand . . . but on the other hand . . ." are a classic way to respond to client conflict and incongruity. Amplifying the alternatives with hand and arm movements helps illustrate conflict even more clearly.

Thus, one of the most powerful influencing skills is based on careful listening. Paraphrasing is particularly useful when the conflict or incongruity revolves around a decision and the pluses and minuses of the decision need to be outlined. Reflection of feeling is important for emotional issues, particularly when clients have mixed feelings ("on one hand, you feel . . ., but on the other, you also feel . . ."). A summary is a good choice for bringing together many strands of thoughts, feelings, and behaviors.

▶ Discerning The Basic Skills of Empathic Confrontation

The combination skill of confrontation is divided here into three steps. The first is listening and observing; the second is providing feedback on internal and external conflict; the third is evaluating how the client received the confrontation and whether it led to change. The third step includes the **Client Change Scale (CCS)**, a systematic way to determine how the client heard you. This CCS is also a way to measure how your client changes and develops over a series of sessions. Thus we can examine the immediate impact of our confrontations, and we can use the same concepts over time to evaluate client success.

EMPATHIC CONFRONTATION, STEP 1: IDENTIFY CONFLICT BY OBSERVING MIXED MESSAGES, DISCREPANCIES, AND INCONGRUITY

This session, while abbreviated and edited, illustrates the fundamentals of confrontation—supporting while challenging.

INTERVIEWER AND CLIENT CONVERSATION	PROCESS COMMENTS
Shavon: (the body language shows excitement) I found this wonderful friend on the Internet. We're emailing at least four times a day. It feels great. I think I'd like to meet him. But it means I may have to go out of town. (Her body language becomes more hesitant and she breaks eye contact.) I wonder what my partner would think if he found out. I'm a little bit anxious, but I really want to meet this guy.	The client demonstrates conflict or incongruity between desire and excitement about meeting the Internet friend and internal anxiety and hesitation. There is the inevitable external conflict of being involved with two people at once. Which discrepancy might you discuss first? Clients who discuss mixed feelings and conflicts usually show them nonverbally as well, through vocal tone and/or body language.
Luis: You really want to meet him, but you're a little bit anxious.	This paraphrase combined with reflection of feeling confronts the mixed feelings in the client. It catches both verbal and nonverbal observations. Note that a confrontation always occurs as part of other microskills. (Interchangeable empathy)
Shavon: Yes, but what would happen if my partner found out? It scares me. I've been so much involved with him over the past two years. But, wow, this guy on the Internet. . . .	The client responds and turns her focus to the discrepancy with her partner. There are at least two issues in this situation.

Your ability to observe verbal and nonverbal incongruities and mixed messages is fundamental to effective confrontation. It will be helpful if you review the brief exchange above and once again identify the internal and external conflicts that Shavon faces.

Shavon faces what for her is a major conflict—the external attraction of the new person and the potential external conflict with her present partner. Internally, this results in mixed and torn feelings. Drawing out those emotions via the basic listening sequence will be important in confronting the conflict. The client has relationship issues, and your relationship is important to her working things through.

Discrepancies between you and the client can be challenging. Many counselor/client relationships will have implicit conflict when the counselor does not agree with a client's choice or has differing values. Counseling is for the client, not you. This is a time to maintain professional behavior and accept and understand where the client is "coming from." If you sense difference or feel a conflict between you and your client, support the client by listening. If you listen carefully, most discrepancies between you and your clients will disappear as you understand how they came to think and behave as they do. Note your own or the client's discomfort with differences between you, and question yourself silently. Keeping one's thoughts and feelings to oneself is part of being nonjudgmental. However, there are times when it is wise to use the influencing skills of feedback or self-disclosure (Chapter 11). This may provide a useful new perspective.

For example, Shavon may be making an unwise decision that you anticipate will lead to even more difficulty in her life. Summarizing the client's point of view and then sharing your alternative thoughts via feedback or self-disclosure may help prevent difficulties. Furthermore, Shavon may fail to see her own contribution to the problems with her present partner by blaming him. Or she may take the blame herself and feel guilty while failing to see the Internet friend as the real intruder.

The distinction between internal and external conflict needs constant attention. All too often, external issues result in internal conflict and anxiety for the client. For example, a woman may think that her partner's abuse is her fault and may stay in a dysfunctional relationship while blaming herself for the situation. Discrimination in the form of racism, ableism, ageism, or other forms of oppression can result in depression and self-blame. Narrative therapy terms this *internalization of the problem*. In such situations, interviewers and counselors need to work toward *externalization of the problem*. When clients blame themselves for the difficult challenges they face, help them discover that the real location of the difficulty is not in them but in others. In case of the several types of oppression, it is the system and the culture that is the real issue. In these situations, we may need to take an action stance and inform clients of what is really happening. These clients often benefit from the new perspective you provide through feedback and other influencing skills.

EMPATHIC CONFRONTATION, STEP 2: POINT OUT ISSUES OF INCONGRUITY AND WORK TO RESOLVE THEM

INTERVIEWER AND CLIENT CONVERSATION	PROCESS COMMENTS
Luis: Could I review where we've been so far? I know you have been having some difficulties with your partner that you've described over the last two sessions. I also hear that you want to work things out even though you're angry with him. You have a lot of positive history together that you hate to give up. On the other hand, you've found this man on the Internet that you're excited about, and he doesn't live that far away. In the middle of all this, I sense you feel pretty conflicted. Have I got the issues right?	This summary indicates that the counselor has been listening. The counselor communicates respect and a nonjudgmental attitude, both verbally and nonverbally. (Interchangeable empathy) The counselor summarizes the major discrepancies that lead to internal and external conflict and checks out with the client to see if the listening has been accurate. (Additive empathy)
Shavon: Yes, I think you've got it. I hear what you're saying; I think I've got to work a little harder on the relationship with my partner, but—wow—I sure would like to meet that guy.	Resolution of conflict and discrepancy occurs best after the situation is fully understood. Through having thoughts and feelings said back, the client starts some movement.

After 10 minutes, Shavon's thoughts and feelings evolve to a new perspective.

INTERVIEWER AND CLIENT CONVERSATION	PROCESS COMMENTS
Shavon: (said with conviction) I've got so much time invested in my partner; I've really got to try harder. (The nonverbals again show hesitancy.) How am I going to work this out with my Internet friend?	Even though the conflict is moving and the client is starting to show evidence of new ways of thinking that weren't there in the first two sessions, conflict remains.
Luis: (solid supportive body language and vocal tone) It looks like you really want to work it out with your partner. You sound and look very sure of yourself. Let's consider what the possibilities are with your Internet friend.	You can confront and help clients face discrepancies, incongruity, and conflict if you are able to listen and be fully supportive. (Additive empathy)

Labeling the incongruity and saying it back through nonjudgmental confrontation may be enough to resolve a situation. Focus on the elements of incongruity rather than on the person. Confrontation is too often thought of as blaming a person for his or her faults; rather, the issue is intentionally facing the incongruity and helping the client think it through. The following summarizes key dimensions of effective empathic confrontation— supporting while challenging:

▶ *Clearly identify the incongruity or conflict in the story or comment.* Using reflective listening, summarize it for the client. The simple question "How do you put these two together?" may lead a client to self-confrontation and resolution.

▶ *Draw out the specifics of the conflict or mixed messages using questions and other listening skills.* If necessary, share your observations. Aim for facts; avoid being judgmental or evaluative. Address each part of the mixed message, contradiction, or conflict one at a time. If two people are involved, attempt to have the client examine both points of view.

▶ *Periodically summarize each dimension of the incongruity.* Variations of basic confrontation include "On the one hand . . . but on the other hand . . .," "You say . . . but you do . . .," "I see . . . at one time, and at another time I see . . .," and "Your words say . . . but your actions say. . . ." Follow with a checkout: "How does that sound to you?" When incongruities are pointed out, the client is confronted with facts.

Many clients are unaware of their mixed messages and discrepancies; point these out gently, but firmly. Notice how they react to your challenge, and assess your impact with the Client Change Scale.

EMPATHIC CONFRONTATION, STEP 3: EVALUATE CHANGE USING THE CLIENT CHANGE SCALE*

The effectiveness of a confrontation is measured by how the client responds. If you observe closely in the here and now of the session, you can rate how effective your interventions have been. You will see the client change (or not change) language and behavior in the interview. When you see immediate or gradual changes over several sessions, you know that your intervention has been successful. When you don't see the change you anticipate or think is needed, this is time for intentionality, flexing, and having another response, skill, or strategy available.

---

*A paper-and-pencil measure of the Client Change Scale was developed by Heesacker and Pritchard and was later replicated by Rigazio-Digilio (cited in Ivey et al., 2005). Factor analytic study of more than 500 students and a second study of 1,200 revealed that the five CCS levels are identifiable and measurable.

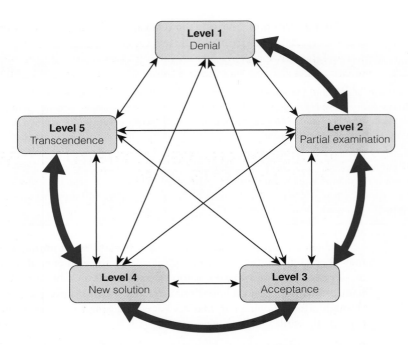

FIGURE 9.1 The Client Change Scale (CCS). The five stages of change may move in a step-by-step order, but the arrows between all the stages show that clients often move backward and forward, up and down, as they encounter new thoughts, feelings, and meanings.

Source: Cengage Learning

Figure 9.1 presents a visual summary of the Client Change Scale (CCS). While the progression from denial through acceptance through significant change can be linear and in a step-by-step order, this does not always happen. Think of a client working through the expected death of a loved one. One possibility is for the client to move through the CCS stages one at a time. But often, the client may seem to be moving forward and then suddenly drop back a level or two. At one interview the client may seem acceptant of what is come, but in the next session move back to partial examination or even denial. Then we might see a temporary jump to transcendence, followed by a return to acceptance.

The Client Change Scale is also useful as a broad measure of success of several interviews. The CCS provides a useful framework for accountability and measuring client growth. The clearest example is that of a client with substance abuse issues. The client may come voluntarily or be referred by the court for cocaine or alcohol abuse (often both). The client may also be depressed and use the drugs to alleviate pain. Often these clients start by denying that they really have a problem (we call this Level 1 on the Client Change Scale). If we are successful in challenging, supporting, and confronting, we will see the client move to Level 2, admitting that there may be a problem. Some call this a "bargaining" stage in which the client moves back and forth between denial and recognizing that something needs to be done.

Acceptance and recognition of the issue occur at Level 3, where the alcoholic admits that he or she is indeed an alcoholic and the cocaine abuser recognizes addiction. But acceptance/recognition is not resolution. The client may be less depressed but continue to drink and use drugs. Some things cannot be changed—for example, a financial loss, a friend who no longer involves you, a divorce in later life, the loss of a limb. New solutions are not always possible, and existential acceptance will be as far as some clients can reach. Resilience can come from acceptance, from a new solution, but best of all from transcendence.

Change occurs when the client reaches Level 4 and actually stops substance use and depression lifts. This is real success in counseling and therapy and not easy to achieve, but clearly it is measureable. Finally, Level 5, transcendence and the development of new ways of being and thinking, may occur. This level will not be achieved by all clients. It is represented by the user who becomes fully active in support groups and helps others move away from addiction. We see changes in life's meaning and a much more positive view of self and the world. This is far more than "getting by."

A MODEL FOR CHANGE IN THE INTERVIEW ITSELF OR OVER THE LONG TERM

Let's move the CCS to a more general level as it might appear in counseling interview with virtually any topic. The client tells us a story and we, of course, listen. If the client is in the denial stage, the story may be distorted, others blamed unfairly, and the client's part in the story denied. In effect, the client in *denial* (Level 1) does not deal with reality. When the client is confronted effectively, the story becomes a discussion of inconsistencies and incongruity, and we see *bargaining and partial acceptance* (Level 2)—the story is changing. At *acceptance* (Level 3), the reality of the story is acknowledged and storytelling is more accurate and complete. Moreover, it is possible to move to *new solutions* and *transcendence* (Levels 4 and 5). When changes in thoughts, feelings, and behaviors are integrated into a new story, we see the client move into major new ways of thinking accompanied by action after the session is completed.

Virtually any problem a client presents may be assessed at one of the five levels. If your client starts with you at *denial* or *partial acceptance* (Level 1 or 2) and then moves with your help to *acceptance* and *generating new solutions* (Level 3 or 4), you have clear evidence of the effectiveness of your interviewing process. The five levels may be seen as a general way to view the change process in interviewing, counseling, and therapy. Each confrontation or other interview intervention in the here and now may lead to identifiable changes in client awareness.

Small changes in the interview will result in larger client change over a session or series of sessions. Not only can you measure these changes over time, but you can also contract with the client in a partnership that seeks to resolve conflict, integrate discrepancies, and work through issues and problems. Specifying concrete goals often helps the client deal more effectively with empathic confrontation.

Box 9.1 presents a narrative of the Client Change Scale, with examples of each of the five levels as the client responds to issues around a divorce. The CCS provides you with a systematic way to evaluate the effectiveness of each intervention and to track how clients change in the here and now of the interview.

If you practice assessing client responses with the CCS, eventually you will be able to make decisions automatically "on the spot" as you see how the client is responding to you. For example, if the client appears to be in denial despite your confrontation, you can intentionally shift to another microskill or approach that may be more successful.

When you confront clients, ask them a key question, or provide any intervention, they may have a variety of responses. Ideally, they will actively generate new ideas and move forward, but much more likely they will move back and forth with varying levels of response. The idea is to note how clients respond (at what level they answer) and then intentionally provide another lead or comment that may help them grow. Clients do not work through the five levels in a linear, straightforward pattern; they will jump from place to place and often change topic on you. It is your task to help them move forward to significant change.

BOX 9.1 The Client Change Scale

The CCS is presented here using five examples of reactions to divorce. Anytime clients are working through change, they talk about their issues with varying levels of awareness. The client may be in denial one moment, then talk as if he or she accepts the problem, and then return to bargaining to avoid change.

CCS STAGES	CLIENT EXAMPLE
Level 1. Denial. The individual denies or fails to hear that an incongruity or mixed message exists.	"I'm not angry about the divorce. These things happen. I do feel sad and hurt, but definitely not angry."
Level 2. Partial examination. The client may work on a part of the discrepancy but fails to consider the other dimensions of the mixed message.	"Yes, I hurt, and perhaps I should be angry, but I can't really feel it."
Level 3. Acceptance and recognition, but no change. The client may engage the empathic confrontation but make no resolution. Until the client can examine incongruity, stuckness, and mixed messages accurately, real change in thoughts, feelings, and behavior is difficult. Coming to terms with anger or some other denied emotion is an important breakthrough and often is a sufficient solution for the client.	"I guess I do have mixed feelings about it. I certainly do feel hurt about the marriage. I hurt, but now I realize how really angry I am."
Level 4. Generation of a new solution. The client moves beyond recognition of the incongruity or conflict and puts things together in a new and productive way. Needless to say, this usually does not happen immediately. It can take several interviews or sessions over a period of weeks and sometimes months.	"Yes, I've been avoiding my anger, and I think it's getting in my way. If I'm going to move on, I will accept the anger and deal with it."
Level 5. Transcendence—development of new, larger, and more inclusive constructs, patterns, or behaviors. *Many* clients will never reach this stage. A confrontation is most successful when the client recognizes the discrepancy and generates new thought patterns or behaviors to cope with and resolve the incongruity.	"You helped me see that mixed feelings and thoughts are part of every relationship. I've been expecting too much. If I expressed both my hurt and my anger more effectively, perhaps I wouldn't be facing a divorce."

Depending on the issue, change may be slow. For some clients, movement to partial acceptance (Level 2) or acceptance but no change (Level 3) is a real triumph. There are a variety of issues where acceptance represents highly successful counseling and therapy. For example, the client has been in a situation where change may be impossible or really difficult. Thus, accepting the situation "as it is" is a good result. A client facing death is perhaps the best example and does not always have to reach new solutions and transcendence. Simply accepting the situation as it is may be enough. There are some things that cannot be changed and need to be lived with. "Easy does it." "Life is not fair." "There is need to accept the inevitable." For a client whose partner or parent is an alcoholic, it is a major step to realize that the situation cannot be changed; acceptance is the major breakthrough that will lead later to new solutions.

Interactive Exercise: Practice With the Client Change Scale
Interactive Exercise: Practicing Confrontation

▶ Observing: Empathic Confrontation in Action

The example interview that follows demonstrates the integration of the three main points presented in this chapter: (1) Listening skills are used to obtain client data; (2) confrontations of client discrepancies are noted; and (3) the effectiveness of the confrontations is considered using the Client Change Scale (CCS). You may want to study the segment carefully to fully understand how confrontation is used in this example. With more experience and practice, these concepts will be useful and important in your interviewing practice.

The following interview presents a conflict that is common to many working couples—balancing home tasks. Male attitudes and behaviors are changing, but many working women are still burdened with the major responsibility. In this example, we have both internal and external incongruity. Dominic is struggling internally with the discovery that things have changed and externally with the realization that his wife is behaving differently. Arguments may be particularly intense as two tired people come home from a hard day's work to face needy children and undone housework. Couples may blame each other rather than attributing a major portion of their issues to external causes at work. An exhausted partner or spouse often has little energy left to deal with the concrete issues at home.

INTERVIEWER AND CLIENT CONVERSATION	PROCESS COMMENTS
1. *Dominic:* I'm having a terrible time with my wife right now. She's working for the first time and we're having lots of arguments. She isn't fixing meals like she used to or watching the kids. I don't know what to do. I'm confused and mixed up.	On the CCS, this client response is rated Level 2; he is partially aware of the problem. At the same time, he denies (Level 1) his role in the problem and is attributing the difficulties in the home to husband-wife issues, failing to see how demands at work play into the system.
2. *Ryan:* (holds out one hand to the right) So, on one hand, your wife is working outside the home, but (holds out the left hand) on the other hand you expect her to continue with all the housework, too. You feel confused and don't like what's going on right now.	Confrontation presented as paraphrase and reflection of feeling. The use of hands with the words helps strengthen and clarify the conflict. Here the counselor shows imbalance or conflict concretely. (Potential additive empathy)
3. *Dominic:* You damn betcha she's expected to do what she's always done—I'm not confused about that.	CCS Level 1—denial. (Obviously, potential was not reached.)
4. *Ryan:* I see. You're not confused; you're really angry and don't like what's going on. Could you give me a specific example of what's happening—something that goes on between the two of you when you both get home?	Paraphrase, reflection of feeling, open question oriented to concreteness. (Return to interchangeable empathy)
5. *Dominic:* (sighs, pauses) Yeah, that's right, I don't like what's going on. Like last night, Sara was so tired that she didn't get around to fixing dinner till half an hour late. I was hungry and tired myself. We had a big argument. This type of thing has been going on for 3 weeks.	While Dominic is able to identify and talk about the conflict (CCS Level 2), he lacks awareness of how his wife feels. He seems insensitive to the fact that his wife is working and he expects her to do everything she did in the past for him (CCS Level 1).
6. *Ryan:* I hear you, Dominic; you're pretty tired and you get angry. Let's change focus for a minute. When dealing with conflict, it helps to concentrate on positive things. Could we search for some positive things that have worked for you and Sara in the past?	This summary introduces an incongruity between the difficult present situation and the good things of the past through the positive asset search. (Additive empathy)

(continued)

INTERVIEWER AND CLIENT CONVERSATION	PROCESS COMMENTS
7. *Dominic:* Yes, I am angry and discouraged. Sara and I were doing pretty well until the baby came. Somehow things just got off kilter. (5-second pause and silence) . . . Well, let me try it your way. We sure had fun times together over the 3 years we've been together. We both like outdoor activity and doing things together. We never seem to have time for that now.	In couples work, it is useful to remind them of positive stories from the past. This has been edited for brevity. It is important to explore positives in the relationship in much more depth than presented here.
8. *Ryan:* It's important to remember that you have a good history and have enjoyed each other. I'm wondering if part of the solution isn't finding time just to be together doing fun things. Let's make that part of our discussion later. But for the moment, let's go back to your main issues. Could you give me a specific example of what goes on between you when she gets home?	Paraphrase, suggestion, open question oriented to concreteness. Getting specifics helps clarify the situation. (Additive empathy) When counseling around couples issues, search for strengths. What brought the couple together, and what maintains them now? This helps clients center themselves in positives as they struggle with the difficult negatives in a relationship. Helping couples recapture the good things in their relationship is important. We often recommend a weekly date night for couples.
9. *Dominic:* . . . (pause) Well, lately, I've had a lot of pressure at work. They're downsizing and morale is bad. I worry I'll be next. I try really hard, but when I get home, I just want to sit. Sara's got the same thing. Her new boss just wants more all the time. I guess when she comes home, she's about as exhausted and confused as I am.	It is common for partners to get angry at each other when external stressors hit one or both individuals in the relationship. One can't argue with the boss or colleagues easily, so the partner is the scapegoat. The external issue has become internalized.
10. *Ryan:* I see. You both work all day and come home exhausted. Dominic, do you really think Sara can work that hard and still take care of you like she used to?	Paraphrase, closed question, with confrontation. Ryan's implicit confrontation is now more concrete. (Potential additive empathy)
11. *Dominic:* I guess I hadn't thought of it that way before. If Sara is working, she isn't going to be physically able to do what she did. But where does that leave me?	For the first time, Dominic is able to see that Sara can't continue as she has in the past (CCS Level 3). Note that he is still thinking primarily of himself. To move to higher levels on the CCS, Dominic would have to be able to take Sara's perspective and articulate what she likely thinks and feels when she gets home.
12. *Ryan:* Yes, where does that leave you? *Dominic,* let me tell you about my experience. My wife started working and I, too, expected her to continue to do the housework, take care of the kids, do the shopping, and fix the meals. She went to work because we needed the money to make a down payment on a small house. I decided to share some of the household work with her and also shop, and I pick the kids up at day care. How does that sound to you? Do you think you want to continue to expect Sara to do it all?	Self-disclosure followed by a checkout. Ryan is speaking up directly with his ideas, but he is also allowing the client to react to them. Clearly, Ryan is trying to get Dominic's thinking and behavior to move. His last statement contains an important implicit confrontation: "On the one hand, your wife is working and if she continues, she'll need some help; on the other hand, perhaps you don't want her to work—what is your reaction to this?" (Additive empathy)

INTERVIEWER AND CLIENT CONVERSATION	PROCESS COMMENTS
13. *Dominic:* Uhhh . . . we need the money. We've missed the last car payment. It was my idea that Sara go to work. But isn't housework "women's work"?	Dominic is now facing up to the contradiction he is posing (CCS Level 3). He has not yet synthesized his desire for his wife to work with the need for his sharing the workload at home, but at least he is moving toward a more open attitude. As he acknowledges his part in the situation and the need for more money, he is beginning to come to a new understanding, or synthesis, of the problem—he is less incongruent and is taking beginning steps toward resolving his discrepancies.

For each developmental task completed in the interviewing process, it often seems that a new issue arises. Just as the counselor is beginning to facilitate client movement, a new obstacle ("women's work") arises. To make the progress shown took half the session. Changing the concept from "women's work" to "work in the home that must be shared if the car payments are to be met" took the rest of the session. Achieving a new and likely more lasting level of male/female cultural differences would require a major transformation in thinking and behavior (CCS Level 5). Major change may require several interviews, group sessions, and time for Dominic, or any other client, to internalize new ideas.

Interactive Exercise: Identification and Classification
Interactive Exercise: Identifying Interviewer Responses

▶ Practicing Becoming Competent in Empathic Confrontation

Exercise 9.1 Identifying Internal and External Conflict
This exercise is designed to reinforce the importance of the distinctions between internal and external conflict and how the two relate to each other.

▶ Identify specific areas of external conflict that are distressing Shavon. In addition, imagine that her partner is abusing her. She may have financial challenges, difficulty with friends, and demanding parents.

▶ What are some areas of internal conflict that affect her emotionally with each of the different challenging problems?

▶ How does the external become internal? How does what happens in our outside world result in emotional and decisional conflict in our minds?

▶ After you have imagined an even more complex situation for Shavon as she faces multiple issues, how might we "unwind" all this?

Exercise 9.2 Thinking About Your Thinking
Observing your own thoughts, feelings, and behaviors in the interview promotes self-examination. Looking at our work in the interview helps us understand our own decisions, enabling us to become more effective.

First, review one of your video or audio recorded interviews. Ask yourself what you were thinking during the interviewing process. Then think about what is going on in your mind as you observe.

Second, pause the recording when a contradiction or discrepancy is noted or when one would be desirable. The following questions will facilitate your recollections:

▶ What were the conflictual issues presented in the interview?
▶ What your thoughts about the client's conflict?
▶ What were your feelings about the conflict?
▶ How well do you believe that you helped the client deal with the conflict?
▶ Can you use the Client Change Scale to evaluate your effectiveness?
▶ How have your views about confrontation changed after learning about empathic confrontation and viewing yourself?

Exercise 9.3 Group Practice
Meet with a group of three or four, taking turns as interviewer and client. One member can run the recording equipment and observe. Particularly important is that the volunteer client discuss a conflictual issue. Examples of internal conflicts could be a difficult choice that needs to be made soon or remembering a time when one was bullied or put down and how one still feels inside. External conflicts could be the same issue of bullying, or conflict with friends, family, or at work. If a second observer is present, he or she takes notes on both listening skills and empathic confrontation. Did the client show any movement on the Client Change Scale? For this exercise, you may find it helpful to use the Microskills Interview Analysis Form (MIAF), Box 7.1 in Chapter 7 and also available online.

Interactive Exercise: Review the CCS and Write Client Responses
Case Study: Confronting a Difficult Client in Romania

▶ Refining Nonempathic Confrontation, Dealing With Resistance, Multicultural Issues, and Working With Groups

NONEMPATHIC CONFRONTATION AND DEALING WITH RESISTANCE

You will find that many clients are not comfortable with coercive or aggressive confrontational and challenging approaches that are provided too quickly. We label these efforts *nonempathic confrontation*. They represent interviewer efforts to work "on the client," instead of working collaboratively "with the client." At best, they represent a genuine effort to get clients to do something about their issues; at worst, they represent ill attempts towards manipulation. Either way, the interviewers' actions disrespect their clients' capacity to resolve their own issues.

Telling the client what the issue is and how to resolve it is not an effective method. This is particularly so with changes the client may not be ready to make. Many new interviewers feel a responsibility to help the client understand what the concern is and what the benefit of change is, only to encounter more client resistance and less change. Direct instructions, persuasion, argumentation, and proselytization are other ways of describing nonempathic confrontation. More extreme forms of nonempathic confrontation can occur with particularly resistant clients. In these cases, interviewers may find themselves engaging

BOX 9.2 The Stages of Change Model

Identify strengths and wellness qualities in clients and their relationships and contexts. When clients are aware of their wellness strengths and positive qualities, they may face difficult issues more easily. Use the Client Change Scale to determine progress.

Pre-contemplation: At this stage, the client is not thinking about changing his or her behavior. Clients may even be unaware that their behavior is problematic.

Contemplation: At this stage, the client is aware that a problem exists but feels ambivalent about doing something to create change. The problematic behavior continues, but the client is now aware that it exists.

Preparation: At this stage, the client has made a decision to change in the very near future. The problematic

behavior continues, but the client has made plans to change soon and begins to set goals.

Action: At this stage, the client is actively taking steps to modify his/her behaviors. S/he is making required lifestyle changes, often with a mix of confidence, pride, and anxiety.

Maintenance: At this stage, the client has successfully changed his/her behavior and has sustained the change for a minimum of six months. In many cases, clients in this stage can experience a relapse and return to an earlier stage of change. Relapse is part of the learning process. Knowledge of the factors that may trigger relapse can serve as useful information for future change attempts.

in active arguments, threatening punishment, or making other efforts to force the client make changes.

Think of resistance as a defense and protective mechanism for your client. Resistance often starts as an effort to keep your client safe. It is better to "go with the resistance" and listen empathically. Resistance to talking about something often relates to some issues from the immediate or more distant past, or the client may simply be uncomfortable with certain topics, ranging from sexuality to past errors they may have made. Respect the resistance, listen, and learn!

It is challenging to work with resistant acting-out children or teens (sometimes called oppositional defiant) and adults who have been diagnosed as antisocial. Understanding others empathically is often difficult or impossible for these individuals; neuroscience research has shown that their brains' mirror neurons, essential for empathy, fire either very little or not at all. In fact, these clients even can take joy in seeing others' pain (Decety & Jackson, 2004). We need to remain empathic in our challenges and confrontation, but we cannot seem weak. Firmness and clarity on your part is often required. This is still an empathic confrontation, but it does look different! Over time, a relationship can be built.

Your personal support can help clients build new behaviors, thoughts, and meanings in a way that recognizes their strengths and current readiness to change and respects their capacity for self-direction and freedom to make decisions that are meaningful to them. See Box 9.2 for a good description of clients' readiness to change.

Here is a partial list of behaviors to avoid in interviewing and counseling; they indicate a lack of empathic confrontation.

▶ *A judgmental vocal tone.* Maintaining respect and warmth with empathy and working with the client where he or she is at the moment lead to a nonjudgmental style. Using the empathic confrontation "on the one hand . . . but on the other hand . . ." and relying on clients' strengths and competencies produce more positive effects.

▶ *Blaming, hostile, and pejorative statements.* Albert Ellis warned us not to attack the person. He disputed negative beliefs, ineffective behaviors, and dysfunctional relations. Listening intentionally and empathically and relating in positive ways produce better outcomes. Carl Rogers called this unconditional positive regard.

- ▶ *Assuming.* Many interviewers believe they know what the client's perceptions of the relationship and level of satisfaction are. Using checkouts and asking for feedback will help you better understand the client's stand on these issues. It can be helpful to provide the client with an evaluation form for the interview just completed.
- ▶ *Pushing.* Forcing your understanding and applicable solutions on your client may backfire. Blunt pointing out of conflicts observed, conflictive views or behaviors, abrupt interpretations, or in-your-face challenging of "denial" only increases the risk of relationship breakdowns. The importance of collaborative work that meets the client where she or he is, in an ethical and culturally sensitive way, is key to counseling success and client change.

INDIVIDUAL AND MULTICULTURAL CAUTIONS

Celebrate diversity, practice acceptance and may we all choose peaceful options to conflict.

—Donzella Michele Malone

Confrontation of discrepancies can be highly challenging to any client, but particularly to one who is culturally different from you. The empathic confrontation process may be made acceptable if the helper takes time to establish a solid relationship of trust and rapport before engaging in confrontation. If you have a fragile client or if the relationship is not solid, confrontation skills need to be used with sensitivity, ethics, and care.

A wide variety of attending and influencing skills may be used to follow up and elaborate on empathic confrontations. Another very different possibility for empathic confrontation is to listen in silence for a short time while the client struggles with internal or external contradiction. Silence is a value within much of the Native American, Dene, or Inuit tradition. You may sit with these clients for several minutes as they sort out issues. One good practice is to ask yourself, how comfortable are you with silence?

This type of empathic confrontation is especially challenging to clients who may ask your opinion—"Don't you agree with me that my partner is wrong?" Most often such questions are answered by throwing the question back to the client (e.g., "Tell me more"). If you say nothing, clients will have to encounter your silence and may more readily find their own answers. But be aware that some clients could interpret your silence as disapproval.

You will find that many clients are not comfortable with confrontational and challenging approaches that are provided too quickly. Your personal support can help clients build new behaviors, thoughts, and meanings. Identify strengths and wellness qualities in clients and their relationships. When clients are aware of their wellness strengths and positive qualities, they may face difficult confrontations more easily.

Within the wide diversity that is multiculturalism, expect a considerable amount of incongruity and conflict between the individual and the external world, as well as internal hurt and anger, even self-blame. Directly, but with as much support as possible, confront conscious and unconscious sexism, racism, heterosexism, and discrimination of all types. For example, a woman may be depressed, but is this depression "her problem" or is her sadness the result of sexual harassment on the job? A White male client who has not been promoted may be angry, believing that his being passed over for promotion may be the result of what he terms "reverse discrimination." Your ability to understand and identify the many conflicts and discrepancies related to multiculturalism is essential. It is also critical that you, the interviewer, work through these issues in your own mind. Seek supervision when necessary.

You will find that direct, aggressive confrontations are not necessary if the client contradiction is stated kindly and with a sense of warmth and caring. Direct, blunt confrontations are likely to be culturally inappropriate for traditional Asian, Latina/Latino, and Native

American clients. But even here, if good rapport and understanding exist between you and the client, empathic confrontations can be helpful. When issues of oppression need to be confronted, your support will be especially essential.

Empathic confrontation must be personally authentic and should be tailored to the specific client's characteristics, style, concern, and context in which the issue occurs. If it is not meaningful for the client, it is likely to fail.

CONFLICT, NEUROSCIENCE, AND SOCIAL JUSTICE

Give peace a chance.

—John Lennon

Confrontational and violent approaches are showcased in the media as effective ways to promote change. What it is usually omitted is the harm done to those who are at the receiving end of such acts. From the developing baby to the older adult, we are all affected by some acts of violence and confrontations. Babies exposed to parental arguments, diverse individuals affronted by overt racism or subtle microaggressions, older adults abused by others, and adolescents treated at "boot" camps are all affected. Their neuronal, emotional, behavioral, and interpersonal development suffers. Even during sleep, babies exposed to the voice of parents arguing show signs of emotional and neuronal distress (Graham, Fisher, & Pfeifer, 2013). Research on tough love or nonempathic confrontational strategies for drug addiction failed to show a single clinical trial demonstrating the efficacy of this approach. To the contrary, several of these studies have documented harmful effects, especially among susceptible groups (White & Miller, 2007).

Racism also acts as a chronic stressor, triggering similar physiological responses and promoting negative consequences such as job strain, interpersonal conflict, smoking, eating disorders, high blood pressure, elevated heart rate, increases in cortisol, and suppressed immunity. Our brains are social, and through special sets of neurons such as the "mirror" neurons, it helps us establish human connections. How we act in a conflictive situation is affected by our brain signals. If something frightens us or we feel danger, the brain signals stress, which triggers the release of cortisol and the fight-or-flight response. Our body is taxed, our mind gets stuck, we develop negative emotions and narrow views, and we become aggressive or defensive. We attack or withdraw while failing to see possible solutions or seize opportunities for success.

Empathic confrontation can be valuable here. It may take some of your own thoughts and awareness if the client is hurting. Survivors of abuse, bullying, or even trauma may blame themselves and internalize the issues. Your goal is to externalize and help them discover or rediscover that they are not at fault. What is really external has become internal and part of the person. It is a matter of social justice to bring awareness to these issues, help clients name and externalize oppressive situations, promote empathic listening and positive relationships, and give peace a chance.

The concluding section of this chapter takes empathic confrontation in a new direction and reviews the mediation process. If you are comfortable and skilled in supportive confrontation, you have the beginnings of an accomplished mediator, whether with couples, groups, or community agencies.

EMPATHIC CONFRONTATION AND MEDIATION OF GROUP CONFLICT

Nobody can get along with everybody all of the time. The more we interact with friends, family, clients, or organizations, the greater the chances we will run into potential conflict. Furthermore, we work in increasingly diverse environments with people who differ in gender, race, ethnicity, age, nationality, immigration status, and conflict resolution style. Conflict can be a creative force or a disruptive one, in spite of all this complexity. It will depend on how you manage it or mediate it.

Empathic confrontation serves as a solid base for mediation. In conflict resolution and mediation, whether between children, adolescents, or adults, the microskills and the five-stage counseling and psychotherapy model provide a useful framework.

Empathic relationship. Develop rapport and outline the structure of your session. Pay equal, neutral attention to each participant or group. Encourage each individual or group member to agree to seek resolution of the conflict. Participants need honesty and not interruption, even when in disagreement. Listening to learn the perspective of the other individual or group is of central importance.

Story and strengths. Define the problem (concern). Use the basic listening sequence to clearly and *concretely* draw out the point of view of each person or group. To avoid emotional outbursts, acknowledge feelings and seek to return to listening and respect. Summarize differing points of view, and check out your accuracy. You might ask each disputant to state the opponent's point of view. Outline and summarize the points of agreement and disagreement, perhaps in written form, if the conflict is complex.

Goals. Use the basic listening sequence to draw out wants and desires for satisfactory resolution. Focus primarily on concrete facts rather than emotions and abstract intangibles. Issues and concerns may be redefined and clarified. Summarize the goals for each, with attention to possible points of agreement. Respectfully outline areas of disagreement.

Restory. Begin negotiation in earnest. Rely on your listening skills to see whether the parties can generate their own satisfactory solutions. When a level of concreteness and clarity has been achieved, the parties involved may be close to agreement. If the parties are very conflicted, meet each separately as you brainstorm alternative solutions. With touchy issues, summarize them in writing.

Action. Contract and generalize. Summarize the agreed-upon solution (or parts of the solution if negotiations are still in progress). Make the solution as concrete as possible, and write down touchy main issues to make sure each party understands the agreement. Obtain agreement about subsequent steps. Congratulate and thank them for their hard work.

INDIVIDUAL AND CULTURAL DIFFERENCES IN MEDIATION AND CONFLICT MANAGEMENT

Skills for communicating with members of different cultural groups are essential. The RESPECTFUL model will help achieve this goal. Participants have different histories and beliefs leading to varying levels of comfort. Some will bring stories of social trauma, discrimination, and victimization. During mediation, some will present with persistent silence or strong suspicion.

The following model, provided by the King Center (2014), will help you enter the world of others who differ from you and reach consensus on a nonviolent and respectful resolution. The five key elements of this model are (1) personal commitment, (2) information gathering and education, (3) negotiation, (4) direct action, and (5) reconciliation. When you work on complex issues of institutional or community change, a review of Dr. King's model may help you in thinking through your approach to major challenges.

1. Personal commitment. Embrace yourself as a cultural being. Learn about your family background and cultural beliefs. Know yourself and your own beliefs, values, biases, and prejudices. How do you behave toward those who are different from you? Encourage the RESPECTFUL model in participants.

Learn about different cultures and the importance of respecting differences. Seek to maintain that essential empathic relationship even though sharp differences may appear. Use multicultural counseling skills. A basic understanding of each participant's cultural worldview and sensitivity to his or her uniqueness will facilitate mediation.

2. Information gathering and education: Observe participants' style and expectations. Draw out the differing stories while continuing to search for underlying issues and observing emotional reactions. Acknowledge emotional difference, but do not go into complications and emotions. Education is important; we cannot assume that participants have any real sense of how the other person(s) sees the situation or and how someone else resolves conflicts. Dialogue with them about the nature of conflict. Explore best ways, and work based on participants' preferred approach to resolution. In short, involve them in the educational process of planning the session.

Use checkouts. Check your assumptions as you mediate. Checking your perceptions increases understanding and trust. Moving forward without checking leads to inaccurate propositions and fosters hostility and resistance.

3. Negotiation: Goals and new solutions; conflict as opportunity. Accepting conflict as part of life creates an opportunity for resolution and progress. The question is not if conflict will occur, but rather what are you going to do when it happens. Without awareness of alternative worldviews, the situation worsens. Different perspectives and opinions help us all reflect and learn anew. It can spur creativity; if a positive resolution is reached, all may come out more creative and satisfied.

Define and redefine goals. Conflict is not usually resolved all at once. If you search out where the interests of both parties coincide or are close, start with the easier issues first. Approach the more difficult problems later. Relax and allow parties to generate as many ideas as they can through brainstorming.

Work on parts of the larger concerns. Through listening and feedback on issues, systematically list and discuss all concerns, perhaps even writing them down. Constant feedback and empathic listening is critical to this process. Start with areas of agreements or small differences, then move to the more challenging areas with creativity and brainstorming.

4. Direct action. Will the conflicting parties agree to take the negotiation process into the real world and implement change? Explicit contracts can be helpful, and provision for follow-up on the action is essential.

5. Reconciliation: Apply the Platinum Rule. The Golden Rule suggests treating people the same way we would like to be treated. The Platinum Rule suggests treating others in the way *they* would like to be treated. Understanding how another person would like to be treated represents an advanced level of multicultural understanding. Mediating to achieve a consensus or compromise among arguments or disagreements represents advance mastery of multicultural competence.

Summarizing Key Points of Empathic Confrontation: Supporting While Addressing Client Conflict

Defining Empathic Confrontation

▶ Confrontation is a supportive challenge to client conflict and incongruity that helps bring a clearer understanding of core issues.

▶ Confrontation is based on effective listening and observation of discrepancies, incongruity, and mixed feelings and thoughts.

Discerning The Basic Skills of Empathic Confrontation

▶ Confrontation does not stand alone as a skill but is a combination of skills.

▶ When confrontation is used clearly and concisely, it encourages clients to look at their situation from a new perspective.

▶ There are three steps of confrontation, each using different skills: Step 1, identify conflict; Step 2, point out issues of incongruity and work to resolve them; and Step 3, evaluate the change.

▶ The Client Change Scale can be applied to any client statement in the interview, or you can use these concepts to evaluate client change over several sessions. The five dimensions are (1) denial; (2) partial examination; (3) acceptance and recognition (but no change); (4) generation of a new solution; and (5) development of new, larger, and more inclusive constructs, patterns, or behaviors—transcendence.

Observing Empathic Confrontation in Action

▶ Three main points are covered in the example interview: (1) Listening skills are used to obtain client data, (2) confrontations of client discrepancies are noted, and (3) the effectiveness of the confrontations can be assessed using the Client Change Scale.

▶ Major client change is an extended process; it may require several interviews, group sessions, and some time for the client to internalize and act on new ideas.

Practicing Becoming Competent in Empathic Confrontation

▶ Practicing what you have read is essential for skill mastery. Each of the exercises including in this section will help you increase your mastery of empathic confrontation.

Refining Nonempathic Confrontation, Dealing With Resistance, Multicultural Issues, and Working With Groups

▶ Nonempathic confrontation is an ineffective attempt to force clients to change. It often increases clients' argumentation and resistance.

▶ Respect client resistance, listen, and maintain awareness that the resistance often has important protective functions for the client.

▶ Be aware of individual and cultural differences, particularly if you are culturally different from the client.

▶ Conflicts are part of our lives and create opportunities for resolution and progress. Develop skills to mediate and manage individual and groups tensions.

▶ Treat people as they would like to be treated.

Assess your current level of knowledge and competence as you complete the chapter:

1. Flashcards: Use the flashcards to check your understanding of key concepts and facilitate memorization of key information.

2. Self-Assessment Quiz: The quiz will help you assess your current knowledge and prepare for course examinations.

3. Portfolio of Competencies: Evaluate your present level of competence on the ideas and concepts presented in this chapter using the Self-Evaluation Checklist. Self-assessment of your competencies demonstrates what you can do in the real world.

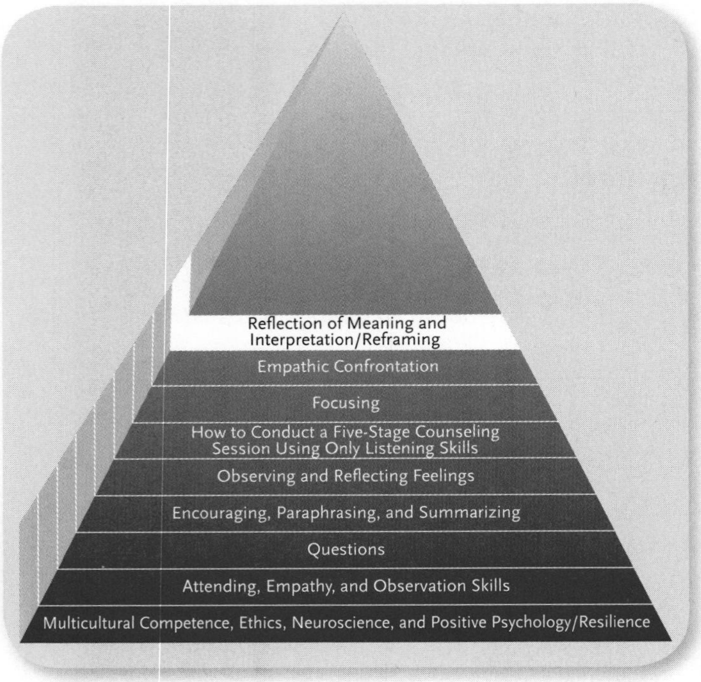
Reflection of Meaning and Interpretation/Reframing
Empathic Confrontation
Focusing
How to Conduct a Five-Stage Counseling Session Using Only Listening Skills
Observing and Reflecting Feelings
Encouraging, Paraphrasing, and Summarizing
Questions
Attending, Empathy, and Observation Skills
Multicultural Competence, Ethics, Neuroscience, and Positive Psychology/Resilience

Chapter 10 Reflection of Meaning and Interpretation/ Reframing

Restorying Client Lives Through Meaning Making

Ever more people today have the means to live, but no meaning to live for.

Challenging the meaning of life is the truest expression of the state of being human.

We who lived in concentration camps can remember the men who walked through the huts comforting others, giving away their last piece of bread. They may have been few in number, but they offer sufficient proof that everything can be taken from a man but one thing: the last of the human freedoms—to choose one's attitude in any given set of circumstances, to choose one's own way.

—Viktor Frankl (who helped many Jews survive and find meaning while imprisoned at Auschwitz during the Holocaust)

Many believe that reflection of meaning and interpretation/reframing are the most important influencing skills of all. You and your clients share a strong need to find meaning in your life experiences and establish personal and collective values. Furthermore, we aim to address basic existential questions: Who am I? Why am I here? What is the purpose or significance of it all? What sense does anything make? How much control do I have over my life? What happens when life ends? Earlier philosophers such as Plato, Socrates, and Aristotle, as well as current counselors and

This chapter is dedicated to the memory of Viktor Frankl. The initial stimulus for the skill of reflection of meaning came from a 2-hour meeting with him in Vienna shortly after we visited the German concentration camp, Auschwitz, where he had been imprisoned in World War II. He impressed on us the central value of **meaning** in counseling and therapy—a topic to which most theories give insufficient attention. It was his unusual ability to find positive meaning in the face of impossible trauma that impressed us most. His thoughts also influenced our wellness and positive strengths orientation. His theoretical and practical approach to counseling and therapy deserves far more attention than it receives. We often recommend his gripping, short book, *Man's Search for Meaning* (1959), to our clients who face serious life crises. It remains fully alive today.

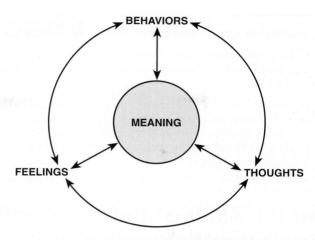

FIGURE 10.1 Meaning as the core of human experience.

therapists, have studied how people understand their life, experiences, and relationships and how this understanding affects their lives. Thus, the pursuit of meaning influences most of what we do.

Some of our understanding leads to positive views of life, while others may be negative enough to create distress, dysfunction, or despair. Interviewing, counseling, and therapy focus on helping clients change their thoughts, feelings, and behaviors to enable them to think in new and more productive ways. Underlying this basic triad is the issue of meaning and how to interpret life experience (Figure 10.1). Two related skills essential to achieve these goals are reflection of meaning and interpretation/reframe.

Reflection of meaning is concerned with helping clients find deeper meanings underlying their thoughts, feelings, and behaviors. In turn, finding a deeper meaning leads to new interpretations of life. **Interpretation/reframing** seeks to provide a new way of understanding these thoughts, feelings, and behaviors and often also results in perspectives on making meaning. Interpretation often comes from a specific theoretical perspective, such as decisional, psychodynamic, or multicultural. Clients generate their own meanings, whereas reframes usually come from the interviewer.

Chapter Goals

Awareness, knowledge, skills, and actions developed through the concepts of this chapter will enable you to:

▶ Understand reflection of meaning and interpretation/reframing and their similarities and differences.

▶ Assist clients, through reflection of meaning, to explore their deeper meanings and values and to discern their goals and life purpose or mission.

▶ Help clients, through interpretation/reframing, find alternative ways of thinking that facilitate personal development.

▶ Apply the Client Change Scale to measure change as you use these skills.

► Defining the Skills of Reflection of Meaning and Interpretation/Reframing

Reflection of meaning and interpretation/reframing explore in-depth issues below the surface of client conversation. Reflecting meaning encourages clients to find new and/or clearer visions for understanding themselves and others, as well as clarifying their purpose in life. Interpretation/reframing is the art of supplying the client with new perspectives and ideas to create new ways of thinking, feeling, and behaving. If you use reflection of meaning and interpretation/reframing skills as defined here, you can anticipate how clients may respond.

REFLECTION OF MEANING	ANTICIPATED CLIENT RESPONSE
Meanings are close to core experiencing. Encourage clients to explore their own meanings and values in more depth from their own perspective, but also the perspectives of others. Questions eliciting meaning are often a vital first step. A reflection of meaning looks very much like a paraphrase but focuses beyond what the client says. Appearing often are the words *meaning, values, vision,* and *goals.*	The client discusses stories, issues, and concerns in more depth with a special emphasis on deeper meanings, values, and understandings. Clients may be enabled to discern their life goals and vision for the future.

INTERPRETATION/REFRAME	ANTICIPATED CLIENT RESPONSE
Provide the client with a new perspective, frame of reference, or way of thinking about issues. Interpretations/reframes may come from your observations; they may be based on varying theoretical orientations to the helping field; or they may link critical ideas together.	The client may find another perspective or way of thinking about a story, issue, or problem. Their new perspective may have been generated by a theory used by the interviewer, by linking ideas or information, or by simply looking at the situation afresh.

► Discerning the Skills of Reflection of Meaning and Interpretation/Reframing

He [or she] who has a why to live can bear almost any how.

—Friedrich Nietzsche

The process of eliciting and reflecting meaning is both a skill and a strategy. As a skill, it is fairly straightforward. To **elicit meaning**, ask the client some variation of the basic question "What does . . . mean to you?" At the same time, effective exploration of meaning becomes a major strategy in which you bring out client stories from the past, in the present, or to anticipate the future. You use all the listening, focusing, and empathic confrontation skills to facilitate this self-examination, yet the focus remains on finding meaning and purpose. Review of life goals and purpose leading toward a more meaningful life can result from this exploration. A decision may be made to spend time helping others and/or in social justice advocacy.

The case of Charlis will serve as a way to illustrate similarities and differences between reflection of meaning and interpretation.

> Charlis, a workaholic 45-year-old middle manager, has a heart attack. After several days of intensive care, she is moved to the floor where you, as the hospital social worker, work with the heart attack aftercare team. Charlis is motivated; she is following physician directives and progressing as rapidly as possible. She listens carefully to diet and exercise suggestions and seems the ideal patient with an excellent prognosis. However, she wants to return to her high-pressure job and continue moving up through the company. You observe some fear and puzzlement about what's happened to her and where she might want to go next.

REFLECTION OF MEANING

Charlis has had a heart attack, and you first need to hear her story, with special attention to the many sad, worried, and anxious feelings that she experiences. There is a need to review what happened before the heart attack and her stressed lifestyle, which likely contributed to hospitalization. She will need help and support in her rehabilitation program. Understanding the meaning of the whole event will be critical. You may prescribe lifestyle changes, support her through an exercise program, review nutrition plans, and counsel her on relationships at work.

You recognize that Charlis is reevaluating the meaning of her life. She asks questions that are hard to answer: "Why me?" "What is the meaning of my life? What is God saying to me? Am I on the wrong track? What should I *really* be doing?" You sense that she feels that something is missing in her life, and she wants to reevaluate where she is going and what she is doing.

To elicit meaning, we may ask Charlis some variation of a basic meaning question: "What does the heart attack mean to you, your past, and—perhaps most of all—your future life?" We may also ask Charlis if she would like to examine the meaning of her life through the process of discernment, a more systematic approach to meaning and purpose defined in some detail in this chapter.

We'd share some of the specific questions below as a beginning; sometimes that is enough to help the client move to new life directions. We'd ask her to think of questions and issues that are particularly important to her as we work to help her discern the meaning of her life, her work, her goals, and her mission. These questions often bring out emotions, and they certainly bring out meaning in the client's thoughts and cognitions. When clients explore meaning issues, the interview becomes less precise as the client struggles with defining the almost indefinable. As appropriate to the situation, questions such as the following can address the general issue of meaning in more detail:

- ▶ "What has given you most satisfaction in your job?"
- ▶ "What's *been missing* for you in your present life?"
- ▶ "What do you *value* in your life?
- ▶ "What *sense* do you make of this heart attack and the future?"
- ▶ "What things in the future will be most *meaningful* to you?"
- ▶ "What is the *purpose* of your working so hard?"
- ▶ "You've said that you wonder what God is saying to you with this trial. Could you share some of your thoughts?"
- ▶ "What gift would you like to leave the world?"

Eliciting meaning often precedes reflection. Reflection of meaning as a skill looks very much like a reflection of feeling or paraphrase, but the key words *meaning, sense, deeper understanding, purpose, vision*, or some related concept will be present explicitly or implicitly. "Charlis, I sense that the heart attack has led you to question some basic understandings in your life. Is that close? If so, tell me more." Eliciting and reflecting meaning provide an opening for the client to explore issues that lead not to a final answer but rather to a deeper awareness of the possibilities of life.

COMPARING REFLECTION OF MEANING AND INTERPRETATION/REFRAMING

Both reflecting meaning and interpretation/reframing are designed to help clients look deeper, first by careful listening and then by helping clients examine themselves from a new perspective. Reflecting meaning involves *client* direction; the interpretation/reframe implies *interviewer* direction. In reflection of meaning, the client provides a new and more comprehensive perspective; in an interpretation/reframe, the interviewer or counselor suggests a new way of seeing issues, self, others, and life.

Here are some brief examples of how reflection of meaning and interpretation may work for Charlis as she attempts to understand some underlying issues around her heart attack.

Charlis: My job has been so challenging and I really feel that pressure all the time, but I just ignored it. I'm wondering why I didn't figure out what was going on until I got this heart attack. But I just kept going on, no matter what.

Eliciting and Reflecting Meaning

Counselor: I hear you—you just kept going. Could you share what it feels like to *keep going on* and what it *means* to you? (Encourager focusing on the key words "*keep going on*"; open question oriented to meaning)

Charlis: I was raised to keep going. My mother always prided herself on doing a good job, even in the worst of times. Grandma did the same thing.

Counselor: Charlis, I hear that keeping going and persistence have been a key family value that remains very important to you. (Reflection of meaning) "Hanging in" is what you are good at. (Positive asset leading to wellness) Could we focus now on how that value around persistence and *keeping going on* relates to your rehab? (Open question that seeks to use the wellness dimensions to help her plan for the future)

Interpretation

Counselor: You could say that you *keep going* until you drop. How does that sound to you? (Mild reframe/interpretation followed by checkout)

Charlis: I was raised to keep going. My mother always prided herself on doing a good job, even in the worst of times. Grandma did the same thing.

Counselor: Many of us become who we are because of family history. It sounds as if several generations have taught you to keep on *no matter what*. You're full of pride as you see the strength in your family that keeps you keeping on. (Interpretation/reframe, includes reflection of feeling)

Both reflection of feeling and interpretation ended up in nearly the same place, but Charlis is more in control of the process with reflection of meaning. Whichever approach is used, we are closer to helping Charlis work on the difficult questions of the meaning and direction of her future life. If the client does not respond to reflective strategies, move to the more active interpretation.

We need to give clients power and control of the session whenever possible. They can often generate new interpretations/reframes and new ways of thinking about their issues.

Interpretations and reframes vary with theoretical orientation. The joint term *interpretation/reframe* is used because they both focus on providing a new way of thinking or a new frame of reference for the client, but the word *reframe* is a gentler construct. When you use influencing skills, keep in mind that interpretive statements are more directive than reflection of meaning. When we use interpretation/reframing, we are working primarily from the interviewer's frame of reference. This is neither good nor bad; rather, it is something we need to be aware of when we use influencing skills.

Linking is an important part of interpretation, although it often appears in an effective reflection of meaning as well. In linking, two or more ideas are brought together, providing the client with a new insight. The insight comes primarily from the client in reflection of meaning, but almost all from the interviewer in interpretation/ reframing. Consider the following four examples.

Interpretation/reframe 1: Charlis, we are all reflections of our family, and it is clear that family history emphasizing success and hard work has deeply affected you—perhaps even to the point of having a heart attack. (Links family history to the heart attack. A family counselor might use this approach.)

Interpretation/reframe 2: Charlis, you seem to have a pattern of thinking that goes back a long way—we could call it an "automatic thought." You seem to have a bit of perfectionism there, and you keep saying to yourself (self-talk), "Keep going no matter what." (Links the past to the present perfectionism from a cognitive behavioral perspective.)

Interpretation/reframe 3. It sounds as if you are using hard work as a way to avoid looking at yourself. The avoidance is similar to the way you avoid dealing with what you think you need to change in the future to keep yourself healthier. (Combines empathic confrontation with linking with what is occurring in the interview series. This is close to a person-centered approach.)

Interpretation/reframe 4. The heart attack almost sounds like unconscious self-punishment, as if you wanted it to happen to give you time off from the job and a chance to reassess your life. (Linking interpretation from a psychodynamic perspective.)

ELICITING AND REFLECTING CLIENT MEANING: SPECIFIC SKILLS

Understanding the client is the essential first step. Consider storytelling as a useful way to discover the background of a client's meaning making. If a major life event is critical, illustrative stories can form the basis for exploration of meaning. Clients do not often volunteer meaning issues, even though these may be central to the clients' concerns. Critical life events such as illness, loss of a parent or loved one, accident, or divorce often force people to encounter deeper meaning issues. Blonna, Loschiavo, and Watter (2011) effectively applied this model to health counseling. If spiritual issues come to the fore, draw out one or two stories that provide concrete example of the client's religious heritage. Through the basic listening sequence and careful attending, you may observe the behaviors, thoughts, and feelings that express client meaning.

Eliciting client meanings. Fukuyama (1990, p. 9) outlined some useful questions for eliciting stories and client meaning systems. Adapted for this chapter, they include the following:

▶ "When in your life did you have existential or meaning questions? How have you resolved these issues so far?"
▶ "What significant life events have shaped your beliefs about life?"

- ▶ "What are your earliest childhood memories as you first identified your ethnic/cultural background? Your spirituality?"
- ▶ "What are your earliest memories of church, synagogue, mosque, a higher power, or of discovering your parents' vital life values?"
- ▶ "Where are you now in your life journey? Your spiritual journey?"

Reflecting client meanings. Say back to clients their exact key meaning and value words. Reflect their own unique meaning system, not yours. Implicit meanings will become clear through careful listening and questions designed to elicit meaning issues from the client. Using the client's key words is preferable, but occasionally you may supply the needed meaning word yourself. When you do so, carefully check that the word(s) you use feel right to the client. Simply change "You feel . . ." to "You mean. . . ." A reflection of meaning is structured similarly to a paraphrase or reflection of feeling. "You value . . .," "You care . . .," "Your reasons are . . .," or "Your intention was. . . ." Distinguishing among a reflection of meaning, a paraphrase, and a reflection of feeling can be difficult. Often the skilled counselor will blend the three skills together. For practice, however, it is useful to separate out meaning responses and develop an understanding of their import and power in the interview. Noting the key words that relate to meaning (*meaning, value, reasons, intent, cause,* and the like) will help distinguish reflection of meaning from other skills.

Reflection of meaning becomes more complicated when meanings or values conflict. Here concepts of empathic confrontation (Chapter 9) may be useful. Conflicting values, either explicit or implicit, often underlie mixed and confused feelings expressed by the client. For instance, a client may feel forced to choose between loyalty to family and loyalty to spouse. Underlying love for both may be complicated by a value of dependence fostered by the family and the independence represented by the spouse. When clients make important decisions, sorting out key meaning issues may be crucial.

For example, a young person may be experiencing a value conflict over career choice. Spiritual meanings may conflict with the work setting. The facts may be paraphrased accurately and the feelings about each choice duly noted, yet the underlying *meaning* of the choice may be most important. The counselor can ask, "What does each choice *mean* for you? What sense do you make of each?" The client's answers provide the opportunity for the counselor to reflect back the meaning, eventually leading to a decision that involves not only facts and feelings but also values and meaning. As with empathic confrontation, you can evaluate client change in meaning systems using the Client Change Scale (Chapter 9).

 Case Study: Finding Family Balance in Greece: A Universal Problem. How would you use reflection of meaning and interpretation/reframing to help Katerina?

▶ Observing the Skills of Reflection of Meaning and Interpretation/Reframing

In the following session, Travis is reflecting on his recent divorce. When relationships end, the thoughts, feelings, and underlying meaning of the other person and the time together often remain unresolved. Moreover, some clients are likely to repeat the same mistakes in their relationships when they meet a new person.

Both the reflection of meaning and the interpretation/reframe can help clients draw meaning from their ordeals and gain new perspectives on themselves and their world.

Andreas, the interviewer, seeks to help Travis think about the word *relationship* and its meaning. Note that Travis stresses the importance of connectedness with intimacy and caring. The issue of self-in-relation to others will play itself out very differently among individuals in varying cultural contexts. Many clients will focus on their need for independence.

INTERVIEWER AND CLIENT CONVERSATION	PROCESS COMMENTS
1. *Andreas:* So, Travis, you're thinking about the divorce again . . .	Encourager/restatement.
2. *Travis:* Yeah, that divorce has really thrown me for a loop. I really cared a lot about Ashley and . . . ah . . . we got along well together. But there was something missing.	(Travis is at Level 2, partial examination, on the Client Change Scale.)
3. *Andreas:* Uh-huh . . . something missing?	Encouragers appear to be closely related to meaning. Clients often supply the meaning of their key words if you repeat them back exactly. (Interchangeable empathy)
4. *Travis:* Uh-huh, we just never really shared something very basic. The relationship didn't have enough depth to go anywhere. We liked each other, we amused one another, but beyond that . . . I don't know . . .	Travis elaborates on the meaning of a closer, more significant relationship than he had with Ashley. (CCS Level 2)
5. *Andreas:* You amused each other, but you wanted more depth. What sense do you make of it?	Paraphrase using Travis's key words followed by a question to elicit meaning. (The paraphrase is interchangeable, the question potentially additive.)
6. *Travis:* Well, in a way, it seems like the relationship was shallow. When we got married, there just wasn't enough depth for a meaningful relationship. The sex was good, but after a while, I even got bored with that. We just didn't talk much. I needed more . . .	Note that Travis's personal constructs for discussing his past relationship center on the word *shallow* and the contrast *meaningful*. This polarity is probably one of Travis's significant meanings around which he organizes much of his experience. (CCS Level 2)
7. *Andreas:* Mm-hmmm . . . you seem to be talking in terms of shallow versus meaningful relationships. What does a meaningful relationship feel like to you?	Reflection of meaning followed by a question designed to elicit further exploration of meaning. (Interchangeable, potentially additive—please note again that questions have potential, but we don't know whether or not they are additive until we see what the client says next.)
8. *Travis:* Well, I guess . . . ah . . . that's a good question. I guess for me, there has to be some real, you know, some real caring beyond just on a daily basis. It has to be something that goes right to the soul. You know, you're really connected to your partner in a very powerful way.	Connection appears to be a central dimension of meaning. We often believe that connectedness is a female construct, but many men also see it as central. (It was additive, CCS Level 2, but moving toward 3.)
9. *Andreas:* So, connections, soul, deeper aspects strike you as really important.	Reflection of meaning. Note that this reflection is also very close to a paraphrase, and Andreas uses Travis's main words. The distinction centers on issues of meaning. A reflection of meaning could be described as a special type of paraphrase. (Interchangeable)

(continued)

INTERVIEWER AND CLIENT CONVERSATION	PROCESS COMMENTS
10. *Travis:* That's right. There has to be some reason for me to really want to stay married, and I think with her . . . ah . . . those connections and that depth were missing. We liked each other, you know, but when one of us was gone, it just didn't seem to matter whether we were here or there.	(CCS Level 2–3 as he is building better understanding)
11. *Andreas:* So the relationship did not feel meaningful to either of you. And I hear that closeness and a meaning-ful relationship is what you missed and what you value.	Reflection of meaning plus some reflection of feeling. Note that Andreas has added the word *values* to the discussion. In reflection of meaning, it is likely that the counselor or interviewer will add words such as *meaning, understanding, sense,* and *value.* Such words lead clients to make sense of their experience from the their own frame of reference. (Interchangeable)
12. *Travis:* Uh-huh. That is really so. (quiet pause, looking puzzled)	
13. *Andreas:* Ah . . . could you fantasize how you might play out those thoughts, feelings, and meanings in another relationship?	Open question oriented to meaning. (Potentially additive)
14. *Travis:* Well, I guess it's important for me to have some independence from a person, but when we were apart, we'd still be thinking of one another. Depth and a soul mate is what I want.	Travis's meaning and desire for a relationship are now being more fully explored. (CCS Level 3+)
15. *Andreas:* Um-hum.	
16. *Travis:* In other words, I don't want a relationship where we always tag along together. The opposite of that is where you don't care enough whether you are together or not. That isn't intimate enough. I really want intimacy in a marriage. My fantasy is to have a very independent partner I care about and who cares about me. We can both be individuals but still have bonding and connectedness.	Connectedness is an important meaning issue for Travis. With other clients, independence and autonomy may be the issue. With still others, the meaning in a relationship may be a balance of the two. (CCS Level 3, moving to 4)
17. *Andreas:* Let's see if I can put together what you're saying. The key words seem to be *independence* with intimacy and caring. It's these concepts that can produce bonding and connectedness, as you say, whether you are together or not.	This reflection of meaning becomes almost a summarization of meaning. Note that the key words and constructs have come from the client in response to questions about meaning and value. (Additive)

Further counseling would aim to bring behavior or action into accord with thoughts. Other past or current relationships could be explored further to see how well the client's behaviors or actions illustrate or do not illustrate expressed meaning.

Interactive Exercise: Identifying Skills. Practice identifying counselor responses.
Interactive Exercise: Identifying/Classifying Meaning. Practice how to help a person clarify the meaning of his or her situation.
Interactive Exercise: Client Issues With Meaning. Help a senior in college explore

whether or not he wants to go on to graduate school in school counseling.

Interactive Exercise: Questions Eliciting Meaning. List questions that might be useful in bringing out the meaning of an event.

Interactive Exercise: Discernment Exercise: Your Future in Interviewing, Counseling, or Therapy. The fields of professional helping are challenging. If you have a sense of mission and purpose, this exercise may help you organize your thinking in a new way and help you get through rough times.

Group Practice Exercise: Group Practice With Reflection of Meaning

---

▶ Practicing the Skills of Eliciting and Reflecting Meaning and Interpretation/Reframing

The following exercises will help you achieve skill mastery and competence.

EXERCISE 10.1 Reflection of Meaning

Go back and review the case of Travis.

- ▶ What thoughts occur to you as you think about his life issues?
- ▶ What do you see as the key issues that relate to the meaning and purpose of his life?
- ▶ What about meaning and purpose in your own life? What are your goals and vision? Take a little time with this as it will you prepare for the discernment exercise coming in the next section.

EXERCISE 10.2 A Client With Marital Problems

A client has been coming to you for therapy and is reviewing his life as it relates to present problems in his second marriage. Last week you reviewed sexual problems in the marriage, and you noted that he tended to focus on himself with little attention to his wife. This week he speaks of a dream in which he is trying to break into a room that contains a golden icon, but is frustrated in the process.

(with agitation) My wife and I had another fight over sex last night. We went to a sexy movie and I was really turned on. I tried to make love and she rejected me again.

How might you paraphrase what the client just said? Reflect this client's feelings? Interpret the client's statement? Generate several possibilities. Perhaps one might relate to how this situation compares to his first marriage, another to his insensitivity to his partner, and the third to the dream.

See the end of the chapter for our suggestions.

EXERCISE 10.3 Group Practice With Reflection of Meaning

For practice with this reflection of meaning, it will be most helpful if the session starts with the client completing one of the following model sentences. The session will then follow along, exploring the attitudes, values, and meanings to the client underlying the sentence.

"My thoughts about spirituality are . . ."

"My thoughts about moving from this area to another are . . ."

"The most important event of my life was . . ."

"The legacy or contribution that I would like to leave the world is . . ."

"The center of my life is . . ."

"My thoughts about divorce/abortion/gay marriage are . . ."

A few alternative topics are "My closest friend," "Someone who made me feel very angry (or happy)," and "A place where I feel very comfortable and happy."

The task of the counselor in this case is to elicit meaning from the model sentence and help the client find underlying meanings and values. The counselor should search for key words in the client response and use those key words in questioning, encouraging, and reflecting. A useful sequence of microskills for eliciting meaning from the model sentence follows:

1. An open question, such as "Could you tell me more about that?" "What does that mean to you?" or "How do you make sense of that?"
2. Encouragers and paraphrases focusing on key words to help the client continue
3. Reflections of feeling to ensure that you are in touch with the client's emotions
4. Questions that relate specifically to meaning
5. Reflecting the meaning of the event back to the client, using the framework outlined in this chapter

It is quite acceptable to have key questions in your lap to refer to during the practice session.

EXERCISE 10.4 Group Practice With Interpretation/Reframe

This is a continuation of the previous exercise, but this time the whole group discusses the meaning issue raised by the client. What different interpretations and/or reframes of the issue can each of you add to the discussion? Continue as time permits, moving between the practice interview and your own personal thoughts about meaning, vision, and the many interpretations that are possible.

For your practice exercise in reflecting meaning, you may find it helpful to use the Microskills Interview Analysis Form (MIAF) (see Chapter 7, Box 7.1).

 Group Practice Exercise: Group Practice With Interpretation/Reframe

---

▶ Refining the Skills of Eliciting and Reflecting Meaning and Interpretation/Reframing

MULTICULTURAL ISSUES AND REFLECTION OF MEANING

For practical multicultural interviewing and counseling, recall the concept of focus (Chapter 8). When helping clients make meaning, focus exploration of meaning not just on the individual but also on the broader life context. In much of Western society, we tend to assume that it is the individual person who makes meaning. But in many other cultures—for example, the traditional Muslim world—the individual will make meaning in accord with the extended family, the neighborhood, and religion. Individuals do not make meaning by themselves; *they make meaning in a multicultural context*. In truth, Western society also draws meaning from family and culture. However, individualism rather than collectivism is generally the focus.

Cultural, ethnic, religious, and gender groups all have systems of meaning that give an individual a sense of coherence and connection with others. Muslims draw on the teachings of the Qur'an. Similarly, Jewish, Buddhist, Christian, and other religious groups will draw on their writings, scriptures, and traditions. African Americans may draw on the meaning strengths of Malcolm X, Martin Luther King, Jr., or on support they receive from Black churches as they deal with difficult situations. Women may make meaning out of relationships, whereas men may focus more on issues of personal autonomy and tasks. Witness the conversations in a mixed social group. Often we find women on one side of the room talking about relationships. Men are more likely to be talking about sports, politics, and their accomplishments.

Viktor Frankl, a German concentration camp survivor, could not change his life situation, but he was able draw on important strengths of his Jewish tradition to change the meaning he made of it. The Jewish tradition of serving others facilitated his survival and enabled him to help fellow sufferers. When times were particularly bad, when prisoners had been whipped and were not being given food, Frankl (1959, pp. 131–133) counseled his entire barracks, helping them reframe their terrors and difficulties, pointing out that they were developing strengths for the future.

> I quoted from Nietzsche, "That which does not kill me, makes me stronger." I spoke to the future. I said that . . . the future must seem hopeless. I agreed that each of us could guess . . . how small were chances for survival. . . . I estimated my chances at about one in twenty. But I also told them that, in spite of this, I had no intention of losing hope and giving up. . . . I also mentioned the past; all its joys and how its light shone even in the present darkness. . . . Then I spoke of the many opportunities of giving life a meaning. I told my comrades . . . that human life, under any circumstances has meaning. . . . I said that someone looks down on each of us in difficult hours—a friend, a wife, somebody alive or dead, or a God—and He would not expect us to disappoint him. . . . I saw the miserable figures of my friends limping toward me to thank me with tears in their eyes.

You may counsel clients who have experienced some form of religious bias or persecution. As religion plays such an important part in many people's lives, members of dominant religions in a region or a nation may have different experiences from those who follow minority religions. For example, Schlosser (2003) and Banks (2012) talk of Christian privilege in North America, where people of Jewish and other faiths may feel uncomfortable, even unwelcome, during Christian holidays. Anti-Semitism, anti-Islamism, anti–liberal Christianity, and anti–evangelical Christianity are all possible examples of spiritual and/or religious intolerance. We also recall that when Christians and other religious groups find themselves in countries where they are a minority, they can suffer serious religious persecution, to the point of death.

RESILIENCE, PURPOSE, AND MEANING

Resilience is the ability to recover from a wide variety of difficulties. Two examples of resilience are the child who is teased, but bounces back cheerfully the next day, and the adolescent who undergoes being "dumped" by a boyfriend or girlfriend, but soon gets over it and moves on. What we see here is the ability to not let bad experiences get one down.

Resilience occurs at a deeper level when the individual suffers a serious or life-threatening trauma. A child is born in poverty experiences abuse and somehow manages to put together a successful life. A businessman goes bankrupt, but within two years has put together a successful business. Two women lose their jobs; after both spend six months searching for employment, one goes into a deep depression, while the other continues on and eventually finds work. Two soldiers on patrol experience an ambush in which one of their comrades is killed. One soldier leaves the warfront and ends up being treated for post-traumatic stress over a period of months; the resilient soldier has a short period of mourning, grief, and recovery and soon is back on the front lines.

Meaning and purpose are key to resilience. Viktor Frankl was a survivor of the German death camps. Many around him gave up and died. Frankl kept his focus on the positive and looked for moments of meaning. He enjoyed a beautiful sunset even though hungry, he thought of his beloved wife, and he wrote a book in his mind while doing painful work. Frankl is the theorist who truly brought the importance of meaning to our field. His personal example and his writings remain relevant for us today.

Teaching resilience results in less childhood depression (American Psychological Association, 2009). Martin Seligman (2009), the founder of positive psychology, has shown that teaching purpose and meaning can make a difference in children. One of his major goals was to encourage children to find purpose and positive expectations.

One exercise involved the students' writing down three good things that happened each day for a week. Examples were: "I answered a really hard question in Spanish class," "I helped my mom shop for groceries" or, "The guy I've liked for months asked me out." Next to each positive event, the students answered the following questions: "What does this mean to you?" and "How can you increase the likelihood of having more of this good thing in the future?" (p. 84)

"What does this mean to you?" is, of course, the basic question to elicit meaning. Meaning is then reflected and synthesized and becomes an important part of the child's cognitive/emotional processing.

African Americans who have suffered trauma but have a sense of purpose and meaning have better mental and physical health (Alim et al., 2008). At the other end of life's spectrum, older people who have a clear sense of meaning and purpose have better mental and physical health. Those with a sense of meaning and purpose are 2½ times less likely to suffer the ravages of Alzheimer's disease (Boyle, Buchman, Barnes, & Bennett, 2010). Meaning is not just found in thoughts and feelings; it also affects the body.

The following discussion of discernment is another way to facilitate your clients' finding deeper meanings and visions for their lives.

DISCERNMENT: IDENTIFYING LIFE MISSION AND GOALS

Listen. Listen, with intention, with love, with the "ear of the heart." Listen not only cerebrally with the intellect, but with the whole of feelings, our emotions, imaginations, and ourselves.

—Esther de Wall

Discernment is "sifting through our interior and exterior experiences to determine their origin" (Farnham, Gill, McLean, & Ward, 1991). The word *discernment* comes from the Latin *discernere*, which means "to separate," "to determine," "to sort out." In a spiritual or religious sense, discernment means identifying when the spirit is at work in a situation—the spirit of God or some other spirit. The discernment process is important for all clients, regardless of their spiritual or religious orientation or lack thereof. Discernment has broad applications to interviewing and counseling; it describes what we do when we work with clients at deeper levels of meaning. Discernment is also a process whereby clients can focus on envisioning their future as a journey into meaning. (See Box 10.1.)

"There is but one truly serious problem, and that is . . . judging whether or not life is worth living" (Camus, 1955, p. 3). Viktor Frankl (1978) talks about the "unheard cry for meaning." Frankl claimed that 85% of people who successfully committed suicide saw life as meaningless, and he blamed an excessive focus on self. He said that people have a need for transcendence and living beyond the self.

The vision quest, often associated with the Native American Indian, Dene, and Australian Aboriginal traditions, is oriented to helping youth and others find purpose and meaning in their lives. These individuals often undertake a serious outdoor experience to find or envision their central life goals. Meditation is used in some cultures to help members find meaning and direction.

Visioning and finding meaning can often be facilitated if issues are explored with a guide, counselor, or interviewer. This can be a spiritual or religious quest for some clients,

BOX 10.1 Questions Leading Toward Discernment of Life's Purpose and Meaning

You may find it helpful to share this list with the client before you begin the discernment process and identify together the most helpful questions to explore. Add topics and questions that occur to you and the client. Discernment is a very personal exploration of meaning. The more the client participates, the more useful it is likely to be. Questions that focus on the here and now and on intuition may facilitate deeper discovery.

Following is a systematic approach to discernment. First, you or your client may wish to begin by thinking quietly about what might give life purpose, meaning, and vision.

Here-and-now body experience and imaging can serve as a physical foundation for intuition and discernment.

▶ Relax, explore your body, find a positive feeling of strength to serve as an anchor for your search. Build on that feeling and see where it goes.
▶ Sit quietly and allow an image (visual, auditory, kinesthetic) to build.
▶ What is your gut feeling? What are your instincts? Get in touch with your body.
▶ Discerning one's mission cannot be found solely through the intellect. What feelings and thoughts occur to you at this moment?
▶ Can you recall feelings and thoughts from your childhood that might lead to a sense of direction now?
▶ What is your felt body sense of spirituality, mission, and life goals?

Concrete questions leading to telling stories can be helpful.

▶ Tell me a story about that image above. Or a story about any of the here-and-now experiences listed there.
▶ Can you tell me a story that relates to your goals/vision/mission?
▶ Can you name the feelings you have in relation to your desires?
▶ What have you done in the past or what are doing presently that feels especially satisfying and close to your mission?
▶ What are some blocks and impediments to your mission? What holds you back?
▶ Can you tell about spiritual stories that have influenced you?

For self-reflective exploration, the following are often useful.

▶ Let's go back to that original image and/or the story that goes with it. As you reflect on that experience or story, what occurs for you?
▶ Looking back on your life, what have been some of the major satisfactions? Dissatisfactions?
▶ What have you done right?

▶ What have been the peak moments and experiences of your life?
▶ What might you change if you were to face that situation again?
▶ Do you have a sense of obligation that impels you toward this vision?
▶ Most of us have multiple emotions as we face major challenges such as this. What are some of these feelings, and what impact are they having on you?
▶ Are you motivated by love/zeal/a sense of morality?
▶ What are your life goals?
▶ What do you see as your mission in life?
▶ What does spirituality mean to you?

The following questions place the client in larger systems and relationships—the self-in-relation. They may also bring multicultural issues into the discussion of meaning.

▶ Place your previously presented experiences and images in a broader context. How have various systems (family, friends, community, culture, spirituality, and significant others) related to these experiences? Think of yourself as a self-in-relation, a person-in-community.
▶ *Family.* What do you learn from your parents, grandparents, and siblings that might be helpful in your discernment process? Are they models for you that you might want to follow, or even oppose? If you now have your own family, what do you learn from them, and what is the implication of your discernment for them?
▶ *Friends.* What do you learn from friends? How important are relationships to you? Recall important developmental experiences you have had with peer groups. What do you learn from them?
▶ *Community.* What people have influenced you and perhaps serve as role models? What group activities in your community may have influenced you? What would you like to do to improve your community? What important school experiences do you recall?
▶ *Cultural groupings.* What is the place of your ethnicity/race in discernment? Gender? Sexual orientation? Physical ability? Language? Socioeconomic background? Age? Life experience with trauma?
▶ *Significant other(s).* Who is your significant other? What does he or she mean to you? How does this person relate to the discernment process? What occurs to you as the gifts of relationship? The challenges?
▶ *Spiritual.* How might you want to serve? How committed are you? What is your relationship to spirituality and religion? What does your holy book say to you about this process?

Source: From Ivey, Developmental Counseling and Therapy, 1E. © 2005 Cengage Learning.

but the discernment process will be useful for all. (Review Ivey, Ivey, Myers, & Sweeney, 2005, for additional information.) In Box 10.1, the specific discernment questions lead to further examination of goals, values, and meaning. Share the list of questions and encourage clients to participate with you in deciding which questions and issues are most important.

THEORIES OF COUNSELING AND INTERPRETATION/ REFRAMING

Theoretical interpretations can be extremely valuable as they provide the interviewer with a tested conceptual framework for thinking about the client. Each theory is itself a story— a story told about what is happening in interviewing, counseling, and therapy and what the story means. Integrative theories find that each theoretical story has some value. Most likely, as you generate your own natural style, you will develop your own integrative theory, drawing from those approaches that make most sense to you.

Imagine now that you have worked for some time with Charlis, who has worked through many of her fears but still faces some real challenges. She comments:

> Yesterday, my manager gave me a new assignment. I sensed that something was wrong, that he wanted something from me, but wouldn't say it. It made me feel very anxious as now I'm not sure what to do. Where do I go next?

First, keep in mind that the interviewer could reflect back the anxiety concerns and turn the issue back to Charlis, or ask her the meaning of the situation. In interpretation, the counselor supplies an alternative perspective. Below are several examples of how counselors with different theoretical orientations might interpret the same information. Before each interpretation is a brief theoretical paragraph that provides a background for the theory-oriented interpretation that follows. Each interpretation is slightly exaggerated for clarity.

Decisional theory. A major issue in interviewing for all clients is making appropriate decisions and understanding alternatives for action. Decisions need to be made with awareness of cultural/environmental context. Interpretation/reframing helps clients find new ways of thinking about their decisions. Linking ideas together is particularly important.

> Charlis, it sounds as if the manager is again giving you a double message, and that causes real anxiety. It's a repetition of some of the things that led to your heart attack. We've spent some time dealing with your tension. This seems a good place to go over breathing and relaxation again.

Person-centered. Clients are ultimately self-actualizing, and our goal is to help them find the story that builds on their strengths and helps them find deeper meanings and purpose. Reflection of meaning helps clients find alternative ways of viewing the situation; interpretation/reframing is not used. Linking can occur through effective summarization.

> Charlis, you are really feeling anxious again, this seems be giving you a difficult again. You're what this means. (Reflection of feeling and eliciting meaning. The conversation has returned to the client.)

Brief counseling. Brief methods seek to help clients find quick ways to reach their central goals. The interview is conceived first as a goal-setting process; then methods are found to reach goals through time-efficient methods. Interpretation/reframing will be rare except for linking of key ideas.

> You're facing some of the old familiar challenges since the heart attack. Let's look at this and think back on what you've done in the past that works in this situation. What comes to mind? (Linking with a mild interpretation and a focus on finding what was effective for Charlis in the past, thus reminding her of her strengths.)

Cognitive behavioral theory. The emphasis is on sequences of behavior and thinking and what happens to the client, internally and externally, as a result. Often interpretation/reframing is useful in understanding what is going on in the client's mind and/or linking the client to how the environment affects cognition and behavior.

> OK, you came in to talk and suddenly the manager seemed almost to be attacking you (antecedent), and then you become anxious (emotional consequence), and this has left you hanging and wondering what to do (behavioral consequence). Now let's analyze the situation. Let's look at what he is doing and how he seems to make you feel. Then we can develop a new alternative and more effective way to deal with this in the future.

Psychodynamic/interpersonal theory. Individuals are dependent on unconscious forces. Interpretation/reframing is used to help link ideas and enable the client to understand how the unconscious past and long-term, deeply seated thoughts, feelings, and behaviors frame the here and now of daily experience. Freudian, Adlerian, Jungian, and several other psychodynamic theories each tell different stories.

> Charlis, this seems to go back to the discussions we've had in the past about your issues with authority, particularly with your father. We've even noticed that sometimes you treat me as an authority figure. We see the here and now with me, the situation with your boss, and the long-term issues with your father. What sense do you make of this possible pattern in your life?

Multicultural counseling and therapy (MCT). Everyone is always situated in a cultural/environmental context, and we need to help clients interpret and reframe their issues, concerns, and problems in relation to their multicultural background (see the RESPECTFUL model). MCT is an integrative theory that uses all of the methods above, as appropriate, to help clients understand themselves and how the cultural/environmental context affects them personally. The following is from a feminist therapy frame of reference.

> Sounds to me like simple sexism once again. Charlis, we've got to work on how you can deal with a work environment that seems continually to be to hassling you. Perhaps a complaint is in order. But at least we have to engage in more stress management to help you as a woman deal with this productively so that it does not destroy your health again.

All of the above provide the client with a new, alternative way to consider the situation. In short, interpretation renames or redefines "reality" from a new point of view. Sometimes just a new way of looking at an issue is enough to produce change. Which is the correct interpretation? Depending on the situation and context, any of these interpretations could be helpful or harmful. The first two responses deal with here-and-now reality, whereas psychodynamic interpretation deals with the past. The feminist interpretation links the heart attack with sexual harassment on the job.

SCIENCE AND ETHICAL DECISION MAKING

To change your life, master your brain.

—Dotti Dixon Schmeling

Many readers think this chapter on meaning and interpretation is the most important because thoughts, feelings, and behaviors often stem from central meaning issues. We think that Viktor Frankl is correct when he points out that the person who has a *why* for living will discover *how* to live. Ethics are the moral principles that define our lives. No matter what we do, our ethics and morals determine our decisions and our behavior. This can be for good or ill. Many people would say that ethics, values, and meaning determine who we really are. We may vocalize one thing but do the opposite. Do we "walk the talk"? Or do we do what is comfortable, or even opposite to what we say?

Neuroscience demonstrates that moral and ethical decision making emerges from a complex interaction among multiple neural systems distributed across brain regions (Greene, 2009; Roskies, 2012). As you might anticipate, brain scans (fMRI) reveal that decision making in the area of values and ethics depends on the executive prefrontal cortex (PFC) in interaction with limbic emotional systems.

The energizing amygdala and the emotional system are fast, automatic, and influenced by the here and now. The executive system is deliberative and both enhanced and limited by memory of past experience (Xu et al., 2013). Value decisions that lead to action rest on the simplistic idea of "good" and "bad." The "good" of the here and now may well conflict with what is considered valuable in the longer term, in which meaning and personal values are more deeply considered.

Serious exploration of the meaning of client stories, as well as the discernment exercise presented here, will strengthen the executive meaning-making system and prepare it for emotional challenges in the future. But the executive cannot and must not be separated from the emotional. Unless meaning and discernment have an emotional base, the cognitive meaning is unlikely to hold and lead to action. This is exemplified when we recall that the amygdala has been activated in relation to personal moral dilemmas (Greene et al., 2004). This emotional component is thought to function as an alarm that directly influences behavior or to add motivational energy to different alternatives considered during the reasoning process.

While reflection of meaning, interpretation, and discernment are often central in ethical, value, and life vision decisions, drawing out clients' stories from multiple perspectives (focusing) and confronting decisional conflict is essential. Exploring emotions is also key to cognitive meaning decisions.

Esther de Wall's quote that begins the discernment discussion is perhaps a good way to end this chapter.

Listen. Listen, with intention, with love, with the "ear of the heart." Listen not only cerebrally with the intellect, but with the whole of feelings, our emotions, imaginations, and ourselves.

Summarizing Key Points of Reflection of Meaning and Interpretation/Reframing: Restorying Client Lives Through Meaning Making

Defining the Skills of Reflection of Meaning and Interpretation/Reframing

▶ A reflection of meaning looks very much like a paraphrase but focuses beyond what the client says. Often the words *meaning*, *values*, and *goals* will appear in the discussion. Clients are encouraged to explore their own meanings in more depth from their own perspective. Questioning and eliciting meaning are often vital as first steps.

▶ Interpretations/reframes provide the client with a new perspective, frame of reference, or way of thinking about issues. They may come from observations of the counselor; they may be based on varying theoretical orientations to the helping field; or they may link critical ideas together.

Discerning the Skills of Reflection of Meaning and Interpretation/Reframing

▶ The two skills are similar in helping clients generate a new and potentially more helpful way of looking at things. Reflection of meaning focuses on the client's worldview and seeks to understand what motivates the client; it provides more clarity on values and deeper life meanings.

▶ An interpretation results from interviewer observation and seeks new and more useful ways of thinking.

▶ Eliciting meaning is accomplished by carefully listening to the client for meaning words. Questions related to values, meaning, life goals, and ultimate causal issues bring out client meanings. "What meaning does that have to you?" "What sense do you make of that?"

Observing the Skills of Reflection of Meaning and Interpretation/Reframing

▶ The counselor uses listening skills and key word encouragers to focus on meaning issues.

▶ Open questions oriented to values and to meaning are often effective in eliciting client talk about meaning issues.

▶ A reflection of meaning looks very much like a paraphrase except that the focus is on implicit deeper issues, often not expressed fully in the surface language or behavior of the client.

Practicing the Skills of Eliciting and Reflecting Meaning and Interpretation/Reframing

▶ Completing the exercises including in this section and on the website will help you achieve mastery of these skills and achieve interviewing competence.

Refining the Skills of Eliciting and Reflecting Meaning and Interpretation/Reframing

▶ Discernment is a form of listening that goes beyond our usual descriptions and could be termed "listening with the heart." Both you and the client seriously search for deeper life goals and direction. Specific discernment questions are in Box 10.1.

▶ Multicultural and family issues and stories may be key in helping clients discover personal meaning. Eliciting meaning and focusing reflection on contextual issues beyond the individual will enhance and broaden the client's understanding of life's deeper concerns.

▶ Focusing and multicultural counseling and therapy are the most certain ways to bring multicultural issues into the interview. A woman, a gay or lesbian, or a Person of Color may be depressed over what is considered a personal failure. Helping the client see the cultural/environmental context of the issue may offer a new perspective, providing a totally new and more workable meaning.

▶ Neuroscience reminds us that finding life's meaning and vision involves both the prefrontal executive and the faster moving emotional limbic system and amygdala.

Assess your current level of knowledge and competence as you complete the chapter:

1. Flashcards: Use the flashcards to check your understanding of key concepts and facilitate memorization of key information.

2. Self-Assessment Quiz: The quiz will help you assess your current knowledge and prepare for course examinations.

3. Portfolio of Competencies: Evaluate your present level of competence on the ideas and concepts presented in this chapter using the Self-Evaluation Checklist. Self-assessment of your competencies demonstrates what you can do in the real world.

Possible Counselor Responses to the Client for Exercise 10.2

You're upset and angry. (reflection of feeling)

You had a fight after the "turn-on" movie and were rejected again. (paraphrase/restatement)

As I've listened to you over the past two sessions, I hear your deep caring for your wife, and when she rejects you, it means you fear losing her. (reflection of meaning)

Sounds like you didn't take it as slowly and easily as we talked about last week and she felt forced once again. Am I close? (interpretation #1 with checkout)

Your anger with her seems parallel to the anger you used to feel toward your first wife. I wonder what sense you make of that? (interpretation #2 with checkout)

The feelings of rejection really bother you. Those sad and angry feelings sound like the dream last week. Does that make sense? (linking interpretation #3 with checkout)

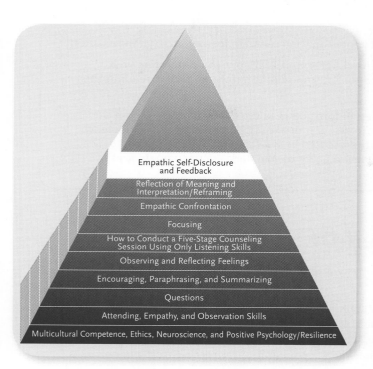

Chapter 11
Empathic Self-Disclosure and Feedback
Relationship, Immediacy, and Genuineness

Before I open up to you (self-disclose), I want to know where you are coming from. . . . In other words, a culturally different client may not open up (self-disclose) until you, the helping professional, self-disclose first. Thus, to many minority clients, a therapist who expresses his/her thoughts and feelings may be better received in a counseling situation.
—Derald Wing Sue and Stanley Sue

If used carefully and sensitively, the two skills of self-disclosure and feedback can bring here-and-now immediacy to the session. When counselors share their experience of the client, or briefly comment on something from their own lives, the client will listen. Counselor **feedback** from observations of client behaviors, thoughts, or feelings can be highly beneficial. However, the use of these two skills remains controversial, as the **power** of the counselor may be used inappropriately.

Chapter Goals

Awareness, knowledge, skills, and actions developed through the concepts of this chapter and this book will enable you to:

▶ Use appropriate self-disclosure, which builds a sense of equality and encourages client trust and openness. Disclosure may be about immediate here-and-now feelings, history of the counselor, or thoughts and feelings that the counselor may have.

▶ Offer accurate feedback on how the client is experienced by the counselor in the here and now of the session; how the client is progressing on issues; how others view the client; and thoughts, feelings, and behaviors observed in the session.

▶ Apply these skills empathically with a sense of personal genuineness and authenticity and an awareness of the counselor's power in the session.

▶ Work effectively with the empathic qualities of immediacy.

► Defining Empathic Self-Disclosure

First and foremost: strict honesty is required.

—James Bugenthal

Self-disclosure is ever present and unavoidable (Barnett, 2011) and takes one of these three forms: deliberate, unavoidable, and accidental (Zur, Williams, Lehavot, & Knapp, 2009). Counselors reveal multiple aspects of themselves through their office décor, clothes, reactions, and nonverbal communication. Because trust and openness are essential if you are to establish a working alliance, many effective interviewers and counselors allow clients to inquire about them as persons and encourage asking questions early in the process. This can help the power balance of the session and build trust if used appropriately. At the same time, as a counselor, you are in a position of power and influence, and the client can overreact to what you say. Ethical sensitivity and awareness of client needs are central.

Self-disclosure by the counselor has been a controversial topic. Some theorists and clinicians argue against counselors' sharing themselves openly, preferring a more distant, objective persona. However, humanistically oriented and feminist counselors have demonstrated the value of sparingly used and appropriate self-disclosure. Multicultural theory considers early self-disclosure as key to trust building in the long run. This seems particularly so if your background is substantially different from that of your client. For example, a young person counseling an older person needs to discuss this issue early in the session. Moreover, research reveals that clients of counselors who self-disclose report lower levels of symptom distress and like the counselor more (Barrett & Berman, 2001).

Following is a brief definition of self-disclosure and how clients might respond when it is used with sensitivity.

SELF-DISCLOSURE	ANTICIPATED CLIENT RESPONSE
As the counselor, share your own related past personal life experience, here-and-now observations or feelings toward the client, or opinions about the future. Self-disclosure often starts with an "I" statement. Here-and-now feelings toward the client can be powerful and should be used carefully.	When used sparingly, appropriately, and empathically, the client is encouraged to self-disclose in more depth and may develop a more egalitarian relationship in the session. The client may feel more comfortable in the relationship and find a new solution relating to the counselor's self-disclosure.

Imagine that you are counseling an older person coping with a forthcoming divorce after 35 years of marriage. Your background is very different, and truly empathizing with the client may be a challenge. Moreover, the older client may have difficulty trusting you and your expertise. Open discussion, some self-disclosure on your part, and exploration of differences may be essential. Self-disclosure of who you really are can be helpful in those situations in which you have not "been there."

Similar issues will occur if you are counseling an alcoholic or a drug abuser and have not been in their shoes. Women or men may not feel comfortable with those of the opposite gender. If you are significantly different from the client on any dimension of the RESPECTFUL model (Chapter 2), the differences may need to be discussed.

Inappropriate and too extensive self-disclosure takes the focus away from the client. There are valid concerns about counselors' monopolizing conversation or abusing the client's rights by encouraging openness too early; critics also point out that counseling and therapy can operate successfully without any self-disclosure at all. The boundaries separating you from the client may be violated, resulting in harm to the client and even ethical violations.

 Interactive Exercise: Using Influencing Skills in the Interview. What self-disclosure would be more appropriate for each segment of the counselor-client interaction? Compare your responses to ours.

▶ Discerning the Skill of Empathic Self-Disclosure

A large majority (93.3%) use self-disclosure. More specifically, over half tell clients that they are angry with them (89.7%), cry in the presence of a client (56.5%), and tell clients that they are disappointed in them (51.9%). The most questioned of these was telling clients of disappointment: 56.8% viewed it as unethical or unethical under most circumstances.
—Ken Pope, Barbara Tabachnick, and Patricia Keith-Spiegel

This quote raises issues we all need to consider. While it is clear that sharing yourself, your thoughts, and your feelings can be facilitative, self-disclosure can go too far. If we are truly empathic, our clients will affect us deeply. You may find yourself feeling in your stomach what you imagine the client is experiencing. You may clench your fists in tension. Furthermore, you will often have experiences and feelings that are remembered (and felt in your body) when you hear a client's parallel story.

You may cry in sympathy, but sympathy and feeling sorry for your client are not empathy. Occasional crying is mostly OK, but beware of losing where the client is as you may start focusing on yourself, resulting in ineffective, nonempathic self-disclosure.

Here are five key aspects of self-disclosure:

1. *Listen.* As with all influencing skills, first attend to the client's story carefully and then assess the appropriateness of any self-disclosure you might make. Be sure that it is relevant to the client's issue. Be sure that the client is prepared to hear what you say.
2. *Self-disclose and share briefly.* Immediately return focus to the client (e.g., "How does that relate to you?"), while noting how the self-disclosure is received.
3. *Use "I" statements.* Self-disclosure almost always involves "I" statements or self-reference using the pronouns *I*, *me*, or *my*—or the self-reference may be implied.
4. *Share and describe briefly your thoughts, feelings, or behaviors.* "I can imagine how much pain you feel." "I feel happy to hear you talk of that wonderful experience—it

was a real change!" "My experience of divorce was hurtful too." "I also grew up in an alcoholic family and understand some of the confusion you feel."

5. *Be empathically genuine and use appropriate immediacy and tense.* Are your self-disclosures real or faked? Don't make up things. The most beneficial self-disclosures are usually made in the here and now, the present tense. ("Right now I feel." "I am hurting for you at this moment—I care.") However, variations in tense are used to strengthen or soften the power of a self-disclosure.

 Interactive Exercise: Writing Self-Disclosure Statements. Practice writing effective self-disclosures.

▶ Refining Empathic Self-Disclosure

Self-disclosure is one way the therapist authentically represents her- or himself in the therapeutic relationship to foster relational movement and growth.
—Mary Tantillo

Can you self-disclose in a way that is meaningful to the client, but not intrusive? Making self-disclosures relevant to the client is a complex task involving the following issues, among others.

GENUINENESS

To demonstrate genuineness, the counselor must first truly and honestly have the feelings, thoughts, or experiences that are shared. Second, self-disclosure must be genuine and appropriate in relation to the client. For example, if you are working with a client who grew up in an alcoholic family and you have had experience in your own family with alcohol, a brief sharing of your own story can be helpful. The danger of storytelling, of course, is that you can end up spending too much time on your own issues and neglect the client. Here the "1-2-3" influencing pattern outlined in Box 11.1 is particularly important.

| **BOX 11.1** | The 1-2-3 Pattern of Listening, Influencing, and Observing Client Reaction |

1. Listen
Use attending, observation, and listening skills to discover the client's view of the world. How does the client see, hear, feel, and represent the world through "I" statements and key descriptors for content (paraphrasing), feelings (reflection of feeling), and meaning (encouragers and reflection of meaning)?

2. Assess and Influence
An influencing skill is best used *after* you hear and understand the client's story. Timing is central—when is the client ready to hear or learn a new way to think

about what has been said? An influencing skill such as feedback, logical consequences, or a directive can be an abrupt change from the listening style unless the client is ready.

3. Check Out and Observe Client Response
Use a checkout as you offer an influencing skill ("How does that seem to you?"), then listen and observe carefully. If client verbal or nonverbal behavior seems incongruent or conflicted, return to the use of listening skills. Influencing skills can remove you from being totally "with" clients.

IMMEDIACY

The following examples show how the use of *here and now* brings immediacy to the session. This may be compared to the *there and then* of the past and future tenses. But recall that all three approaches will be useful in the session—we need to know some past, and we also need to anticipate the future. Your self-disclosure or feedback, given in the here and now, can show clients that you are with them in the moment.

Madison:	I am feeling really angry about the way the way I'm treated by men in power positions.
Onawumi:	(present tense) You're coming across as really angry right now. I like that you finally are in touch with your feelings.
Onawumi:	(past tense) It makes me angry too. I've had the same difficulty. I recall when I would just sit there and take it.
Onawumi:	(future tense) This awareness of emotion can help us all be more in touch in the future. I know it will continue to aid me.

Be careful when clients say, "What would you do if you were in my place?" Clients will sometimes ask you directly for opinions and advice on what you think they should do. "What do you think I should major in?" "If you were me, would you leave this relationship?" "Should I indeed have and keep the child?" Effective self-disclosure and advice can potentially be helpful, but it is *not* the first thing you need to do. Your task is to help clients make their own decisions. Involving yourself too early can foster dependency and lead the client in the wrong way. Note the following exchange.

Madison:	How do you think I ought to tell my boss what I think of him hitting on me?
Onawumi:	I'm not in your position, and I haven't heard too much about your boss yet nor enough about the job. First, could we explore your relationship in more detail? (past tense)
Onawumi:	(If you feel forced to share your thoughts when you prefer not to, keep your comments brief and ask the client for her or his reflections.) I sense your anger right now, Madison (here and now, present tense). My own thought would be to share with the boss what he is doing. He may not be fully aware of how he's affecting you. But I'm not you. He may not hear that. What might be the outcome if you told him? (future tense)
Onawumi:	(after drawing out more information on the relationship) From what I've heard, it sounds wise to bring up what's going on at a safe time. I think it is important to bring it up directly, and I admire the way you are thinking ahead. How does that sound to you? (future tense moving to the here and now as the issue is explored further)

TIMELINESS

If a client is talking smoothly about something, counselor self-disclosure is not necessary. However, if the client seems to want to talk about a topic but is having trouble, a slight leading self-disclosure by the counselor may be helpful. Too deep and involved a self-disclosure may frighten or distance the client.

CULTURAL IMPLICATIONS AND EXPLORING DIFFERENCES

You cannot expect to have experienced all that your clients bring to you. After an appropriate time for introducing the session, when a man works with a woman, for example, it may be useful to say, "Men don't always understand women's issues. The things you are talking about clearly relate to gender experience. I'll do my best, but if I miss something, let me know. Do you have any questions for me?" If you are White or African American working with a person of the other race, frank disclosure that you recognize the differences in cultures at the initiation of the session can be helpful in developing trust (see Box 11.2). Also, encourage the client to ask you further questions at any point in the interview.

When you are culturally different from your client, self-disclosure requires more forethought, and it is vital that you are personally comfortable with differences. Open discussion and some self-disclosure on your part may be essential. The frank disclosure of differences can be helpful in establishing rapport. Many alcoholics are dubious about the

---

BOX 11.2 National and International Perspectives on Counseling Skills

When Is Self-Disclosure Appropriate?
Weijun Zhang

My good friend Carol, a European American, has had lots of experience counseling minority clients. She once told me that one of the first questions she asks her minority clients is "Do you have any questions to ask me?" which often results in a lot of self-disclosure on her part. She would answer questions not only about her attitudes toward racism, sexism, religion, and so forth, but also about her physical health and family problems. During the initial interview, as much as half of the time available could be spent on her self-disclosure.

"But is so much self-disclosure appropriate?" asked a fellow student in class after I mentioned Carol's experience.

"Absolutely," I replied. We know that many minority clients come to counseling with suspicion. They tend to regard the counselor as a secret agent of society and doubt whether the counselor can really help them. Some even fear that the information they disclose might be used against them. Some questions they often have in mind about counselors are "Where are you coming from?" "What makes you different from those racists I have encountered?" and "Do you really understand what it means to be a minority person in this society?" If you think about how widespread racism is, you might consider these questions legitimate and healthy. And unless these questions are properly answered, which requires a considerable amount of counselor self-disclosure, it is hard to expect most minority clients to trust and open up willingly.

Some cultural values held by minority clients necessitate self-disclosure from the counselor, too. Asians, for example, have a long tradition of not telling personal and family matters to "strangers" or "outsiders" in order to avoid "losing face." Thus, relative to European Americans, we tend to reveal much less of ourselves in public, especially our inner experience. A mainstream counselor may well regard openness in disclosing as a criterion for judging a person's mental health, treating those who do not display this quality as "guarded," "passive," or "paranoid." However, traditional Asians believe that the more self-disclosure you make to a stranger, the less mature and wise you are. I have learned that many Hispanics and Native Americans feel the same way.

Because counseling cannot proceed without some revelation of intimate details of a client's life, what can we do about these clients who are not accustomed to self-disclosure? I have found that the most effective way is not to preach or to ask, but to model. Self-disclosure begets self-disclosure. We can't expect our minority clients to do well what we are not doing ourselves in the counseling relationship, can we?

According to the guidance found in most counseling textbooks here, much of self-disclosure by the counselor is considered unprofessional; but if we are truly aware of the different cultural styles, it seems that a different approach may be needed.

The classmate who first questioned the practice asked with a smile, "Why are you so eloquent on this topic?" I said, "Perhaps it is because I have learned this not just from textbooks, but mainly from my own experience as both a counselor and a minority person."

---

ability of nonalcoholics to understand what is occurring for them. A client with cancer or heart disease may feel that no one can really understand if that person has not experienced this particular illness. On the other hand, if you have had a difficult life experience with alcohol, sharing your story briefly can be very helpful. It often helps to say to the client, "Please feel free to ask questions about me if you wish." Just remember to keep your responses brief! Be sure that what you do or do not do is authentic to your own knowledge and comfort level, but also be sure that the self-disclosure is reasonable and comfortable for the client.

Ziv-Beiman (2013, p. 65) summarizes relevant research findings regarding self-disclosure as follows:

1. Counselor or interviewer self-disclosure is widely used.
2. Its value is conceptualized differently according to a range of theories.
3. Immediate disclosure is strongly established as beneficial to the therapeutic alliance and serves to advance a range of therapeutic goals.
4. Despite being theoretically conceptualized and frequently evaluated as useful by patients, nonimmediate counselor self-disclosure remains a controversial technique.
5. Moderate use of self-disclosure demonstrably contributes to the working alliance.
6. Additional research is required to deepen our understanding of the value of feedback

▶ Defining Empathic Feedback

To see ourselves as others see us,
To hear how others hear us,
And to be touched as we touch others . . .
These are the goals of effective feedback.

—Allen Ivey

When you use feedback skillfully as described below, you can anticipate the client's response.

EMPATHIC FEEDBACK	ANTICIPATED CLIENT RESPONSE
Present clients with clear, nonjudgmental information on how the counselor believes they are thinking, feeling, or behaving and how significant others may view them or their performance.	Clients may improve or change their thoughts, feelings, and behaviors based on the interviewers feedback.

▶ Discerning the Skill of Empathic Feedback

Feedback is the breakfast of champions.

—Kenneth Blanchard

Immediacy, genuineness, timeliness, and cultural awareness are critical aspects of effective empathic feedback. These concepts, described in the previous section, have important contributions if you seek to influence the client positively.

Knowing how others see us is a powerful dimension in human change, and it is most helpful if the client solicits feedback. Feedback is an important influencing strategy if you have developed good rapport and enough experience with the client to know that he or she trusts you. The following feedback guidelines are critical.

1. *The client receiving feedback should be in charge.* Listen first to determine whether the client is ready for feedback. Feedback is more successful if the client solicits it. (Listen—provide feedback—use checkout.)

2. *Feedback is best received when you focus on strengths and/or something the client can do something about.* It is more effective to give feedback on positive qualities and build on strengths. Corrective feedback focuses on areas in which the client can improve thinking, feeling, and behaving. Corrective feedback needs to be about something the client can change, or a situation the client needs to recognize and accept as something that can't be changed.

3. *Feedback needs to be concrete and specific.* "You had two recent arguments with Chris that upset both of you. In each case, I hear you giving in almost immediately. You seem to have a pattern of giving up, even before you have a chance to give your own thoughts. How do you react to that?" Avoid abstractions and generalizations, such as "That seems typical of you." "It would be better if you stopped that way of thinking/behaving."

4. *Feedback is best when it is empathic, relatively nonjudgmental, and interactive. A central dimension of empathy is to be with the client in a nonjudgmental fashion.* Stick to the facts and specifics. Facts are friendly; judgments may or may not be. Demonstrate your nonjudgmental attitude through your vocal qualities and body language. "I do see you trying very hard. You have a real desire to accept the way Chris is and learn to live with what you can't change. How does that sound?"—not "You give in too easily, I wish you'd try harder" or the all-too-common "That was a *good* job."

5. *Feedback should be lean and precise.* Don't overwhelm the client; keep corrective feedback brief. Most of us can hear only so much and can change only one thing at a time. Select one or two things for feedback, and save the rest for later.

6. *Check out how your feedback was received.* Involve clients in feedback through the checkout. Their response indicates whether you were heard and how useful your feedback was. "How do you react to that?" "Does that sound close?" "What does that feedback mean to you?"

 Interactive Exercise: Personal Experience With Feedback. An exercise to help you reflect on your personal experience receiving feedback.

▶ Refining the Skill of Empathic Feedback

True intuitive expertise is learned from prolonged experience with good feedback on mistakes.

—Daniel Kahneman, Nobel Prize Winner

Everyone has a story that makes me stronger. I know that the work I do is important and I enjoy it, but it is nice to hear the feedback of what we do to inspire others.

—Richard Simmons

Authenticity and sensitivity to the here-and-now needs of the client is central. Authentic feedback can be negative, which provides little help. Or it can be corrective and clarify the situation, as Kahneman suggests. Overall, Simmons is right—feedback, regardless of type, needs to be inspiring and help the client build a stronger, more meaningful story.

Following are some examples of negative, positive, and corrective feedback.

Vague, judgmental, negative feedback	Madison, I really don't think the way you are dealing with your boss is effective. You come across as just letting things go and accepting what's happening.
Concrete, nonjudgmental, positive feedback	Madison, you can do it; I've seen what you can do in other settings. Your ability to be assertive will be important. May I suggest some specific things that might be helpful the next time you face that situation?
Corrective feedback	Your effort was in the right direction. You can do even more if we set up a role-play to practice new ways to cope with these challenges. After that, I can offer more suggestions to help you change behavior. (See Chapter 12 for ideas on directives, psychoeducation, and logical consequences.)

Positive feedback has been described as the "the breakfast of champions." "You can do it, and I'll be there to help." Your positive, concrete feedback helps clients restory their problems and concerns. Whenever possible, find things right about your client. Even when you have to provide challenging feedback, try to include positive assets of the client. Help clients discover their wellness strengths, positive assets, and useful resources.

Corrective feedback is a delicate balance between negative feedback and positive suggestions for the future. When clients need to examine themselves, corrective feedback may need to focus on things they are doing wrong or behavior that may hurt them in the future. Management settings, correctional institutions, and schools and universities often require the counselor to provide corrective feedback in the form of reprimands and certain types of punishments. When you must give negative corrective feedback, keep your vocal tone and body language nonjudgmental and stick to the facts, even though the issues may be painful.

Negative feedback may be necessary when the client has not been willing to hear corrective feedback. For example, in cases of abuse, planned behavior that hurts self or others, and criminal behavior, negative feedback—including the negative consequences the client's actions can bring—is necessary and can be beneficial. It is our responsibility to act in these situations, but listening to the client's point of view, even if the client is a perpetrator, remains important.

Feedback from the client. Most theorists, counselors, and clinicians believe it is wise to check out and ask the client how things are going. Usually, this is done with a question. "We've been talking for some time now. Do you mind if I ask how being here and sharing yourself feels?" "I'd appreciate your taking a moment to tell me about how helpful working with me has been—and where might I have been 'off-base' or missed something?" Your follow-up self-disclosure, of course, depends on what the client has said. And, at the close of a session, ask, "What have we missed today?" or "How did today feel for you? Any suggestions?"

Remember your own positive and negative experiences with feedback. These may have involved feedback from friends, family, teachers, or a work supervisor. What do you notice personally about effective and ineffective feedback? Now that you have read this section of the chapter, what would you do better, or what would you avoid, when giving feedback?

 Interactive Exercise: Writing Feedback Statements. Practice writing positive and negative feedback statements to Alisia.
Group Practice Exercise: Group Practice With Self-Disclosure and Feedback

► Observing Empathic Self-Disclosure and Empathic Feedback

The following is an excerpt from the later stages of another session with Madison in which the counselor, Onawumi, uses self-disclosure and feedback. Madison is speaking of a difficult situation at work, where her supervisor has started "hitting" on her.

INTERVIEWER AND CLIENT CONVERSATION	PROCESS COMMENTS
1. *Onawumi:* Madison, as we've talked these last few sessions, you come across as very able and sure of yourself. In this situation, you are not so sure. Now your boss, Jackson, is hitting on you and you're scared for your job, but still angry at what he is doing. The anger makes a lot of sense.	Onawumi starts here with brief feedback on Madison's strengths, followed by contrasting her abilities with her current fears and anger. Two major discrepancies are confronted here: (1) internal strengths shown in the past, contrasted with current fears and indecision; (2) the conflict with her boss, including her present silence and the need to keep her job. Onawumi also provides feedback on the validity of Madison's anger, giving less attention to the fear. (Additive)
2. *Madison:* That's right, what am I doing? What can I do?	As Madison talked earlier about this issue, it was clear that she was moving between Levels 1 and 2 on the Client Change Scale. At times she denied that anything was happening, but then she was able to partially examine the issue. Finally, she moved to Level 3, showing clear awareness of what was happening.
3. *Onawumi:* You've got considerable strength. Women go through this too often, but something can be done. What occurs to you?	Feedback again on strength, but also provides feedback on other women. She then tosses the decision back to Madison, thus providing both respect and potential empowerment. (Additive)
4. *Madison:* Well, talking with you is helping. I think I see what's going on more clearly. I think I can pull myself together after all. I know that Jackson does like the work I am doing. I think I got that raise because my work is good, not because he had his eye on me.	Madison is drawing on the positive feedback and showing more certainty about herself and her abilities. She has moved to Level 3 on the CCS.
5. *Onawumi:* So you know he respects your work; and I gather from what you say that you are holding an important project together for him and the company. Clearly, you have some leverage in that. Go on, I think you are on the right track.	Paraphrase, feedback. (Additive)
6. *Madison:* I wonder what would happen if I sat down with him and reviewed the present status on the project and how well things are going. He has never directly really forced or embarrassed me, but the hints are really there. Perhaps, after we review the project, I could simply say that I like working with him, but feel that we must keep our relationship on a professional level. I've not been comfortable with a few things he has said, but I still respect him. Would he mind if we kept our focus on the project?	This did not happen this fast; only in the last 15 minutes of the session did Madison show real movement. Here she is at CCS Level 3, but moving toward Level 4.

INTERVIEWER AND CLIENT CONVERSATION	PROCESS COMMENTS
7. *Onawumi:* I'm impressed with your view of the situation. Well done. If he is a reasonable person, that likely could work. You are giving him respect, but still standing up for yourself. Let's try your idea in a role-play. I'll be Jackson, and you go through and practice what you might say.	Positive and appropriate self-disclosure followed by feedback to Madison on her ideas, followed by a directive for testing them out in a role-play (see the next chapter).

The role-play then follows, and Madison clearly summarizes the project and her feelings about the sometimes tense relationship. As she presents the situation, she firmly maintains her "cool," but also shows respect for the boss. She feel good about the way she handled the role-play.

8. *Onawumi:* That was great, Madison. I think that is about what you can try. Let's hope that it works. On the other hand, I can say that I've been in a similar situation, and I tried to do what we just role-played. The boss there just ignored what I said, but he did start leaving me alone. Nonetheless, I wonder if I lost some power. You've got some ideas that might well work, but—what if they don't?	Feedback followed by a self-disclosure. The self-disclosure of a parallel situation in Onawumi's life is accurate, but it can also give Madison some additional fear. Nonetheless, it is also important that Madison explore the possible logical negative consequences of speaking up. (Potentially additive)
9. *Madison:* Oh, I know that he might become angry, but it's good to know that you had to deal with the same thing. Onawumi, women run into this all too often and we have to start standing up for our rights. I think this boss can take it, and he may have to. We just had corporate training on sexual harassment and company policy. He might even appreciate my feedback. I suppose it is possible that he doesn't know what he is doing.	More indication of Level 4 responding. Note that the feedback and self-disclosures have still left Madison in control of decisions and future actions.
10. *Onawumi:* So, what if it doesn't go well?	Open question confronting what might happen. (Potentially additive)
11. *Madison:* Actually, I'm not as scared for my job as I was. I think I'll get through this. I have to stand up. He needs me for this project. And if it doesn't work out, I'm ready to leave—I know that I can find something else.	Level 4 on CCS.
12. *Onawumi:* That makes me feel good that you are able to take that risk. And my sense is that you have the power and wisdom to make it work, no matter what the result. I'll be waiting here to learn what happened. You are doing the right thing for you,	Self-disclosure and feedback.

► Practicing the Skills of Empathic Self-Disclosure and Empathic Feedback

EXERCISE 11.1 What Are Your Thoughts About Self-Disclosure?
How comfortable are you with sharing personal information? What do you think about revealing yourself to your interviewee or client? When do you think it is appropriate, and when is it not? What kind of personal information is more appropriate to share? What would be the purpose of self-disclosure? Write down your thoughts.

Now read the following online articles on self-disclosure:

Ofer Zur. (2011). Self-disclosure & transparency in psychotherapy and counseling. *Zur Institute*. http://www.zurinstitute.com/selfdisclosure1.html#top

Thomas G. Gutheil. (2010). Ethical aspects of self-disclosure in psychotherapy. *Psychiatric Times*. http://www.psychiatrictimes.com/articles/ethical-aspects-self-disclosure-psychotherapy

EXERCISE 11.2 Writing Self-Disclosure Statements
Individually or in pairs, write one effective and one ineffective self-disclosure for each of the following statements that might come from a client.

"I'm feeling very vulnerable talking about my stuff right now. I wonder if you're really understanding my story?"

"I have to do a class presentation, and I feel very afraid speaking in front of my classmates. I don't like this type of course assignment. What should I do?"

"I feel like my boss is hovering over me all the time. You've heard my story. I feel so closed up. Have you ever experienced that?"

(in tears) "My mother died last weekend. You have been so helpful and understanding, but I hurt so much."

"I passed that exam (or got the job, made a new friend, etc.). I was afraid that it wouldn't happen."

EXERCISE 11.3 Learning About Feedback From Your Own Experience
Remember a positive and a negative experience with feedback. These may have involved feedback from friends, family, teachers, or a work supervisor. What do you notice personally about effective and ineffective feedback? Now that you have read this section of the chapter, what would you do better, or what would you avoid, when giving feedback? How do the concepts of genuineness and immediacy relate to the feedback that you received?

EXERCISE 11.4 Writing Feedback Statements
Individually or in pairs, write one effective and one ineffective feedback comment for each of the following statements that might come from a client.

"I'm afraid that I may fail that next exam. I've always been careful not to cheat before, but a friend hacked the professor and now has the exam that he is willing to share with me."

"As you know, I've never had sex, but this (boy/girl) is so attractive and nice. I think our relationship is turning into something important. What should I do?"

Summarizing Key Points of Self-Disclosure and Feedback: Relationship, Immediacy, and Genuineness

Defining Empathic Self-Disclosure

▶ Self-disclosures are basically "I" statements made by the interviewer to share her or his own personal experience, thoughts, or feelings. Here-and-now self-disclosures tend to be the most powerful.

▶ Self-disclosure is ever present and takes one of three forms: deliberate, unavoidable, and accidental.

▶ When used appropriately and empathically, self-disclosure encourages more in-depth client disclosure, may result in a more egalitarian relationship in the session, and may help the client find a new solution related to the counselor's self-disclosure.

Discerning the Skill of Empathic Self-Disclosure

▶ The self-disclosure needs to be relevant to the client's worldview. It needs to be brief, genuine, authentic, and timed appropriately to client needs.

▶ Use the 1-2-3 pattern: First attend to and determine the client's frame of reference; then assess her or his likely reaction before using an influencing skill; finally, check out the client reaction to your use of the skill.

▶ Empathic self-disclosure involves indicating your thoughts and feelings to a client as follows:

1. Listen.
2. Self-disclose and share briefly.
3. Use "I" statements.
4. Briefly describe your thoughts, feelings, or behaviors.
5. Be empathically genuine and use appropriate immediacy and tense.

▶ Self-disclosure tends to be most effective if it is genuine, timely, and phrased in the present tense. Keep your self-disclosure brief. At times, consider sharing short stories from your own life.

Refining Empathic Self-Disclosure

▶ Demonstrate genuineness. Share truly and honestly felt feelings, thoughts, or experiences, as appropriate in relation to the client.

▶ Put clients first. When clients ask you what you would do in their place, your task is to help them make their own decisions. Involving yourself too early can foster dependency and potentially mislead clients.

▶ It is important to discuss multicultural differences openly and with respect. Be culturally intentional and diversity respectful.

Defining Empathic Feedback

▶ Use the 1-2-3 pattern to ensure appropriateness of feedback. Involve the client as much as possible.

▶ Feed back accurate data on how you or others view the client. Remember the following:

1. The client should be in charge.
2. Focus on strengths.
3. Be concrete and specific.
4. Be nonjudgmental.
5. Keep feedback lean and precise.
6. Check out how your feedback was received.

Discerning the Skill of Empathic Feedback

▶ Effective feedback is empathic, timely, genuine, and culturally sensitive.

Refining the Skill of Empathic Feedback

▶ Positive feedback is "the breakfast of champions." Use this skill relatively frequently; it will balance more challenging necessary corrective feedback.

▶ Corrective feedback will be necessary at times. It needs to focus on things that the client can actually change. Seek to avoid criticism. Be specific and clear and, where possible, supplement with client strengths that support the change.

▶ In some cases (such as abusive or criminal behavior), negative feedback—including the negative consequences the client's actions can bring—is necessary and can be beneficial. It is our responsibility to act in these situations. But listening to the client's point of view, even if the client is a perpetrator, remains important. Use this skill sparingly.

▶ Feedback from the client can help guide your interventions, increase relationship satisfaction, and reduce breakdowns. When asking clients how things are going, the use of formal methods of seeking feedback provides the best benefits.

Observing Empathic Self-Disclosure and Empathic Feedback

▶ The case of Madison and her difficult situation at work illustrates how feedback can be used to focus the interview, reduce digressions or avoidance, and confront challenging issues.

Practicing the Skills of Empathic Self-Disclosure and Empathic Feedback

▶ The exercises offered in this section will help you critically think about self-disclosure and practice to increase relevant and effective use of self-disclosure.

▶ Bank on your personal experiences with feedback to guide your understanding of this skill. Practice in a group to get immediate and constructive feedback and master the skill of feedback.

Assess your current level of knowledge and competence as you complete the chapter:

1. **Flashcards:** Use the flashcards to check your understanding of key concepts and facilitate memorization of key information.

2. **Self-Assessment Quiz:** The quiz will help you assess your current knowledge and prepare for course examinations.

3. **Portfolio of Competencies:** Evaluate your present level of competence on the ideas and concepts presented in this chapter using the Self-Evaluation Checklist. Self-assessment of your competencies demonstrates what you can do in the real world.

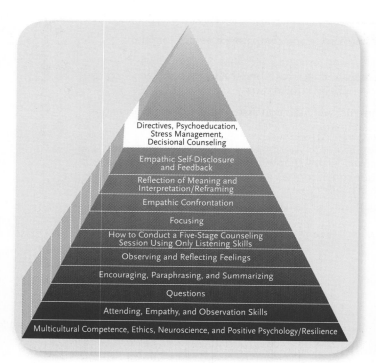

Directives, Psychoeducation,
Stress Management,
Decisional Counseling

Empathic Self-Disclosure
and Feedback

Reflection of Meaning and
Interpretation/Reframing

Empathic Confrontation

Focusing

How to Conduct a Five-Stage Counseling
Session Using Only Listening Skills

Observing and Reflecting Feelings

Encouraging, Paraphrasing, and Summarizing

Questions

Attending, Empathy, and Observation Skills

Multicultural Competence, Ethics, Neuroscience, and Positive Psychology/Resilience

Chapter 12
Influencing Client Actions and Decisions
Directives, Psychoeducation, Logical Consequences, and Decisional Counseling

Blessed is the influence of one true, loving human soul on another.

—George Eliot

This chapter presents four major highly related skill and strategy areas. As such the chapter is organized somewhat differently from most of the others. In particular, please note that the important practice exercises are at the end of the chapter.

You can facilitate clients' examining of new possibilities for behaviors, thoughts, and feelings by providing them with **information** or skill development, thus helping them see the **logical consequences** of decisions to act or not to act.

Chapter Goals

Awareness, knowledge, skills, and actions developed through the concepts of this chapter and this book will enable you to:

▶ Continue your awareness that listening comes first when using any of the influencing skills.

▶ Provide information and psychoeducation to clients in an appropriate and timely manner.

▶ Help clients examine the natural and logical consequences of past, present, and future decisions through decisional counseling.

▶ Utilize several types of directives and psychoeducation in interviewing practice and stress management.

▶ Develop a beginning understanding of how you can share information on the brain with clients.

▶ Defining Directives, Providing Information, and Psychoeducation

> If you lose direction, go to a higher ground.
>
> —Toba Beta

If you listen carefully and fully, many clients can generate their own new ways of thinking, feeling, and behaving. However, most will benefit from your expertise through your use of influencing skills. In this chapter, we turn our focus to several counseling and clinical strategies that provide information leading to new possibilities for change.

When you provide information, give advice, or teach clients skills via psychoeducation, you will need to follow up to see whether the new ideas "took" and were useful. Nonetheless, we can anticipate how the client may react to your use of these skills.

DIRECTIVES, INFORMATION, AND PSYCHOEDUCATION	ANTICIPATED CLIENT RESPONSE
Share specific information with the client—e.g., career information, possible college majors, where to go for community assistance and services. Offer advice or opinions on how to resolve issues and provide useful suggestions for personal change. Through psychoeducation, teach clients specifics that may be useful in a variety of life situations.	If information and ideas are given sparingly and effectively, the client will use them to act in new, more positive ways. Psychoeducation that is provided in a timely way and involves the client in the process can be a powerful motivator for change. You may help clients develop a wellness plan, teach them how to use microskills in interpersonal relationships, and educate them on multicultural issues and discrimination.

PROVIDING INFORMATION AND ADVICE

Be aware that this area is fraught with danger. Unless actively sought, it is difficult for the client to hear even the best of ideas that you provide. For example, try offering teens suggestions on how they should dress, drive, and or handle alcohol. Children or teens often resist any direction or advice. Adults may be reminded to stop smoking or drinking, lose or gain weight, get more exercise, or eat more fruits and vegetables. All of us have difficulty in listening to others who seek to help us change our behavior—or even to learn new things that will make our life better.

When listening to information or receiving advice, the client needs to be in charge and actually desire your feedback. Career and college counselors provide students with needed job or admissions information. Here the teen may actually listen without too much prodding. Students facing critical life decisions frequently want to know your opinions and advice. In family counseling, the family may be caring for an older parent and may want advice on how to handle serious challenges. Lifestyle coaching, executive coaching, and family coaching all require the ability to give skillful advice and direction.

Crisis counselors must give information and advice to flood or hurricane survivors. A family with a child who is hyperactive and/or failing school may ask for advice and direction. Cognitive behavioral counselors and therapists often instruct clients diagnosed as depressed or borderline on the causes of their condition, possible treatments, the value of medications, and how to obtain a favorable outcome from counseling and therapy.

The word **directive** refers to sharing information, giving advice, or teaching skills to clients. While listening is always central, there are many times when sharing your expertise and knowledge will be useful to your clients. The most effective directives follow these microskill specifics:

1. *Listen and observe.* Draw out the client's story; search out what clients have already done to resolve their concerns. How aware are they of possible solutions? Through your listening to clients, they will be better prepared to listen to you.

2. *Use specific, concrete language.* Show that you understand what you are saying. Avoid abstractions.

   Not: "Shane, if you really want to get along, I think you need to slow down and stop complaining about your partner when sex is suggested."

   Better: "Sounds like your partner is not responding to the way you're approaching sex. I've listened to what you've said and I know how you feel hurt and rejected. May I share some specifics that I've observed?" (Client nods OK) "Looks like you are so into it that you are rushing and that may be one reason that you get turned down. What do you think? . . ." The session continues with concrete observations of relationship behavior on the part of both partners. Suggestions for Shane to change are specific, with Shane being encouraged to react and participate in the process of dealing with advice and suggestions.

   Not: "Go out and arrange for the secretary to give you a test."

   Better: "Mahaloni, I'll walk out with you and show you the test and then point specifically how and why it may help you decide on your major (or career)."

   (Working with nonverbals) *Not:* "Relax, don't worry!"

   Better: "Ashi, I sense your tension as you tell me about _____. Could we stop for just a moment? (pause) . . . Close your eyes, take a deep breath and hold it (pause). Breathe out easily. Note the tension in your shoulders. Let's tense them up (pause), now relax." The counselor observes tension and senses when it is time to start the challenging topic again—or continues the brief relaxation.

3. *Check out fully* with the client about how he or she thought and felt. Use your listening skills intentionally to see how the directive, advice, or information was received.

4. *Use the Client Change Scale* to determine if the advice, information, directive, or psychoeducational teaching strategy was useful when you meet for follow-up sessions.

REFERRAL TO COMMUNITY SOURCES

We all need to be aware of resources that support mental and physical health. These can range from a gym to a financial consultant to an appropriate physician. Single parents, cancer survivors, parents of children diagnosed with attentional issues, Alzheimer's caregivers, and many others benefit from referral to support groups. If the client is a veteran, explaining community resources for social, emotional, and financial help may turn out to be the pivotal turning point. For teens, youth groups and activities of all types can be encouraged.

SOME DANGERS OF ADVICE

Advice untethered from a strategy for implementation is merely annoying.

—Joe Queenan

Know when to tune out. If you listen to too much advice, you may wind up making other people's mistakes.

—Ann Landers

Seeking advice you have no intention of following is a time-honored American tradition. Once again, listen first and ensure that your client is interested and ready for new ideas. Does what you share meet their goals and life vision? Are you and they ready for that fifth stage of the interview—action in the real world?

Ann Landers has a point. It is possible that providing information causes clients to rely too much on you and fail to make their own decisions. This can result in dependency or, ultimately, a complete rejection of your ideas. Work with the client's goals and life vision—not your own.

▶ Discerning The Specifics of Psychoeducation

Education is the most powerful weapon which you can use to change the world.

—Nelson Mandela

Clients are often unaware of skills and knowledge that could enlarge their intentionality and abilities. Timely and appropriate psychoeducation in the here and now of the session can help change the client's world.

Psychoeducation is future oriented and seeks to teach clients critical life skills. This is obviously information giving, but it is structured, planned, and oriented to teaching. What distinguishes psychoeducation from directives, information, and advice is that these are relatively short comments in the interview. Psychoeducation uses these same skills but integrates listening and many other influencing skills as well. Introducing therapeutic lifestyle changes (TLCs) to your clients is a form of education, drawing on wellness and positive psychology traditions.

Psychoeducation sets specific objectives in connection with client needs and desires. Not only can we provide useful information for clients, we can also teach them skills of daily living, ranging from how to be more assertive to role-plays in which interactions with friends and family are evaluated and improved. Psychoeducation teaches social skills, how to plan for a college application, decision making, developing a better understanding of how culture and cultural identify affect one's life, and many other life skills. Team members in a mental health facility teach those diagnosed with posttraumatic stress, schizophrenia, or depression how to understand their diagnosis and what they can do on their own and with family to live more effectively. Many of the skills of psychoeducation discussed here are used daily in psychotherapy with the severely distressed.

One specific form of psychoeducation is called life skills training; it can involve role-playing and enactment of scenes from the client's life or situations anticipated in the future. This type of psychoeducation may involve several interviews, planned homework, and readings; it is also frequently found in group work. Life skills training is actually the foundation of the microskills approach. Teaching clients the same skills of listening that you have learned has been a foundation of the model since its inception.

An example of the psychoeducational microskills is the teaching of attending behavior and listening skills to clients, thus enabling better interpersonal communication. During the Vietnam War, Allen used listening training with inpatients in a VA hospital. Particularly important here was videotaping the client in a role-play, which was then debriefed, after which the clients themselves chose behaviors that they wished to change. This, coupled with the TLC of relaxation training, enabled them to leave the locked ward for home (Ivey, 1973).

Teaching communication skills is particularly effective with clients who face conflict with their partner, family, or friends. They may have mild depression and find it difficult to speak to others. It is a good social skills training method for all types of clients and patients. You will also find the psychoeducation approach of microskills taught widely in training managers, physicians, agricultural extension officers, ministers and priests, librarians, and even police departments and the CIA, who use them as base interrogation strategies. Obviously, value issues come into play when you teach these skills to certain groups.

Teaching communication skills is similar to what you have encountered in your own practice sessions. The specific psychoeducation steps include the following:

1. Check out your clients' interest in learning listening skills.
2. Role-play ineffective listening, or a real situation that they have encountered.
3. Debrief the specific behaviors that the two of you have observed. It is wise to use the client's name for the behaviors that they select as needing change.
4. Provide information on the content of the attending behavior (visual contact, verbal following, vocal tone, and body language). Or focus on one or two skills, perhaps different, that the client has selected.
5. Role-play positive communication using good listening skills.
6. Work with the client to take action in the real world.

Clients often benefit from role-playing a variety of challenging issues in their lives. A situation with a difficult supervisor, for example, may be assisted through a role-play. The client tells you the general nature of the supervisor's behavior. You play the difficult supervisor, while the client takes her or his usual role. Through this, both you and the client gain a better understanding of what is going on. Then the two of you can discuss alternative behaviors so that the client can deal with the situation more effectively. Then, further role-plays are tested until the client has a good sense of how the situation should be played out. Couple or family conflict, nervousness about a presentation, preparing for a job interview, and many other issues benefit from this approach. Of course, video or audio recording on a computer camera, the client's cell phone, or your own office video equipment makes the process more effective.

Therapeutic lifestyle changes (TLCs) are becoming an important psychoeducational force leading toward mental and physical health. The discussion in Chapter 2 on positive psychology provides a list of key TLC's that need to be integrated into the practice of interviewing, counseling, and psychotherapy. Psychoeducation can help clients develop elements of a healthy lifestyle such as good nutrition, exercise, and meditation.

Provide clients with clear information on the health gains that will result—along with information on how their brain will change with the development of new neural networks. More challenging is a student not doing well in school or an office worker who shows up late for work. Both know that change would be wise and what your advice is likely to be. It is all the more important to hear their stories and points of view before attempting any psychoeducation. Change your approach when you see clients roll their eyes, slump back in their chair, or look at the ceiling.

Never give up on any client. Go with their resistance, listen to them, be supportive, and intentionally search for different ways to reach healthy objectives.

► Defining Natural and Logical Consequences and Decisional Counseling

Not to decide is to decide.

—Harvey Cox

Decisions require creativity, a disciplined freedom, an openness to change, and the capacity to envision the potential consequences of given actions. There is an old Zen fable that goes something like this, updated for today.

> A woman is hiking along a California Sierra trail along the edge of a 15-foot drop. As she rounds a bend, she sees a bear, which starts to charge. Surprised but still able, she grabs a wild vine and swings over the edge. As she thankfully hangs and looks for a safe place to jump, she sees another bear below! There are summer strawberries growing on the vine so she decides to hold on with one hand and reaches for a few berries with the other. How sweet they taste!

Your clients face bears of decisions. One bear promises one thing, while the other may bring something else. We can help clients taste the sweetness of strawberries and the importance of the moment before they jump. Let us hope that their decisions are friendlier than between bears.

THE DECISIONAL STRATEGY OF NATURAL AND LOGICAL CONSEQUENCES

Natural and logical consequences are key to making decisions at the restorying phase of the interview. Like the five-stage interview structure, this strategy is helpful in many theoretical approaches. When you use this strategy, you can anticipate how clients may respond and use that response to help them explore the results of their actions in more detail.

LOGICAL CONSEQUENCES	ANTICIPATED CLIENT RESPONSE
Explore specific alternatives with the client and the concrete positive and negative consequences that would logically follow from each one. "If you do . . ., then . . . will possibly result."	Clients will become less confused and frustrated as they change thoughts, feelings, and behaviors through better anticipation of the consequences of their actions. When you explore the positives and negatives of each possibility, clients will be more involved in the process of decision making.

For most decisions, the various alternatives are apt to have both negative and positive consequences. Potential **negative consequences** of changing jobs, for example, could include leaving a smoothly functioning and friendly workgroup, disrupting long-term friendships, moving children to a new school, and other factors that might result in major difficulties. **Positive consequences** might be a pay raise and the opportunity for further advancement, a better school system, and money for a new home. Decisions often mean leaving the known for an unknown future.

Cognitions are not enough for effective decision making. Becoming aware of natural and logical consequence is first a cognitive reality issue. However, emotions are critical in making decisions. The decision may be cognitively correct, make sense, and even be wise, but the client needs to feel emotional satisfaction or a sense of peace to move forward. There are many examples of tough but wise choices, such as accepting a new job in a new city with its positive consequences, but also with the negative consequences of major up-rooting and leaving behind friends and family. Leaving an abusive partner will make the client safe, but it also has the logical consequence of going it alone and facing financial hardship.

Many good decisions may have negative natural consequences that we need to prepare our clients to face. Similarly, the client may be heading toward a decision that will produce long-term and serious issues, such as continuing alcohol or drug use. A client may be deciding whether to agree to meet that "fascinating person" with the "great" photo and story whom the client has met online. With both the wise and unwise decisions our clients make, we need to help them anticipate how they may feel *emotionally* as a consequence of the decision and work toward being able to live with the anticipated results.

For disciplinary issues or when the client has been required to come to the session, it is important to note that even more power rests with the interviewer. The school, agency, or court may ask the interviewer to recommend actions that the legal system can take. The use of natural and logical consequences may require you to be more firm and direct. Warnings are a form of logical consequences and may center on anticipation of punishment; if used effectively and coupled with client rapport and listening, warnings may reduce dangerous risk taking and produce desired behavior.

DECISIONAL COUNSELING METHODS

If you don't know where you are going, you may end up somewhere else.
—Kenneth Blanchard

Underneath almost every interview you hold will be decisions that the client wishes to make. Often discussions about cognitive decisions may start with the client talking about confusion and/or frustration. A good cognitive decision will not hold unless emotions are worked through. Settling for confusion or frustration as the emotional base is often not enough. You will want to encourage emotional expression of deeper feelings, as relevant to the issues and goals of the session.

It is the rare human behavior that does not have its costs and benefits. By involving the client in examining the pluses and minuses of alternatives, the interviewer gives the decision to the client, or at least shares it more openly. Consider the following suggestions for using the decisional strategy of logical consequences.

1. *Listen and ensure that client goals are clear.* As the restory develops, however, the goals may change.
2. *Generate alternatives leading to a new story.* Use questions and **brainstorming** to help the client generate alternatives for resolving issues. Where useful, provide your own additional alternatives for consideration.
3. *Identify positive and negative natural and logical consequences.* Box 12.1 presents one way to diagram and clarify issues. Also ask the client to generate a story of what the future might be like if a particular choice is made. For example, "Imagine yourself two years from now. What would your life be like if you chose the alternative we just discussed?" In anticipating the future, special attention needs to be paid to likely emotional results of decisions. Emotions are often the ultimate "decider."
4. *Provide a summary.* As appropriate to the situation, provide the client with a nonjudgmental summary of positive and negative natural consequences. Consider asking the client to make the summary.
5. *Encourage client decision making and action.* Work collaboratively with the client to reach a decision and to develop an action plan for implementation.

BOX 12.1 The Cognitive and Emotional Balance Sheet

List below the positive and negative factual and emotional results for each of the possible alternatives. If there is more than one alternative, make a separate Cognitive and Emotional Balance Sheet for each one.

What is the decision? What happens if I leave my abusing partner?

What are the possible positive gains for me?	What are the emotional gains for me?	What are the possible positive gains for others?	What are the emotional gains for others?
Abuse will stop and I won't get hurt.	I won't be so scared.	My mom won't have to talk to me on the phone constantly.	Mom will be so relieved that it's over.
I'll be able to move on with my life.	Perhaps I can return to feeling good about myself.	My mom would like to help.	She'd feel that she is important to me again.
I can be myself.	I used to feel OK, and that would be a relief.		

What are the possible losses for me?	What are the emotional losses I might face?	What are the possible losses for others?	What are the emotional losses for others?
I'll be on my own.	This frightens me as much as staying.	None that I can think of.	Again, none that I can think of. They'll be happy to see him gone.
How can I finance things by myself?	This terrifies me.	I'll have to go home. My parents may have to support me for a while.	They aren't that well off, and they told me not to go out with him. They may be angry, even though they'll help.
I still love that man, despite it all.	I'll be lonely.	My friends will be there for me.	They'll be glad for me and listen.
He might follow me, and that might make it worse.	I'll have no future and be totally alone.	My counselor is there to advise and support me.	I can sense that I'm not as alone as I might think I am. I feel supported and cared for.

Source: Adapted from Leon Mann (Mann, 2001; Mann, Beswick, Allouche, & Ivey, 1989); also see Miller & Rollnick, 2002.

Decisions could be called the most basic issue in interviewing, counseling, and therapy, as all clients will be making decisions whether they decide by themselves or engage in any of the more than 500 theoretical approaches to helping. **Decisional counseling** underlies the five stages of the interview and involves the skills in which you have already become competent. For example:

Empathic relationship. Structuring the decisional counseling interview entails developing a working alliance and sharing what the client might expect in the session. Drawing out the client's preliminary goals for the session gives the working alliance purpose.

Story and strengths. Listen and draw out the background of client issues and decisional concerns. Balance the concerns with stories of strengths and resources. Where have they succeeded in the past, and how can we draw on those positive assets for resolution?

Goals. Spend sufficient time to elaborate and clarify the preliminary goals. Check to see how committed your client is to reaching the goals.

Restory. One way to approach restory is to brainstorm multiple routes toward resolution. Here you may want to carefully provide information, advice, and/or psychoeducation. If the client does not have at least three alternatives for action, continue the search, although doing nothing—not deciding—is always one possibility of the three. An important part of restorying is helping the client to face the natural and logical consequences of decisions, as well as reviewing the natural and logical consequences of past failed and successful decisions. The final step of restorying is actually making a decision and being ready with alternative paths if the decision does not work out as hoped and anticipated.

Action. Commitment to do something immediately is crucial. Can the client do something today or tomorrow to start implementation of the new program or decision? This may involve a single action or a planned program of change, which could involve further counseling sessions, psychoeducation, and/or referral to self-help books, support groups, a gym, or a variety of activities available in the community.

Decisional counseling is sometimes called "problem-solving therapy" (Haley, 1997; Nezu, Nezu, & D'Zurilla, 2012). With the word *problem*, we assume that "the problem will be solved." Decisional counseling views it differently. Yes, the concern, issue, or challenge will be resolved, but we also know that new decisions and challenges await us around the corner. If you and your client are "armed" with this basic framework, you both are ready to engage with many issues over time. Decisional counseling is not just for the moment—it teaches a lifelong process.

Allen and Mary present a complete transcript of a decisional interview on the website. This is accompanied by a pre-interview plan and a discussion of the session, looking at the structure of the interview as well as use of microskills.

Clients come to us with pieces of their lives literally "all over the place." Decision making is a lifelong issue, and decisional counseling's method provides a lifelong practical approach that clients can use to solve later life issues and make decisions. Young adults must choose a college or a career or decide whether to continue a relationship, get married, or have a child. Later they will make decisions about how to succeed in a work setting, deal with difficult colleagues, and plan finances for their children's education and their own retirement. They will find that decisions don't end with retirement; the first question faced by many retirees is "What shall I do with all this time?" Difficult decisions around health issues, wills, and plans for their own funeral often require counseling.

COGNITIVE AND EMOTIONAL BALANCING

The **cognitive and emotional balance sheet**, created by the Australian Leon Mann, clarifies natural and logical consequences. A major strength of the balance sheet is that the client is asked to think through the implications of gains and losses, cognitively and emotionally, for other individuals. Writing down the multiple aspects of a decision clarifies issues and helps ensure that the most critical matters are considered. As an example, Box 12.1 shows a decisional balance sheet created with a woman who has experienced abuse. Notice how powerful the arguments are for staying in an abusive relationship. This is why your support in such cases is significant.

As another example, we can help substance abusers use the emotional balance sheet to look at issues around drinking or using drugs. Adding focus concepts to the balance sheet helps alcoholics see the broader implications of their drinking for the lives of others and makes possible the exploration of long-standing emotional issues and the natural and logical consequences of either continuing or stopping substance abuse. Motivational interviewing uses Leon Mann's strategy and his balance sheet with addiction (Miller & Rollnick, 2013).

▶ Refining Stress Management as Psychoeducation

> Treatment is based on managing stress.
>
> —Patrick McGorry, M.D., Ph.D.

> There's good stress, there's tolerable stress, and there's toxic stress.
>
> —Bruce McEwen

Stress management and therapeutic lifestyle changes support the decision-making process. As stated previously, stress underlies most issues that clients bring to us. Thus, understanding and managing the impact of stress is often necessary for effective, rational, and emotionally satisfying decisions.

Regardless of your approach to interviewing and counseling, you will constantly be working with clients who have some form of stress. This is logical when you think about it: How often do you go through a stress-free day? This section defines stress more precisely and discusses the impact of both positive and negative stressors on key parts of the brain.

Chapter 1, page 13, notes that we can help clients detect and manage stress. However, stress remains endemic in our society. Effective counseling can make a significant difference in helping clients anticipate stress, prepare to deal with it, and see it in a total life perspective.

Stress is a part of living; it helps keep us alert and alive. But continuous day-to-day stress, over and over again, can be seriously damaging to physical and mental health. Or a single traumatic incident can accomplish this in a few seconds or minutes. Severe stress leads to many mental and physical health issues. Figure 12.1 shows the impact on the brain of severely stressful situations. Also recall that the body responds to stress in ways that are personally damaging.

We tend to think of stress as related to issues such as an argument with a family member or employer. More serious and long-term damage to the brain can be caused by being mugged in a dangerous alley, being raped, or serving in a war. Smaller incidents may cumulate over time; seemingly small stressors become large through accumulative stress. These may include child neglect, parental pressure, teasing that turns into bullying, or a constant series of put-downs and negations from a loved one.

All of us may experience some form of common life stressors such as unemployment, divorce, illness or the illness of a close family member, death of a parent, financial reversals,

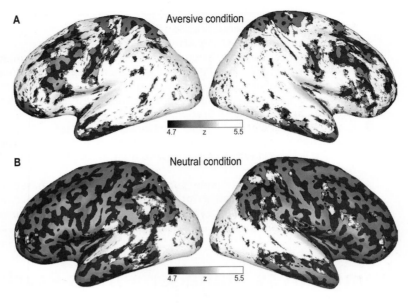

FIGURE 12.1 The brain under aversive stress.

Source: Hermans, E., van Marle, H., Ossewaarde, L., Henckens, A., Qin, S., Kesteren. M.,Schoots, V., Cousijn, H., Rijpkema, M., Oostenveld, R., & Fernández, G. (2012) Stress-related noradrengenic activity prompts large-scale neural network configuration. Science, 334, 1151–1153. Reprinted with permission from AAAS.

a sibling with a mental illness, credit card debt, and many others. All of these stressors have produced the following results:

▶ Forty-three percent of all adults suffer adverse health effects from stress.
▶ Seventy-five percent to 90% of all doctor's office visits are for stress-related ailments and complaints.
▶ Stress can play a part in problems such as headaches, high blood pressure, heart problems, diabetes, skin conditions, asthma, arthritis, depression, and anxiety.
▶ The Occupational Safety and Health Administration (OSHA) declared stress a hazard of the workplace. Stress costs American industry more than $300 billion annually.
▶ The lifetime prevalence of an emotional disorder is more than 50%, often due to chronic, untreated stress reactions (Goldberg, 2012).

As emphasized in Chapter 1, we need some level of stress to get ready for an exam, an athletic event, or a job interview. Stress is necessary for learning. Positive stressors make us happy and joyful in many ways. Examples are planning for a big date or marriage, being deeply involved at an opera or a baseball game, rock climbing, running, or driving fast on a racetrack. Positive stress can provide fulfillment, such as the satisfaction of a job well done, being able to help another person, graduating from a master's degree program, or helping build a Habitat for Humanity house.

Figure 12.2 shows how stress can be either growth producing or, at high levels, destructive. Appropriate levels of stress can "pump us up" to prepare for that exam or other challenge. Normal stress helps the brain grow through neurogenesis, resulting in new neurons and neural connections. But severe stress results in negative neurogenesis, with neural loss and with possible long-term damage to the brain. This loss can occur with one really traumatic event (war, rape, a severe accident) or a continued series of chronic damaging stressors such as bullying, racism, poverty, abuse, neglect, and even so-called "normal" stress in the workplace.

The skills and strategies of this book are all oriented toward enabling clients to reach their full potential. Almost always you will find that stressors of many types lead to client

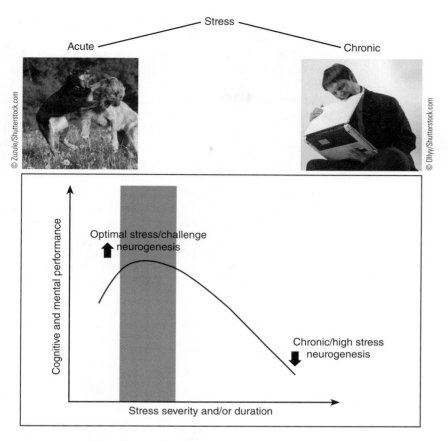

FIGURE 12.2 Optimal levels of stress contrasted with chronic stress.

Source: Mr. Sander's article on research at https://newscenter.berkeley.edu/2013/04/16/researchers-find-out-why-some-stress-is-good-for-you/. Reprinted with permission from Daniela Kaufer, Dept of Integrative Biology, University of California, Berkeley.

issues. Our first avenue to helping our clients deal with stress is to listen to their stories carefully and fully. Usually, there is a logical reason that they are reacting or behaving as they are. Underlying academic failure, family conflict, depression, and even physical illness is stress.

EXERCISE 12.1 Experiencing the Impact of a Simple Stressor

With a partner: Try this exercise with a friend, taking turns as leader and follower. Have your partner close his/her eyes. Then say in a slow, even voice, "NO NO NO NO NO NO." The partner then opens his/her eyes and you debrief the experience. How was the word "NO" felt in the body? What thoughts and feelings occurred for your partner? You will find that repetition of the negative word "NO" has an almost immediate impact.

Now ask your partner to close his/her eyes again. In a slow, even voice, say, "YES YES YES YES YES YES." Open the eyes and debrief this experience, comparing it with "NO." Typically, it takes several "YES's" to relax a person from the multiple "NO's."

On your own: Close your eyes and visualize some error or difficult experience from the past, along with the word "NO." Take a moment to know both body and mind. Then follow this with a positive, joyful memory. Take some time to debrief.

This exercise provides some understanding of what even the simplest stressor can do to the body. Negative stressors over time produce negative neurogenesis.

A child or adult who receives a negative comment or goes through a difficult personal experience easily becomes stressed. These negative events have an impact on the brain and imprint memory at a deeper level than most positives. Some say that it takes 5 to 10 positives to counteract one negative comment. If there is a trauma, even 10 will not be enough, and the negative memory may take over one's life. A single negative—to a child "You're fat and a freak" or "We don't want to play with you"; or something as simple as a friend saying with a quizzical expression, "Your hair looks different today"—can ruin the whole day or more, no matter what else happens. Any number of direct or subtle put-downs that we all experience build a negative self-concept.

It has been suggested that clients who have a negative view of life should start a journal and note both positive and negative events. Usually, they are surprised at how many positives have occurred without their full attention. Similarly, when we anticipate a negative experience, we can protect ourselves partially by deliberately planning at least three positives during the day before it happens. If the negative experience is anticipated, rather than an unpleasant surprise, the impact will be less powerful and damaging.

Critical for us to understand in stress management is the role of hormones in the *stress response system* (Figure 12.3) associated with the emotional limbic system: the hypothalamus, pituitary, and adrenals (HPA; also see Appendix C). HPA hormones are activated by the energizing amygdala, and in times of emergency, fear, or anger they may bypass prefrontal

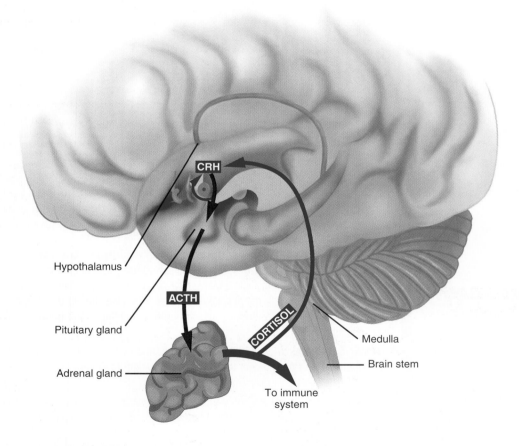

FIGURE 12.3 The stress response system.

executive functioning. This protects us when we have no time to think and make decisions—when we see a snake, a fist coming at us, or a car swerving directly into our lane. We need the fast-acting limbic and stress systems or we as a human species would not be here.

The **hypothalamus**, the H of HPA, is the master gland controlling hormones that affect heart rate, biological body responses to emotions such as hunger, sleep, aggression, and other biological factors. The **pituitary** is another control gland that receives messages from the hypothalamus and influences growth, blood pressure, sexual functioning, the thyroid, and metabolism. The **adrenal glands** produce *corticosteroids*, including *cortisol*.

Under significant negative emotional stress of many kinds, cortisol is potentially damaging. High cortisol levels from prolonged or chronic stress produce side effects such as cognitive and memory impairment, increased blood pressure, blood sugar imbalance, lowered immunity, and inflammatory responses. However, the normal stressors of a challenge, an examination, or a race produce appropriate levels of cortisol, resulting in positive effects such as improved memory, reduced sensitivity to pain, and increased sustained energy (Lee & Hopkins, 2009).

▶ Refining Directives and Psychoeducational Strategies for Building Mental and Physical Health

> Directives and stress management strategies are preferred modes of treatment, useful for the majority of your clients. They also serves an important PREVENTIVE function, thus enabling the client to make better decisions and cope more effectively with present and future stressors.
>
> —Mary Bradford Ivey

The directive strategies discussed here are designed to be used with virtually all theoretical approaches. They can be called on when you know or suspect that a particular intervention will be useful for client growth. While they are also key to stress management, it is important to remember that these can provide the lever for client improvement not only in resolving current issues and concerns but also for developing a resilient lifestyle, the ability to learn from and bounce back from life challenges.

To involve clients in these strategies, help them develop body awareness in the here and now. Note when your own body and mind are tense, where and how it feels. How does your body feel during positive moments, including relaxation and peaceful feelings? How are you breathing? Exercise 12.1 helped you understand the nature of stress and its body impact. The TLCs and the strategies presented in Box 12.2 will help you reduce stress and self-regulate to achieve a more relaxed and effective mode of functioning.

BOX 12.2	Psychoeducational Strategies

These example strategies are presented in very brief form. With further study and some imagination and practice—and client participation in the process—you can successfully use many of them. For more detailed presentations of these and other strategies in highly concrete form, see Ivey, D'Andrea, and Ivey's *Theories of Counseling and Psychotherapy: A Multicultural Perspective* (2012).

Therapeutic lifestyle changes (see pages 39–47)
All of the TLCs are oriented to positive management of stress. Help clients learn and improve mental and physical health through TLCs such as exercise, nutrition, meditation, social relations, cognitive challenge, sleep, social justice action, joy, humor, and zest for living. While sharing the many possibilities, it is best that the client select one or two for action. KEEP IT SIMPLE!

(continued)

BOX 12.2   (continued)

The interview in the following section of this chapter illustrates the use of TLC's within a decisional counseling framework.

Directives, sharing information, and advice

"I suggest you try . . ."

"Vanessa, the next time you go to the garage and they start giving you a bad time, I'd like you to stand at the counter, make direct eye contact, and clearly and firmly tell the manager that you have a meeting at 10:00 and you need prompt service—now! If he says there will be a delay, get him to make a firm time commitment. Then follow up 15 minutes later."

Detail and concreteness are very important when providing a directive.

Spiritual imagery (useful as a TLC)

"You say you gain strength from your spirituality and religion. Could you tell me about an image that comes to your mind related to a spiritual strength?" (Listen to the story and the feelings that go with it.)

"Now, close your eyes and visualize (that symbol, person, experience) and allow it to enfold you completely. Just relax, focus on that image, and note what occurs in your body."

In recent years the counseling and interviewing field has recognized the strengths and power in spirituality and religion. Many clients benefit from spiritual imagery and often find peace in their inner body and strength to move on. Forgiveness of the transgressions and omissions of others can come from spiritual imagery, or your client may find new strengths to deal with a difficult illness or serious loss. A spiritual orientation even helps some clients recover from operations or serious illness.

Role-play enactment

"Now return to that situation and let's play it out."

"Let's role-play it again, only change the one behavior we agreed to."

Role-playing is an especially effective technique to make the abstract concrete. It makes the client behavior clear and specific. This is one of the most basic techniques used in assertiveness training.

Positive reframing (Chapter 10) combined with a directive

"We've identified the problem and how it feels. Now feel that wellness strength in your body. Do it fully, magnify it, and take it to the problem."

Taking real positives to attack problems through the body can be effective. If the positive strength is not able to meet and counteract the negative, add another resource—or just have the strength approach one part of the negative at a time.

Relaxation (like meditation, this is an important TLC)

The simplest way to learn the relaxation response and the basic act is to (1) notice body tension; (2) take a deep breath; and (3) hold it for just a moment and let it go as the body gradually relaxes.

In teaching clients, the following may be useful: "Close your eyes and focus on the moment."

"Tighten your forearm, very tight, now let it go." (Directions for relaxation continue throughout the body parts, ending with full body relaxation.)

Clients may be tight and tense, but once they are able to relax and gain control of their body, they are better able to cope with stressful daily encounters. Teaching the "relaxation response" is an essential skill for all interviewers and counselors (Benson & Proctor, 2010).

Physical exercise and related therapeutic lifestyle changes

A past president of the American Psychiatric Association has stated that any physician who does not recommend exercise to patients is unethical. Interviewing, counseling, and therapy have been very weak in this area. Exercise is a preventive health activity that needs to become part of everyone's practice. Moreover, research is now showing that exercise helps clients deal with stress, which in turn helps with many difficult issues ranging from depression to Alzheimer's (Ratey & Manning, 2014).

A sound body is fundamental to mental health. Beyond encouraging clients to exercise regularly, remind them that proper eating habits and a regime of stretching and meditation can make a significant difference in their lives. Teaching clients how to nourish their bodies is becoming a standard part of counseling. We love and work more effectively if we are comfortable in our bodies.

Imagery focusing on a relaxing scene (useful as a TLC)

This directive may be used with any positive image, person, or situation. All of us have past positive experiences that are important for us—maybe a lakeside or mountain scene, or a snowy setting, or a quiet, special place. The image can become a positive resource to use when we feel challenged or tense. For example, you feel tension in your body when anticipating making a presentation, being interviewed by an employer, or having difficulty going to sleep. When giving a guided imagery directive, time your presentation to your observations of the client.

> Close your eyes and relax. [Pause] Notice your breathing and the general feelings in your body. Focus on that place where you felt safe, comfortable, and relaxed. Allow yourself to enter that scene. What are you seeing? . . . Hearing? . . . Feeling? . . . Allow yourself to enjoy that scene in full relaxation. Notice the

good feelings in your body. Enjoy it now for a moment before coming back to this room. Now, as you come back, notice your breathing [pause] and as you open your eyes, notice the room, the colors, and your surroundings. How was this experience for you?

Thought stopping

"Wind back the tidal wave of the frenetic motion of the mind." (Joseph Ting)

This simple strategy has consistently been found to be one of the most effective interventions we can use. It is useful for all kinds of client problems: perfectionism, excessive culture-based guilt or shame, shyness, and mild depression. Almost everyone engages in internalized negative self-talk. Self-talk is stressful thoughts you say to yourself, perhaps several times a day. For example:

"Why did I do that?"

"I'm always too shy."

"Can't I stop making mistakes?"

"I should have done better."

"Life is so discouraging for me."

"Nobody will listen to me."

The following is the basic process for learning and using thought stopping.

Step 1. Learn the basic process. Relax, close your eyes, and imagine a situation in which you make the negative self-statement. Take your time and let the situation evolve. When the thought comes, observe what happens and how you feel after the negative self-talk. Then tell yourself silently "STOP." If you are alone, put it loudly and firmly.

Step 2. Transfer thought stopping to your daily life. Place a rubber band around your wrist and every time during the day that you find yourself thinking negatively, snap the rubber band and say "STOP!" This simple step almost sounds silly, but it works. (Snapping the rubber band is not a form of punishment but a way to interrupt or interfere with negative thinking. Be kind to yourself; use it with this purpose only.) The client can, of course, just say "Stop," but the rubber band adds extra reinforcement to the effort to change thought patterns.

Step 2. Add positive imaging. Once you have developed some understanding of how often you use negative self-talk, and after you say "STOP" or snap the rubber band, immediately substitute a more positive statement about yourself. You may use positive imagery, or think about an example when you had a positive experience, or use a brief broader statement emphasizing general strengths.

"I can do lots of things right."

"I am lovable and capable."

"I sometimes mess up—no one's perfect."

"I did the best I could."

Journaling

"Alisia, you like to write and think about things. How would it be if you started a journal of your work with me? You might want to reflect on each interview and its impact and how what we discuss relates to what you see happening in your life daily. You can share this with me or not, as you choose."

Keeping a journal is helpful to many clients. This helps them reflect on the interview and its impact on them during the week.

▶ Observing: Introducing Therapeutic Lifestyle Changes Through Decisional Counseling

When Allen asked DNA Noble Prize winner James Watson about his current work in cancer, he replied: Exercise is 30% of preventing cancer, another 30% is not becoming obese (personal communication, February 6, 2014). Thus, diet and nutrition are essential.

Therapeutic lifestyle changes are obviously a "different" way to approach interviewing and counseling. With Dr. Watson's surprisingly strong support, it means that we need to take TLCs rather seriously. But TLCs involve psychoeducation, providing information, and at times even giving advice. Encouraging clients to engage in new behaviors and ways of thinking is never easy.

In the interview that follows, Imani is counseling 48-year-old Bill Jensen, a veteran of the Gulf War who periodically experiences a mild depression. He has been working as a senior mechanic for the last 10 years, was divorced two years ago, shares custody, and sees his two children on weekends. Bill has recently felt more stressed and depressed than usual and has requested help from Imani, a local licensed clinical counselor.

Bill has now seen Imani for two sessions. He has shown some improvement as a result of her empathic listening skills. Now she has decided that therapeutic lifestyle changes would be useful to energize Bill.

Several sources have described exercise as the number one TLC (Ratey & Hagerman, 2013). Exercise not only builds the body, but the increased blood flow increases the possibility of neurogenesis and developing new neural nets. Once a client gets the body moving, it is more difficult to feel down and depressed. In fact, 25 research studies have found that exercise can prevent depression (Bergland, 2013).

Increased oxygen flow makes a difference in mental spirit as well as in developing a more healthy body. At the end of the second interview, Imani explained TLCs to Bill and gave him a handout that expands the discussion you read earlier on pages 37–39 as part of positive psychology and wellness. You may want to photocopy those pages and use them as a handout in your own work.

As he came in for the third session, Bill said that he found the handout interesting and that it "might be helpful" if he did something about TLCs. He said that had stopped exercising two years before, at the time of the divorce, and that he had heard that meditation was "something good."

We enter the session after 10 minutes of reviewing Bill's rapid progress. He and Imani have a solid working alliance, developed over the previous sessions. Bill has identified the general goal of getting out of his depressed rut and doing something. The TLCs are being introduced as part of a larger treatment plan to encourage Bill to get out in the world and find more life satisfaction.

Note that when offering psychoeducation in something like TLCs, at first the interviewer is likely to take much more talk time than usual. Nonetheless, it remains important to involve the client constantly to ensure that what is being suggested is meaningful and might actually be used outside the session.

INTERVIEWER AND CLIENT CONVERSATION	PROCESS COMMENTS
1. *Bill:* Thanks, Imani, I think I've got a handle on spending less time thinking about my problems. (Pause) Let's talk about how I get exercising again. I recall that you said exercise can change and perhaps even strengthen the brain.	He sits up and is clearly ready to move on. We see evidence that her ability to listen to him confront some of his issues has already made a difference.
2. *Imani:* Right Bill, exercise is a real brain builder, not just the body. But first, I'd like to hear more about your past exercise. As I recall, you said that you stopped about two years ago at the time of the divorce.	Imani starts the process by learning a bit more about Bill and his past experience with exercise. The entire session can be classified as "potentially additive" as we won't know how valuable this is until he reports back in later interviews.
3. *Bill:* Yes, I was angry and depressed. And the settlement and having to move out of the house was not fun. I had to take on extra hours at the garage and was lucky to get them. I had no time for exercise. I used to run 2 miles most days, and other days I'd take a walk with my wife. Obviously, that ended.	Divorce is often traumatic and can start a continuing cycle of sadness/anger leading to doing less and thinking about self and one's issues more and more. We want to break into that cycle and move our clients to action, often with less self-reflection.
4. *Imani:* How was running for you? What was the experience?	Open question.

INTERVIEWER AND CLIENT CONVERSATION	PROCESS COMMENTS
5. *Bill:* The first half-mile always was difficult and some-times even painful. But then that "runner's high" seemed to come out and the endorphins were great fun.	
6. *Imani:* It felt real good and it sounds like it was a highlight. I've read that people increase endorphin production, which makes them feel better.	Reflection of feeling, providing information.
7. *Bill:* Amen. I miss that; I was a happier guy when I was running.	
8. *Imani:* I hear that. Sounds like you are ready to start again. You already know how good those endorphins feel, but did you know that 25 research studies have shown that exercise can prevent depression? Just like counseling, exercise changes the brain.	Providing information.
9. *Bill:* No, I thought depression was just in the mind. Tell me more.	

Below we see a longer discussion, which is often part of psychoeducation. Introducing stress management and TLC strategies will (for a short time) result in more interviewer talk.

10. *Imani:* Here I have a picture of the brain. There is too much detail here, but let me share some highlights and how the brain reacts to exercise. (See website for downloadable picture and other information.)

First, notice that word amygdala (points). The amygdala is our energizer, but it is also the seat of those negative emotions you experience. When we are depressed that area is working overtime.

Now look at the prefrontal cortex (points). Easier to say PFC. That's where our thinking and decision making are primarily located. It is also where we manage and control our emotions.

Exercise increases blood flow to the brain and body. When people are depressed, they often stop exercising, they sleep less well, and eat less carefully. This actually increases the depression.

What we want to do is increase the power or the PFC, the top part of the brain, to regulate and control those negative emotions and relieve depression.

How does that sound? (Psychoeducation followed by checkout—potentially additive)

11. *Bill:* OK, I've more or less got it. I think you are tell-ing me that I might be able to control my depression with thinking, but how does that relate to exercise?	When we use influencing skills, the checkout becomes all the more important. It ensures client participation, and the client's response lets us know what is best to do next.

12. *Imani:* Good question. Look again. When we are exercising, several things are happening. Important among them is that we are increasing oxygen and blood flow, not only to the heart but also to the brain. Somehow the stillness of depression almost seems to slow blood flow, or at least it feels like it.

Blood flow to the brain is as important as blood flow is to the heart. In fact, there are important neural connec-tions between the two—and some think that the heart action is what starts the brain moving. I'm not sure of that, but they are certainly connected.

With the increased body action, Bill, we get a positive cycle going. The body feels better, the brain feels better, and likely that is an important part of how exercise fights depression. So . . . (Psychoeducation, potentially additive)

(continued)

INTERVIEWER AND CLIENT CONVERSATION	PROCESS COMMENTS
13. *Bill:* (Interrupts) I get it, and the endorphins come out of that and I feel better. I used to love my runner's high. (Pause and points) So, exercise endorphins likely hit the PFC here and allow it to better regulate or control the negative emotions. Is that right?	Counseling changes the brain, and this can cause clients to take the information you give them more seriously. It also increases the likelihood of clients' doing something outside the interview. Action beyond the session is perhaps the most important part of our work with clients.
14. *Imani:* Yes, exercise enlivens our whole brain. It brings about a better mood. Not only will you feel better with exercise, your brain will operate more effectively.	Summary of the possibilities within the TLC of exercise.
15. *Bill:* Wow, makes exercise a must.	So far Imani's psychoeducational efforts are additive, but that always depends on whether or not something happens after the session.
16. *Imani:* Great, learning how the brain responds to exercise makes you want to return to it even more. Well, let's discuss how you want to start.	Positive feedback and an open invitation to discuss the matter or other topics further.
17. *Bill:* Yes, I'm ready. I do want to learn about meditation, but I think what you told me already is more than enough. I'll start tomorrow with just a mile. I'll walk one telephone pole, and then jog the next. That's enough to start.	Bill gets very specific as to his plans to do something different. Not too many clients will be this enthusiastic, so be ready to work with them for a realistic step-by-step action plan.

▶ Practicing: Integrating the Skills and Strategies of This Chapter

EXERCISE 12.2 Directives and Providing Information

▶ Reflect on your past thoughts and feelings when a teacher, supervisor, or family member told you what they thought you should do.

▶ Close your eyes and recall a situation that did not work out well, perhaps left you angry and frustrated. What behaviors do you recall that led to your resistance? Observe your body for emotions that still may be there.

▶ Close your eyes again. Find a situation in which someone gave you directions on what to do and provided useful information that was facilitative and helped you. What did they do right? What were their behaviors?

EXERCISE 12.3 Natural and Logical Consequences

▶ Identify a decision that you wish to make, one that is important to you.

▶ Fill out the Cognitive and Emotional Balance Sheet (Box 12.1) as it relates to this decision.

▶ With this experience, take a volunteer client through the same exercise.

EXERCISE 12.4 Stress and Your Own Life Experience

▶ Recall when you have been stressed and it felt damaging to you. It may help to close your eyes and seek a specific image of the situation and play it out like a movie. What happened to your body?

▶ What stress management strategy or skill learned in this book might be useful? Spend some time thinking through how it might be helpful.

▶ Can acute stress ever be useful? Can you anticipate how you could handle other stressful situations effectively?

▶ It is agreed that a moderate amount of stress is necessary for learning; otherwise, we are not motivated. Think of a situation in your life to which this applies. It could be academics, a speaking opportunity, athletics, dance, or another situation in which you are expected to perform. Again, close your eyes and visualize like a movie how well you handled this event. How does your body feel with the accomplishment? What goes on in your mind?

▶ Just as we must stress muscles if they are to strengthen, do you agree that some level of stress is necessary for effective learning and living?

EXERCISE 12.5 Stress and a Volunteer Client's Life Experience
With a volunteer client, work through one or more parts of Exercise 12.4.

EXERCISE 12.6 Role-Playing an Interview Providing Directives and/or Psychoeducation

▶ With a volunteer, jointly select a strategy to use, perhaps a TLC or one of the other strategies in Box 12.2.

▶ As preparation for using this strategy, ensure that you establish an empathic relationship, that you hear the story or issue that led to the volunteer's interest in that strategy, and that the two of you agree on a goal for the session (stages 1, 2, and 3).

▶ Sharing information, using directives, or engaging in psychoeducation takes you immediately to stage 4, restory. Share the directive and/or teach the specific lifestyle that was agreed to earlier.

▶ Work with the volunteer client on action. Will he or she actually try what the two of you have discussed?

▶ Obtain feedback on the interview.

For your practice exercises in influencing skills, you may find it helpful to use the Microskills Interview Analysis Form (MIAF) (see Chapter 7, Box 7.1).

Summarizing Key Points of Influencing Client Actions and Decisions: Directives, Psychoeducation, Logical Consequences, and Decisional Counseling

Defining Directives, Providing Information, and Psychoeducation

▶ All these influencing skills are concerned with imparting information to the client. The task is to be clear, specific, and relevant to the client's world.

▶ All directives in the book follow these microskill specifics: (1) listen and observe; (2) use specific and concrete language; (3) check out; (4) use the Client Change Scale.

► Provide specific data, judiciously, that will help clients in decision making; you may follow up with logical consequences. Counselor advice may be taken too seriously, so use it sparingly. Use the 1-2-3 pattern of listening when providing information or advice and check out with the client how you were received.

► Make referrals to community resources. We all need to be aware of and use community resources that support mental and physical health.

► To use these skills effectively, listen first and ensure that your client is interested and ready for your information.

Discerning The Specifics of Psychoeducation

► Different from directives and information, psychoeducation involves relatively short comments in the interview and integrates listening and other influencing skills.

► Psychoeducation sets specific objectives in connection with client needs and desires; it teaches them skills of daily living, ranging from how to be more assertive to life skills training, to help them meet their needs.

► Life skills training usually involves role-playing and enactment of scenes from the client's life or anticipated for the future.

► Specific psychoeducation steps include: (1) Check out your clients' interest in learning listening skills. (2) Role-play ineffective listening. (3) Debrief the specific behaviors that the two of you have observed. (4) Provide information on the content of the attending behavior (eye contact, verbal following, vocal tone, and body language. (5) Role-play positive communication using good listening skills. (6) Work with the client to take action in the real world.

► Homework is a useful strategy to help clients generalize what they have learned in the interview to their daily life in the "real world."

► Therapeutic lifestyle changes (TLCs) is an important positive force leading toward mental and physical health.

► Psychoeducation plays an important role enabling clients develop a healthy lifestyle, including good nutrition, exercise, relaxation, and meditation.

Defining Natural and Logical Consequences and Decisional Counseling

► Decisions are the most basic issue in interviewing, counseling, and therapy, as all clients will be making decisions.

► Decisions have consequences. Even good decisions may have negative natural consequences that we need to prepare our clients to face.

► Explore concrete positive and negative consequences that may logically follow from each decision. Use the "If you do . . ., then . . . will possibly result" approach.

► Making decisions is observed in all of the more than 500 theoretical approaches to helping.

► The steps of decisional strategy of logical consequences are: (1) Listen to clients and clarify issues. (2) Generate alternatives with them to resolve their issues. (3) Carefully sort out the positive and negative consequences of each alternative. (4) Summarize the alternatives using a nonjudgmental and positive stance. (5) Encourage client decision making.

► Decisional counseling underlies the five stages of the interview and involves the skills in which you have already become competent throughout the book.

► The cognitive and emotional balance sheet, created by the Australian Leon Mann, clarifies natural and logical consequences.

Refining Stress Management as Psychoeducation

▶ More than 40% of adults suffer adverse health effects from stress.

▶ Stress accounts for 75% to 90% of all doctor's office visits and plays a part in problems such as headaches, high blood pressure, heart problems, diabetes, skin conditions, asthma, arthritis, depression, and anxiety.

▶ Stress can be positive (eustress), but with high intensity, long duration, or both, it produces damaging mental and physical consequences.

▶ Stress management is an organized series of instructional, psychoeducational, and counseling strategies to enable clients to deal more effectively with stress.

▶ Stress management is based in solid neuroscience research and is used by therapists and counselors from many theoretical orientations.

Refining Directives and Psychoeducational Strategies for Building Mental and Physical Health

▶ Directive strategies are used to tell the client what to do in a timely, individually and culturally appropriate manner. Directives may range from telling the client to seek career information to using complex theoretical strategies.

▶ Used sparingly, directives can be critical in leading to client change and growth, as measured on the Client Change Scale (CCS).

▶ Therapeutic lifestyle changes (TLCs) are a key component of psychoeducation to promote mental and physical health and wellness.

▶ Therapeutic lifestyle changes offer clients a way to improve health, prevent physical and mental decay, manage stress, and promote wellness and well-being.

▶ Other strategies include spiritual imagery, role-play enactment, positive reframing, relaxation, meditation, physical exercise, imagery focusing on a relaxing scene, thought stopping, and journaling.

Observing: Introducing Therapeutic Lifestyle Changes Through Decisional Counseling

▶ Therapeutic lifestyle changes are a "different" way to approach interviewing and counseling.

▶ TLCs involve psychoeducation, providing information, and—at times—even giving advice.

▶ Encouraging clients to engage in new behaviors and ways of thinking is never easy.

Practicing: Integrating the Skills and Strategies of This Chapter

▶ Practice providing directives, information, and natural consequences.

▶ Explore the stress in your own life and in the lives of others: current, past, and future.

▶ Practice pondering alternatives using Leon Mann's cognitive and emotional balance sheet.

▶ Take advantage of website activities, including (1) Writing Logical Consequences Statements, (2) Logical Consequences Using Attending Skills, (3) Group Practice With Logical Consequences, and (4) Natural and Logical Consequences.

▶ Complete interactive exercises offered online: (1) Your Personal Experience With Psychoeducational Information, Advice, Opinions, and Suggestions; (2) Writing Psychoeducational Information, Advice, Opinion, and Suggestion Statements.

▶ Conduct a group practice exercise: Psychoeducation and Information.
▶ Review online video activities: (1) Relaxation Demonstration; (2) Thought-Stopping Technique.

Assess your current level of knowledge and competence as you complete the chapter:

1. Flashcards: Use the flashcards to check your understanding of key concepts and facilitate memorization of key information.

2. Self-Assessment Quiz: The quiz will help you assess your current knowledge and prepare for course examinations.

3. Portfolio of Competencies: Evaluate your present level of competence on the ideas and concepts presented in this chapter using the Self-Evaluation Checklist. Self-assessment of your competencies demonstrates what you can do in the real world.

SECTION IV
Skill Integration, Crisis Counseling, and Developing Your Own Style

> Despite their good intentions, today's businesses are missing an opportunity to integrate social responsibility and day-to-day business objectives—to do good.
>
> —*Cindy Gallop*

The management consulting star Cindy Gallop has an important point that we can apply to our own work in interviewing and counseling. It is not just meeting our needs and our clients' needs; there is also a moral and meaning imperative. We not only want change, we want to do good.

What is your style of interviewing and counseling? This section provides a framework to help you integrate the many skills and concepts of this book. Central to this process is Chapter 13, where we ask you to make a recording of a complete interview.

Chapter 13, Skill Integration Through Examining Your Own Session, Treatment Plans, and Case Management, provides the opportunity for you to examine your interviewing style through a detailed analysis of a recorded interview. You can compare your interview with those earlier in this book. Important in this chapter is information on developing treatment plans and case management. Stanford professor John Krumboltz's "happenstance theory" is introduced to enable awareness of chance in everyday life. We are not always in charge of our decisions.

Chapter 14, Crisis Counseling and Assessing Suicide Potential, shows how to use microskills in a crisis counseling session. Following this is a brief discussion of observations around suicide issues.

Chapter 15, Determining Personal Style in a World of Multiple Theories, begins by discussing how the microskills framework can be used with many different counseling and psychotherapy frameworks. The chapter then reviews the major concepts of the book and suggests ideas for your future plans in the helping fields.

We are nearly at the end of our journey through the basics of interviewing, counseling, and psychotherapy. You have communication skills that can be used in many settings, as these are foundational units of all communication beyond the helping fields—business, sales, law, medicine, child or peer helping, and many others throughout the nation and the world. How will you use and adapt these skills and strategies in your professional and personal life?

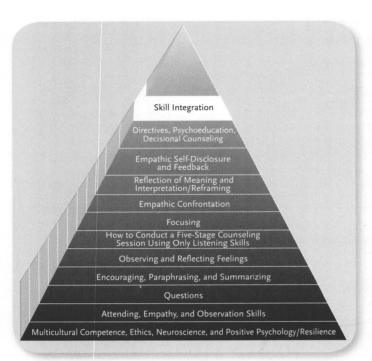

Skill Integration

Directives, Psychoeducation, Decisional Counseling

Empathic Self-Disclosure and Feedback

Reflection of Meaning and Interpretation/Reframing

Empathic Confrontation

Focusing

How to Conduct a Five-Stage Counseling Session Using Only Listening Skills

Observing and Reflecting Feelings

Encouraging, Paraphrasing, and Summarizing

Questions

Attending, Empathy, and Observation Skills

Multicultural Competence, Ethics, Neuroscience, and Positive Psychology/Resilience

Chapter 13
Skill Integration Through Examining Your Own Session, Treatment Plans, and Case Management

When I look at the world I'm pessimistic, but when I look at people I am optimistic.
—Carl Rogers

You and your clients will greatly benefit from a naturally flowing interview and treatment plan using a smooth integration of the skills, strategies, and concepts of intentional interviewing and counseling. This chapter presents several key issues, but the most important of these is that you conduct a complete interview and analyze what happens between you and the client.

Chapter Goals

Awareness, knowledge, skills, and actions developed through the concepts of this chapter and this book will enable you to:

▶ Analyze your own interviewing style in detail and determine the progress you have made. Integrate the concepts, skills, and strategies learned in previous chapters.

▶ If you chose, use the detailed interview evaluation form available at the conclusion of this chapter as part of this analysis.

▶ Utilize a pre-interview checklist that summarizes the many issues you need to be aware of before the interview begins.

▶ Develop a plan for the first interview, keep systematic interview records, and develop long-term treatment plans.

▶ Examine case management as related to the treatment plan.

▶ Consider the chance factors that are present in decision making from Krumboltz's "happenstance theory."

▶ Use relapse prevention as a systematic model to facilitate transfer of learning from the interview to the real world.

▶ Consider two neuroscience-based technologies for client change: biofeedback and neuro-feedback.

▶ Defining Skill Integration

The aim of skill integration is to take the microskills, stages of the interview, and your natural expertise and then examine where you stand now. How might you want to intentionally grow? As you increase interviewing competence, the result will be increasing intentionality and the ability to flow naturally with your clients.

SKILL INTEGRATION	ANTICIPATED CLIENT RESPONSE
Integrate the microskills into a well-formed interview and generalize the skills to situations beyond the training session or classroom.	Developing interviewers and counselors will integrate skills as part of their natural style. Each of us will vary in our choices, but increasingly we will know what we are doing, how to flex when what we are doing is ineffective, and what to expect in the interview as a result of our efforts.

Taking a careful look at your interview behavior, as well as your competence in analyzing the session, is the central focus of this chapter. Specifics for this analysis will be found at the conclusion, where specifics for recording a session with a volunteer client are presented. In preparation for this, several key issues are discussed—among them are an interviewing checklist, treatment planning, case management, referral, and relapse prevention.

It may be helpful to visit the web resources and view a full interview transcript of Allen and Mary in a decisional interview. As a supplement to this chapter, it will enable you to see decisional counseling in action, and it shows another way to analyze the interview in even more detail than discussed here.

Interactive Exercise: Influence the Interview. This exercise provides a good review of the skills learned.

▶ Discerning: Suggestions to Facilitate Skill Integration

PLANNING THE FIRST INTERVIEW USING A CHECKLIST

Given the complexity of relationships, particularly professional relationships, the physician Atul Gawande has written *The Checklist Manifesto* (2009). Focusing first on medicine, he found that a surgical checklist of basic and often obvious factors significantly reduced dangerous errors during operations. He has gone on to point out that thinking ahead about what one is going to do improves performance regardless of the field. Since the original book, Gawande's innovative thinking has entered the methods of many professions. His ideas have profound mean for our practice in interview and counseling.

Box 13.1 provides a checklist for the first interview. You will see that there are many issues that we can consider both before, during, and after we work with clients. Review

---

BOX 13.1 **The Interview Checklist**

Before the Session

_____ Are you familiar with HIPAA, the policies of your agency, and key state laws? Are key policies posted in the agency waiting area? These need to be shared early in the session.

_____ If there is a file, have you read it? Do you need notes from the file as a refresher?

_____ Do the room and setup ensure confidentiality?

_____ Does the room provide adequate silence? Do you need a sound machine working outside your door?

_____ Do you have an inviting atmosphere where you will meet the client? Is it neutral, or do you have interesting art and objects relevant to those who may come to this office? Are chairs placed in a position where the power is relatively equalized?

During the Session

Relationship
How did you plan to establish a relationship and connect with this unique client?
Did you:

_____ 1. Discuss the client's rights and responsibilities? The interviewing, counseling, and therapy relationship works in part because of clearly defined rights and responsibilities of each person involved.

_____ 2. Provide an explanation of what might happen in the session and/or how the interview is likely to be structured?

_____ 3. Review HIPAA, agency policies, and key legal issues? If your agency requires diagnostic labels, did you explain that to the client and offer to share that diagnosis if he or she wishes?

_____ 4. Discuss confidentiality and its limits?

_____ 5. Obtain the client's permission to take notes and/or record the session? Was the client informed that these notes and the recording are available if he or she wishes to review them?

_____ 6. If working with an underage client, obtain the appropriate parental permission as required by your agency and/or state law?

_____ 7. Provide an opportunity for the client to ask you questions before you started? Were issues of multicultural differences addressed?

_____ 8. Work with the client to establish an early preliminary goal or objective for the session?

_____ 9. Come prepared if the client immediately started talking about issues and concerns? Did you listen carefully and return later to cover these important points?

Story and Strengths
Did you:

_____ 1. Allow and encourage the client to present the story fully? Did you reframe the word *problem* into a more positive, change-oriented perspective with words such as *issue, challenge, concern, opening for change*?

_____ 2. Bring out the key facts, thoughts, feelings, and behaviors related to the story? Did you perhaps also look for underlying deeper meanings behind the story?

(continued)

3. Avoid becoming enmeshed in the client's story by becoming a voyeur (endless fascination and searching for details about the client's interesting issues) or by unconsciously putting a "negative spin" on what the client said? Are you part of the problem or part of the solution?

4. Bring out stories and concrete examples of client personal strengths and external resources? Did you search for specific images within these stories and perhaps anchor these positive images in specific areas of the body?

5. Ask the critical questions "What else relates to what we've talked about so far?" "What else is going on in your life?" and "Is there anything else I should have asked you but didn't?"

Goals
Did you:

1. Review the early goals set by the client and revise them in accordance with new information about the story and strengths?

2. Jointly make these goals as specific and observable as possible?

3. When necessary, break down large goals into manageable step-by-step objectives that can be reached over time? Did you prioritize these goals?

4. Remind the client of the strengths and resources that he or she brings to achieve these goals?

Restory
Did you:

1. Include brainstorming without a theoretical orientation? Confront with a supportive challenge, summarizing the goal and the issue? ("On one hand the goal is . . ., but on the other hand the main challenges you face are. . . . Now what occurs to you as a solution?) With your support, clients will often come up with their own unique and workable answers, often ones that you did not think of.

2. Use a variety of listening and influencing strategies to facilitate client reworking and restorying of issues? What were they?

3. Use identified positive strengths and resources to remind clients during low points of their own capabilities?

4. Agree on homework or personal experiments to be completed between sessions?

5. Develop a clear definition of a more workable story that can lead to action and transfer to the real world?

Action
Did you:

1. Work with the client to take specific learning and action from the interview to the "real world"?

2. Agree on a plan for transfer of learning that is clear and doable?

3. Contract with the client to do at least one thing differently during the week, or even tomorrow?

4. Agree to plans to look at this homework during the next session?

5. Check how it was for client? Does the client think he or she could work with you? Did you agree to work together?

6. Set a date for the next session or follow-up interview?

After the Session
Did you:

1. Write notes that are clear to both your agency and your client? Give special attention to client strengths and resources? Sign and date the record?

2. Review the checklist and make plans to include missing items in your next session?

3. As necessary, develop a treatment plan for the future?

4. Anything else?

Source: Based on Gawande physician's checklist, 2009.

this list before you record an interview with a volunteer client, and review it afterward checking to see what might have been missed. For you, personally, what might you add or delete from this checklist? Making this checklist fit for you and your agency is essential.

The first interview presents many challenges as you develop a relationship, structure the session, and get to know a new client. The interview plan and checklist may be helpful as you plan for any session, particularly that first session in which there are so many issues to cover. Experience suggests that even the most seasoned professional forgets and omits important items on this list and in the treatment plan as well. Review Box 13.2 for additional information regarding client and therapist's perceptions of counseling and the counseling process.

TREATMENT PLANNING

Treatment planning, in the form of specific written goals and objectives, is increasingly standard and often required by agencies and insurance companies. When possible, negotiate the specific goals with the client and write them down for periodic joint evaluation. They should be as concrete and clear as possible; indicators of behavioral change and emotional satisfaction are important. The more structured counseling theories, such as cognitive behavioral, strongly urge interview and treatment plans with specific goals developed for each issue. Their interview and treatment plans are often quite specific.

Less structured counseling theories (Gestalt, psychodynamic, person-centered) tend to give less emphasis to treatment plans, preferring to work in the moment with the client. In short-term counseling and interviewing, the interview plan serves as the treatment plan. As you move toward longer-term counseling (5 to 10 sessions), a more detailed treatment plan with specific goals is often required. Many agencies also use case management as part of the treatment plan.

One way to start a treatment plan is to use the five-stage interview structure. It is also a useful framework for taking notes.

STAGE/DIMENSION KEY QUESTIONS	NOTE TAKING AND INTERVIEW REVIEW
Empathic Relationship. Initiate the session, develop rapport, and structure the interview. *What structure do you have for this interview? Do you plan to use a specific theory? What special issues do you anticipate with regard to rapport development?*	Reflect on the session and the nature of the relationship and working alliance. If you have asked the client to complete a feedback form at the conclusion of the session, use that information as a starting point. More and more agencies are including feedback as part of standard practice.
Story and Strengths. Gather data; draw out stories, concerns, problems, or issues. *What are anticipated difficulties? Strengths? How do you plan to define the issues with the client? Will you emphasize behavior, thoughts, feelings, meanings?*	Summarize the key narratives in brief form. Make special note of the emotions associated with the stories. It is very important to develop an early list of client strengths and resources.
Goals. Set goals mutually; establish outcomes. *What is the ideal outcome? How will you elicit the client's idealized self or world?*	Define the identified goals clearly. Again, if you and the client don't know where you are going, you may end up in confusion and indecision. An increasing number of agencies require very specific goals that are accountable, especially if the session is insurance based. Consider the possibility of a specific list of goals and a time frame for them to be addressed.

STAGE/DIMENSION KEY QUESTIONS	NOTE TAKING AND INTERVIEW REVIEW
Restory. Explore and create alternatives; confront client incongruities and conflict; restory. *What theories would you probably use here? What specific incongruities have you noted or do you anticipate in the client? How will you generate alternatives?*	Most likely, there will be minimal restorying at this point, but if there is clear change, describe the change and how it might have occurred. What strategies and skills did you actually use in the session? Which were most effective? Least effective.
Action. Generalize and act on new stories. *What specific plans do you have for transfer of training? What will enable you, the interviewer, to personally feel that the interview was worthwhile?*	Pay special attention to the Maintaining Change Worksheet later in this chapter (Box 13.3). This is the place for an action contract, usually verbal. Writing down specific things for the client to do or try as concrete action can make a big difference in what happens. Outline plans for homework and follow-up that you can review in the next session. Is there a planned date to achieve defined goals?

▶ Refining Supplementary Concepts and Methods

CASE MANAGEMENT AS RELATED TO TREATMENT PLANNING

Management is doing things right; leadership is doing the right things.

—Peter Drucker

Although this book focuses on the interview and counseling skills, treatment planning often needs to be extended to case management. For human service professionals, social workers, and school counselors, case management will be as or more important than treatment planning. Case management requires the professional helper to coordinate community services for the benefit of the client and, very often, the client's family as well. Let's look at the complexity of case management with an example.

A single mother and her 10-year-old boy are referred to a social worker in a family services agency. The family physician thinks that the child's social interaction issues may be a result of Asperger's syndrome. The child has few friends, but is doing satisfactorily in school. The social worker interviews the mother and reports the child's social and academic situation at school. The mother says that she has financial problems and difficulty in finding work. Meeting with the child a few days later, the worker notes good cognitive and language competence, but that the child is unhappy and demonstrates some repetitive, almost compulsive behavioral patterns.

The agency staff meets and starts to initiate a case management treatment plan. It is clear that the mother needs counseling, and that the child needs psychological analysis and likely treatment as well. At the staff meeting, the following plan is developed: The child is to be referred to a psychologist for evaluation, and based on that evaluation, recommendations for treatment are likely to be made. The social worker is assigned to do supportive counseling with the mother and to take overall responsibility for case management. The social worker will eventually need a treatment plan for the mother and a case management plan for the family.

School performance is good, but the social worker contacts the elementary school counselor and finds that the counselor already has the child in counseling. In fact, the counselor has been working with the teacher to help the client develop better classroom and playground relationships. The school counselor and social worker discuss the situation and realize that the boy is often left alone without much to do. Staying after school for an extended day would likely be helpful. The school counselor, with the social worker's backing, contacts a source for funding through a local men's club.

This is but a beginning for many cases—case management involves multiple dimensions beyond this basic scenario. Counseling and therapy are an important part of case management, but only a part. The Division of Youth Services or Family and Children Services may need to be called to review the situation if the counselor, or another mandated reporter, suspects there are signs of abuse or neglect; the mother may need short-term financial assistance and career counseling. If the father has not been making child support payments, legal services may have to be called in.

Through all of the above, the social worker maintains awareness of all aspects of the case. Through individual counseling with the mother, the worker is in constant touch with all that is going on and remains in contact with key figures working toward the positive treatment of the child. The child can only be successfully treated in the family, school, and community situation.

Advocacy action is important throughout, and it does not happen by chance. Throughout this process, the social worker will constantly advocate for the mother and the child by establishing connections with various agencies. The school counselor who goes to the local men's club for funding has to advocate for the child; encouraging a teacher to change teaching style to meet individual child needs also requires advocacy. Elementary counselors spend a good deal of time in various activities advocating for students individually, and often advocating for fair school policies as well.

Social justice issues may appear. In many communities and agencies, all of the above services may not exist—and equally or more likely, they do not work together effectively. This may require organizing to produce change. The child may be bullied, but the school has no policy to prevent bullying. Social justice and fairness demand that each child be safe. Discrimination against those who are poor or are from minority backgrounds may exist. Individual and group education or actual forced change may be necessary.

HAPPENSTANCE LEARNING THEORY: THE ROLE OF CHANCE FACTORS IN INTERVIEWING AND COUNSELING

> Chance events over one's life span can have both positive and negative consequences. Unpredictable social factors, environmental conditions, and chance events over the life span are to be recognized as important influences in clients' lives.
>
> —John Krumboltz

Despite all our planning, establishing goals, and thinking we know where we are going, Krumboltz (2014) points out, we are not always the masters of fate that we think we are. Happenstance theory focuses on chance as a major determinant of who we are and where we are heading. The results of chance or happenstance affect our goals and decisions. Stop for a moment and think of various decision and chance points that might have led you to this moment in your life.

As your clients set goals and generate possibilities for action, there is a need to encourage flexibility and exploration. Expectations for success help the client succeed, but at the same time we all need to know that the way forward is not always a straight line.

Thus, decisions and chance factors that bring us opportunity are always likely to change. In short, flexible planning and awareness of happenstance can be valuable.

"Make lemonade out of lemons," goes the old saying. Positive psychology demonstrates the importance of optimism. Persistence and flexibility are key. As goals are developed, encourage clients to organize broad-based action plans that are doable. Also, help them be flexible and ready to change to meet developing circumstances, unexpected obstacles and setbacks, and unanticipated outcomes.

Change involves risk and the ability to turn adversity to one's advantage—and you, the interviewer or counselor, being ready to help your clients through the challenges. At times, anticipating alternative futures and discussing and/or role-playing new behaviors can help.

BIOFEEDBACK AND NEUROFEEDBACK: TECHNOLOGIES FOR CLIENT CHANGE

The stress response can be defined as changes in the body when we encounter a challenging or threatening situation. Physiological changes that may occur as the result of stressors may include increased blood pressure, tightened muscles, breathing, and heart rate—all representing the brain and body's fight-or-flight reaction.

Biofeedback and neurofeedback training may be defined as the use of technical equipment to provide feedback on immediate here-and-now body and brain responses. This is provided so that clients can become aware of inside-the-skin behaviors and thus learn how to self-regulate their neurobiological responses to challenges that they may face.

—Carlos Zalaquett

It is very likely that many reading this book will become skilled in biofeedback, a counseling and therapeutic technique that helps clients gain control of physiological processes. To accomplish this goal, clients' physiological responses are monitored using specific equipment, and the signals are fed back to the client and therapist. A trained therapist uses different teaching techniques to facilitate the acquisition of self-regulation skills. There is now some research evidence that an empathic relationship will improve client response to these technologies.

As we recommend for interviewing practice, the client assumes responsibility and becomes an active participant in his/her own improvement. In biofeedback, only the client can actually produce the desired changes through following the counselor's instructions and regular practice. In addition to improving the specific physiological target, the client increases his/her feelings of self-control and personal mastery.

At some point in your career, you are likely to want to learn the terms and general areas of five commonly employed feedback technologies:

Electromyography (EMG), used to register muscle movements

Skin conductance level (SCL), also known as galvanic skin response (GSR), which registers sweat gland activity

Thermal, which registers temperature changes

Heart rate (HR), which registers rate and pulse volume

Neurofeedback (EEG), which register brain waves

Biofeedback is an efficacious treatment for anxiety, chronic pain, headaches, hypertension, and Raynaud's disease. Neurofeedback has been found successful for attention deficit disorder and epilepsy. These techniques are also used in the treatment of various physiological diagnoses;

they are used as adjunctive therapy within a variety of treatments, such as CBT, hypnosis, and stress management, and within various stress- and anxiety-producing situations, such as job- or study-related physical and psychological issues, as well as in cases where a person wishes to improve wellness and learn to relax more. In all these cases, structured stress- and anxiety-reducing interventions using biofeedback have proven to be effective. Furthermore, biofeedback and neurofeedback seem to have a very low potential for damage (if used properly, of course.)

Neurofeedback is particularly important given the new advances in neuroscience and the confirmation of neurogenesis. Our brain has billions of neurons that discharge electricity whenever stimulated. These electrical patterns called brain waves can be registered with computer technology and transformed into a video display on a computer monitor. In neurofeedback training, your brain waves are displayed as information you can learn from. You can watch your brain waves in action. Changes in brain wave patterns might show that you are concentrating, suppressing your impulses, indulging in your impulses, experiencing tension, or relaxing.

Neurofeedback training converts the brain waves into gamelike displays in which you learn to control the brain waves. For example, concentrating on an animated airplane on the computer screen enables it to take flight. The longer you stay focused, the longer it flies. As soon as you drift off or lose interest, your airplane falls. Other games might include making a superhero fly or putting together puzzle pieces. You score only when your attention remains focused. By doing this, you gain a sense of what concentration feels like and at the same time learn how to maintain your train of thought for a period of time.

This enhanced brain activity becomes a natural part of your daily routine. If learning how to control your brain waves sounds a little farfetched for you, stop what you are doing and count to 10. You just controlled your brain activity. We have been doing it all of our lives. The procedure is simple; with practice, you can learn to control your brain waves without the help of neurofeedback. It is hard to explain how to concentrate to clients suffering with attention deficit disorder, but once they understand neurofeedback and learn self-regulation, difficult tasks become less stressful and easier to overcome. Clients feel in control and successful while at the same time you are improving your physiological and brain's activity.

In spite of its technological base, neurofeedback practitioners provided early evidence of microskill effectiveness. Edward Taub, as early as 1977, suggested that the most powerful factor influencing whether different forms of biofeedback learning will occur was the quality of the relationship between the practitioner and the client. He called this the "person factor," and the communication skills used by the most effective practitioners closely resembled the microskills.

To learn more about neurofeedback, read the work of Chapin and Russell-Chapin (2014) and Collura (2014). The former has solid chapters on brain structures, their functions and meanings, and the latter offers foundational information on neurofeedback technology. For biofeedback, we suggest Schwartz and Andrasik's *Biofeedback: A Practitioner's Guide* (2005).

| BOX 13.2 | National and International Perspectives on Counseling Skills |

What's Happening With Your Client While You Are Counseling?

Robert Manthei, Christchurch University, New Zealand

There is more going on in interviewing beyond what we see happening during the session. Clients are good observers of what you are doing, and they may not always tell you what they think and feel. Research shows that clients expect counseling to be shorter than do most counselors and therapists. Clients see counselors as more directive than counselors see themselves. And what the counselor sees as a good session may be seen otherwise by clients, and vice versa. Counselors and clients may vary in their perceptions of interviewing effectiveness.

I conducted a study of client and counselor experience of counseling. Among the major findings are the following:

(continued)

Clients often have sought help before

Most people don't come for counseling immediately. Talking with friends and family and trying to work it out on their own were usually tried first. Reading self-help books, prayer, and alcohol and drugs are among other things tried. Some deny that they have problems until these become more serious.

Implications for practice

Ask clients what they have tried before they came to you, and find out what aspects of prior efforts seemed to have helped. You may want to build on past successes. This is an axiom of brief solution-oriented counseling (see Chapter 14).

First impressions are important

That first interview sets the stage for the future, and the familiar words "relationship and rapport" are central. I found that clients generally had favorable impressions of the first sessions and viewed what happened even more positively than counselors. Sometimes sharing experience helps. One client who did not feel positive about the first session commented, "Maybe if the counselor had gone through a similar experience of divorce and children, it would have helped."

Implications for practice

Obviously, be ready for that first session. Cover the critical issues of confidentiality and legal issues in a comfortable way. Structuring and letting the client know what to expect

seem important. Some personal sharing, used carefully, can help. And empathic listening always remains central.

Counseling helps, but so do events outside of the interview

Resolution of their issues was attributed to counseling by 69% of clients, while 31% believed events outside the session made the difference. Among things that helped were talking and socializing more with family and friends, taking up new activities, learning relaxation, and involvement with church.

Implications for practice

The interview is important, but generalization of behavior and thought to daily life is central. Homework and specific ideas for using what is learned in the session are important.

Things that clients liked

Relationship variables such as warmth, understanding, and trust were important. Clients liked being listened to and being involved in making decisions about the course of counseling. Reframes and interpretations helped them see their situations in a new way; also valued were new skills such as imagery, relaxation training, and thought-stopping to eliminate negative self-talk.

All of the above speaks to respecting the client's ability to participate in the change process. I think it is vital that we tell clients what we are doing, but also ask them to share their perceptions of the session(s) with us. We can learn from and learn with more client participation.

▶ Practicing Topics Important to Interviewing Practice and Recommendations for Reviewing Your Own Interviewing Style

REFERRAL

Sometimes the client/interviewer relationship simply doesn't work as well as we all would like. When you sense the relationship isn't doing well, avoid blaming either the client or yourself. Focus on client goals, and seek to hear the story completely and accurately. Ask the client for feedback on how you might be more helpful. Seek consultation and supervision, and most often these "difficult patches" can be resolved to the benefit of all. When an appropriate referral needs to be arranged, we do not want to leave our clients "hanging" with no sense of direction or fearful that their problems are too difficult. Maintain contact with the client as the referral process evolves, sometimes even continuing for a session or two until arrangements are complete.

Another key referral issue is whether this is a case in which interviewer expertise and experience are sufficient to help the client. Even if the counselor thinks that he or she is working effectively, this may not be enough; it may be a case in which supervision and case conferences can be helpful. Opening up your work to others' opinions is an important part of professional practice. Clients, of course, should be made aware that you as counselor or therapist are being supervised. It is critical that the client never feel rejected by you. Your support during the referral process is essential.

Your first interviews with a client are individually unique and give you an opportunity to learn about diversity and the complexity of the world.

MAINTAINING CHANGE: RELAPSE PREVENTION

Once change has been effected, a relapse plan can reduce slips or relapses. This approach, originally developed for use in alcohol or drug abuse treatment, is now used in a variety of settings. Relapse prevention is now a standard part of most cognitive behavioral counseling, and the concepts presented here will enrich all theories of helping described in this chapter.

The Maintaining Change Worksheet (Box 13.3) helps the client plan to avoid relapse or slips into the old behavior. The counselor hands the client the worksheet and they work through it together, giving special emphasis to things that may come up to prevent treatment success. Research and clinical experience in counseling both suggest that this may be the most important thing you can do with clients—help them ensure that they actually *do* something different as a result of their experience in the interview.

BOX 13.3 Maintaining Change Worksheet: Self-Management Strategies for Skill Retention

I. Choose an Appropriate Behavior, Thought, Feeling, or Skill to Increase or Change

Describe in detail what you intend to increase or change.

_____

_____

Why is it important for you to reach this goal?

_____

_____

What will you do specifically to make it happen?

_____

_____

II. Relapse Prevention Strategies

   A. *Strategies to help you anticipate and monitor potential difficulties: regulating stimuli*

Strategy	*Assessing Your Situation*
1. Do you understand that a temporary slip may occur but it need not mean total failure?	_____ _____
2. What are the differences between learning the behavioral skill or thought and using it in a difficult situation?	_____ _____

(continued)

3. Support network: Who can help you maintain the skill? _____

4. High-risk situations: What kinds of people, places, or things will make retention or change especially difficult? _____

B. Strategies to increase rational thinking: regulating thoughts and feelings

5. What might be an unreasonable emotional response to a temporary slip or relapse? _____

6. What can you do to think more effectively in tempting situations or after a relapse? _____

C. Strategies to diagnose and practice related support skills: regulating behaviors

7. What additional support skills do you need to retain the skill? Assertiveness? Relaxation? Microskills? _____

D. Strategies to provide appropriate outcomes for behaviors: regulating consequences

8. Can you identify some probable outcomes of succeeding with your new behavior? _____

9. How can you reward yourself for a job well done? Generate specific rewards and satisfactions. _____

III. Predicting the Circumstance of the First Possible Failure (Lapse)

Describe the details of how the first lapse might occur; include people, places, times, and emotional states.

_____

_____

_____

Permission to use this adaptation of the Relapse Prevention Worksheet was given by Robert Marx.

Remember that every client is unique and the relapse prevention plan needs to be tailored to the individual client and the characteristics of the problem. Every person's prevention plan should look somewhat different.

Include in the plan a list of people who can be counted on for support, the triggers to the unwanted behaviors, possible responses to those triggers, and rewarding alternatives client can do instead of engaging in the negative behavior.

People you can contact by telephone, text, or email: Develop a list of three or more people who can provide immediate support during times of potential relapse or slip.

Triggers to undesired behavior, and possible responses for those triggers: Discuss and write down the biggest temptations or triggers. What are the things or situations that make you want to engage in the negative behavior (frequenting some people, visiting a particular place, the sight, smell, or sound of . . .)? What can you plan to do in response to each of these triggers?

Alternative and rewarding activities that do not involve the unwanted behavior: Inactivity and boredom are some of the greatest threats to change. Working with your client, develop a plan of activities that he or she enjoys and wants to do to prevent voids of time that can facilitate negative behaviors.

Rewards the client can give her- or himself for meeting change targets: What would the client do when reaching a week of progress, two weeks, and longer periods of change?

Remember that change takes time and effort. Maintaining an intentional effort to change will pay off at the end.

▶ Your Turn: The Final Audio or Video Exercise

You have engaged in the systematic study of the counseling process. By now you have experienced many ideas for analyzing your style and skill usage. Furthermore, the microskills learned through this text and the practice exercises have provided you with alternatives for intentional responses. But your responses needs to be authentically your own. If you use a skill or strategy simply because it is recommended, it could be ineffective for both you and your client. Not all parts of the microskills framework are appropriate for everyone. We hope that you will draw on the ideas presented here, but ultimately *you are* the one who will put the science together in your own art form.

You have practiced varying patterns of helping skills with diverse clientele. We hope that you have developed awareness and knowledge of individual and multicultural differences. Study and learn how to "flex" and be intentional when you encounter diversity among clients. Developing trust and a working alliance takes more time with some clients than with others. For example, you may be more comfortable with teenagers than with children or adults, or you may have difficulties with elders.

To help ensure your understanding of your own style, we suggest that you complete another audio or video session. The guidelines provided in Box 13.4 will enable you to study your own work and obtain the best feedback from others.

BOX 13.4   Reviewing and Analyzing Your Interviewing and Counseling Style

1. Find a volunteer client willing to role-play a concern, issue, challenge, or opportunity.
2. Before you conduct the interview, review the Interview Checklist in Box 13.1.
3. Counsel the volunteer client and record for at least 20 minutes. Feel free to spend more time to gain a sense of completion. For the exercise in this section, you don't have to transcribe a full session, although a full transcript likely will be most beneficial to you. Be sure to indicate what happened in the rest of the counseling session so that the context of the transcript is clear.
4. Use regular ethical guidelines as in the past. Ask the volunteer client, "May I record this session?" and obtain written permission. Remind the client that the interview may be stopped at any time.
5. Use your own natural communication style.
6. Select a topic beforehand. The client makes the choice, but you may help by making suggestions, such as interpersonal conflict, choice of courses for the following term, or whether or not to attend graduate school. A good topic

is to explore some significant event from childhood or teen years and how it might influence the client today.

7. Ask for a community genogram. This could be completed the day before or as the first part of the interview. If you do so, we suggest recording that as well.
8. As appropriate, complete a relapse prevention form with the client and discuss how any learning from the interview may be taken home.
9. Obtain client feedback. Use the Client Feedback Form (Box 1.4, page 17). You will find it very helpful to get immediate feedback from your client. You may even watch or listen to the recording together.
10. You may find the Microskills Interview Analysis Form (MIAF) useful as you review your session. A careful analysis of your behavior in the session will aid in identifying your natural style and its special qualities as well as your skill level. Use the ideas presented in this chapter to further examine and analyze your work, referral and consultation, and treatment planning. See Chapter 7, page 148 for the MIAF.

A session transcript may be useful in demonstrating your competence and obtaining an internship or job. Consider the transcript a permanent part of your developing professional life and your portfolio.

Transcribe the session (see Box 13.5), and use what you have learned so far to analyze your work. A careful analysis of your behavior in the session will aid in identifying your natural style and its special qualities, as well as your skill level. Use the ideas presented in this chapter to further examine and analyze your work, referral and consultation, and treatment planning.

 Interactive Exercise: A Full Interview Transcript. Prepare a paper in which you demonstrate your interviewing style, classify your behavior, and comment on your own skill development as an interviewer.
Group Practice Exercise: Your Turn: An Important Audio or Video Exercise

---

| BOX 13.5 | Transcribing and Analyzing Sessions |

Check off the following points to make sure you have included all the necessary information in the transcript. In addition (or as an alternative), you can use the Microskills Interview Analysis Form (MIAF), asking a colleague to review your transcript and provide the form as another summary of your work.

_____ Describe the client briefly. Do not use the client's real name.

_____ Outline your session plan *before* the session begins.

_____ Be sure you obtain the client's permission before recording the session, and include a summary of this agreement in the transcript. The client should be free to withdraw at any time. Rarely, at the end the client might say, "Do not use this session." If that happens, then find another client for your recording. Ethically, we protect the rights of the client.

_____ Organize the transcript in a format similar to the two-column transcripts used in this book.

_____ Number all interactions, and be sure to indicate who is speaking at the beginning of each interaction.

_____ Please note the focus of each interviewer and client comment and the skill classification. Periodically, use the Client Change Scale to look at possible change in the session. Comment briefly on what you observe, as done in the interview transcript throughout the book.

_____ If you confront, note this in the Process Comments column; also note how the client responds, using the Client Change Scale.

_____ Indicate when you feel that you have reached the end of a stage. Do not feel that you must cover all stages; in some cases, you may cycle back to an earlier stage or forward to a later stage.

_____ At the end, write a commentary that summarizes what happened.

_____ Summarize your use of skills through a skill count. What skills do you seem to use most?

_____ Assess your competence levels. What skills have you mastered, and what do you need to do next? This is also a summary of your strengths and the areas that need further development. What did you like and not like about your work? Your ability to understand and process "where you are" and discuss yourself is essential for personal and professional growth.

_____ End the transcript report with a treatment plan for a hypothetical future series of sessions.

_____ Finally, go back to the transcript or recording of the session you recorded earlier, in Chapter 1 and/or Chapter 6. Note how your style has changed and evolved since then. What particular strengths do you note in your own work?

Summarizing Key Points of Skill Integration Through Examining Your Own Session, Treatment Plans, and Case Management

Defining Skill Integration

▶ Self-analysis of your own interviewing style and its impact on the client provides the foundation for mastering skills and integrating these into your own natural style.

▶ Your goal is to integrate the microskills into a well-formed interview and to practice to generalize the skills beyond this course or training.

▶ The five-stage structure of the interview is clinically useful for short- and long-term interviewing planning. Specifying goals and outcomes is particularly important.

▶ In addition, you may decide to use the interview analysis form in Box 13.5.

Discerning: Suggestions to Facilitate Skill Integration

▶ It is possible to use the restorying model structure to plan your interview before you actually meet with a client. The checklist for the first interview (Box 13.1) can help you accomplish this. This same five-stage structure can be used as an outline for note taking after the interview is completed.

▶ A treatment plan is a long-term plan for conducting a course of interviews or counseling sessions. Case conceptualization is your integration of client, issues, challenges, and goals for the future.

Refining Supplementary Concepts and Methods

▶ Case management often leads to alternative change strategies beyond the original interview and may involve efforts with the family, community, and other agencies.

▶ Happenstance theory asserts that chance is a major determinant of who we are and where we are heading. The results of chance or happenstance affect our goals and decisions.

▶ Biofeedback and neurofeedback are therapeutic techniques with the capacity to help clients gain control of physiological processes. New advances in neurosciences call for counselors' use of brain-based technologies. Neurofeedback is particularly suited for counselors' and therapists' work.

Practicing Topics Important to Interviewing Practice and Recommendations for Reviewing Your Own Interviewing Style

▶ When referral is necessary, provide sufficient support during the transfer process.

▶ Relapse prevention and planning are especially effective to help clients maintain change. Work together with your client to identify possible challenges that may prevent success and to establish possible actions to reduce their impact and maintain course.

Your Turn: The Final Audio or Video Exercise

▶ Using the constructs of this book, examine your own interviewing style and your use of the microskills. Pay special attention to your ability to produce a well-formed interview and its effect on a client's cognitive, behavioral, and emotional response.

Assess your current level of knowledge and competence as you complete the chapter:

1. Flashcards: Use the flashcards to check your understanding of key concepts and facilitate memorization of key information.

2. Self-Assessment Quiz: The quiz will help you assess your current knowledge and prepare for course examinations.

3. Portfolio of Competencies: Evaluate your present level of competence on the ideas and concepts presented in this chapter using the Self-Evaluation Checklist. Self-assessment of your competencies demonstrates what you can do in the real world.

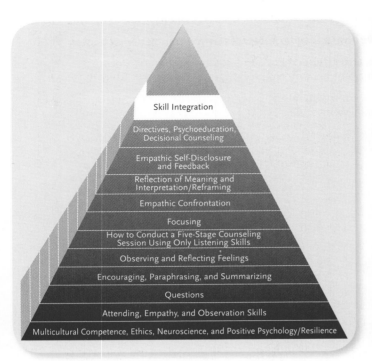

Skill Integration

Directives, Psychoeducation, Decisional Counseling

Empathic Self-Disclosure and Feedback

Reflection of Meaning and Interpretation/Reframing

Empathic Confrontation

Focusing

How to Conduct a Five-Stage Counseling Session Using Only Listening Skills

Observing and Reflecting Feelings

Encouraging, Paraphrasing, and Summarizing

Questions

Attending, Empathy, and Observation Skills

Multicultural Competence, Ethics, Neuroscience, and Positive Psychology/Resilience

Chapter 14
Crisis Counseling and Assessing Suicide Potential

We want people to know that their emotions and reactions are completely normal.

—Anonymous crisis helper

We all face crises, and people cope the best they can. Attend to their uniqueness. Listen, support and respect differences in style and manner. There are underlying reasons and explanations for virtually everything you will encounter. Be empathically with your clients and avoid judging and evaluation. (Of course, take action when needed.)

—Allen Ivey

This chapter addresses two challenging issues that you are likely to encounter in your helping career. Often frontline workers are the first to be called when it comes to crises and the potential of suicide. Experienced professionals often come in later, so it is essential that all of us be ready to help when we are needed.

You will find that empathic listening and observation microskills remain basic and essential. Many times you will find yourself giving information and providing directives.

Chapter Goals

Awareness, knowledge, skills, and actions developed through the concepts of this chapter and this book will enable you to:

▶ Understand the basics of crisis intervention.

▶ Examine key issues related to suicide potential.

▶ Discover key references to continue advancing your knowledge.

> **Assess your current level of knowledge and competence as you begin the chapter:**
>
> 1. Self-Assessment Quiz: The chapter quiz will help you determine your current level of knowledge. You can take it before and after reading the chapter.
>
> 2. Portfolio of Competencies: Before you read the chapter, please fill out the Self-Evaluation Checklist to assess your existing knowledge and competence on the ideas and concepts presented in this chapter. Then, at the end of the chapter, complete the checklist again to summarize your competencies after study and practice.

Virtually everyone who reads this book will at some point become involved in an immediate crisis. It may be as an experienced professional counselor dealing with the aftermath of a hurricane, a beginning professional on a suicide hotline, or a volunteer working with flood survivors. At another level, you will encounter individuals and family members who have just had a heart attack, learned about cancer, or discovered that they cannot become pregnant. War, rape, divorce, abuse, gay bashing, witnessing a street shooting, and an endless array of incredibly difficult issues may come to you.

Whether you enter the helping profession or not, you will almost certainly find yourself involved in crisis in some form or other. All of us can offer support and help at this time, as experienced professionals are often not there at the moment. The purpose of this chapter is to provide some *very beginning* key points so that you are better prepared. We urge you to follow this up with further reading, Internet study, and videos available in the web resources.

▶ Defining Crisis Counseling

> When written in Chinese, the word "crisis" is composed of two characters.
> One represents danger and the other represents opportunity.
> —John Fitzgerald Kennedy

Crisis counseling is the most pragmatic and action oriented form of helping. The word *pragmatism* comes from the Greek word for deed, act, to practice, and to achieve, πράγμα (*pragma*). Even more than decisional counseling, crisis counseling is concerned with action and useful, practical results for the client. Both are provided with a caring attitude and with respect for client response.

WHAT IS A CRISIS?

It has been pointed out that virtually all the world's population experiences one or more crises or traumas in their lifetime. In that sense, crisis is a normal life event; "normalizing" the crisis is one foundational idea to keep in mind. Crisis is closely related to the word *trauma*. Following is a sampling of the various types of crisis that you may encounter.

Immediate here-and-now crises demanding rapid practical action include flood, fire, earthquake, rape, war, refugee status, a school or community shooting, personal assault (including abuse), being held hostage, a serious accident, and the sudden discovery or diagnosis of a major medical problem (e.g., heart attack, cancer, multiple sclerosis). These clients need immediate help, with you offering both a listening ear and routes toward action.

Most people are resilient and recover after a disaster, despite their intense psychological and somatic initial reactions. These distressing but normal reactions to overwhelming events subside within weeks or months, either on their own or with the use of natural helping networks and resources in the family and community. Remember, all of our clients have resilience and capacity to deal with stress. Time can heal. In fact, providing counseling can sometimes be damaging.

The second type of crises is a more "normal" and "typical" part of life but can also be extremely challenging. Many of your clients will see you about divorce or the breakup of a long-term relationship, foreclosure of their home, job and income loss, a home break-in and burglary, or death of a loved one. Examples of children's crises include bullying, dealing with their own or parental illness, or leaving friends and moving to a new location. For some, not getting into the desired college or failing an important exam will become a crisis situation.

However, the more resilient the client, the less the need for counseling. "Loss, Trauma, and Human Resilience: Have We Underestimated the Human Capacity to Thrive After Extremely Aversive Events?" is the arresting title of an article by George Bonanno (2004). The author points out that trauma and crisis response is more complex than usually thought. This would seem to fit with the traditions of positive psychology and searching for client strengths. While we do not want to exaggerate this possibility, it seems logical that reminding clients of their strengths in the midst of crisis may indeed produce resilience, both now and in the future.

With many crisis survivors, appropriate follow-up and further counseling can be helpful. Interviewing and counseling skills are obviously needed in both phases, but the immediate crisis will demand more cognitive and emotional flexibility—and the ability to join the client in the here and now. The second phase usually gives you more time to develop a working relationship and will begin to look like a more typical counseling situation.

THE TEAM APPROACH TO CRISIS

A more comprehensive team approach is required in serious crises such as when a person with a gun approaches and attacks a school, university, bank, or office. The aftermath of a bombing, fire, earthquake, or other disaster will typically need follow-up group and individual work.

As one example of the team approach, Mary Bradford Ivey, as counselor in an elementary school, was with the excited children when the classroom watched the liftoff of the NASA spaceship *Challenger*. Their excitement turned to fear and tears when the ship blew up in front of them. Mary and the teaching staff faced a major here-and-now crisis.

First, the teachers had worked with Mary to plan ahead in case a school crisis should occur. It had been agreed that the teachers would take the children to their homerooms and encourage them to talk and ask questions, with awareness that each child would have her or his own unique reactions. It was important that the teacher maintain composure and a calming atmosphere in the classroom, reassuring the children that they were safe and encouraging them to ask questions. Mary went from classroom to classroom supporting each teacher and the students. If one or more children were particularly upset, Mary took them with her to the school counselor's office.

Debriefing continued the next day and later that week. Some children wanted to express their feelings through art. Mary had extensive experience in group work and set up several sessions for students who wanted or needed to explore issues in more detail. While virtually all children proved resilient in the long run, appropriate short-term calming interventions are important in the early stages of crisis.

The message here: Prepare for crisis. We never know when one will occur and what the crisis will be.

▶ Discerning Specifics of Crisis Counseling

While we are first addressing those who have just survived a major crisis such as a fire or flood, the suggestions here also hold for those you may meet after the event.

NORMALIZING

Normalize, normalize, normalize.

—Allen Ivey

Do we call those who experience trauma "victims," "survivors," or, more simply, just "people who have encountered a disaster"? The latter terms are more empowering for clients and puts them more in control. Thinking of people as victims tends to depersonalize them and put them in a helpless position controlled by external forces.

Disorder is an inappropriate term that ignores that the client has responded in a normal fashion to an "insane" situation. Many crisis workers object to the term *posttraumatic stress disorder* (PTSD), pointing out that virtually any serious encounter with crisis will produce extreme stress and challenges to the whole physical and mental system. An alternative frame of reference is *severe stress reaction* (SSR). Your clients have gone through a far from normal experience. Helping clients see that their "problem" is not inside them but the logical result of external stressors is one step toward normalizing the situation. Thus, the cultural/environmental contextual focus becomes especially important, as we want to avoid attributing any client response as solely "in the person." All survivors of a crisis need to know that however they responded to severe challenges is OK and to be expected (see Box 14.1).

At the same time, posttraumatic stress "disorder" (PTSD) has become part of daily language for both professionals and the lay public. As such, it is best that you do not disagree with others who prefer more traditional terminology.

BOX 14.1 | Normalizing and Treating Severe Stress Reaction (SSR) (aka PTSD)

War, rape, abuse, constant harassment or bullying, even extreme parental pressure on middle-class students can produce severe stress reactions (SSR). The word *disorder* attributes the concern as something problematic in the client, while simultaneously ignoring what happened in the disordered real world. A severe and marked reaction to stress is indeed a logical and natural cognitive and emotional response.

Allen learned this simple point years ago when working with Vietnam veterans who had been sent to a Veterans Administration (VA) hospital. His instructions were to search for malingers, veterans who "made up" their issues and behavior without a rational reason.

Using psychoeducational microskills and video feedback, Allen taught the patients, some of whom had been in the hospital for years, attending behavior and listening skills (Ivey, 1973). He also emphasized the importance of body awareness and introduced relaxation training on the locked ward. Patients viewed themselves and selected the behaviors that they wanted to change. It was impressive to see extremely active vets diagnosed as bipolar sit

back quietly or diagnosed depressed vets look up with interest as they viewed themselves and decided what they wanted to change. This, coupled with some basic narrative and cognitive therapy, led to rapid change and movement out of the hospital.

As Allen listened to the stories of the veterans, he also focused on their strengths and resources. The more he listened, the more he learned that the veteran "patients" were not as "sick" and "pathological" as the hospital staff and VA believed. Given the vivid experience and horror of Vietnam, the fact that they survived at all was impressive. Their stories emotionally paralleled those of Viktor Frankl in the Auschwitz death camp. Many of these veterans felt guilty and ashamed at their "breakdown." It does not take a lot of thought to realize that SSR is a normal and natural logical consequence of severe stress.

Normalize, normalize, normalize has become one of the mottos of the microskills approach. Expect the vast majority of your clients to have a logical and natural explanation for their behavior, even behavior that you find troublesome and challenging.

Regardless of terminology, crises and trauma are typically imprinted deeply in the brain and sooner or later may bring out sudden and unexpected fears, crying, or anger. These expressions may or may not be stimulated by the immediate situation. Watch for other signs of distress, such as flashbacks and intrusive thoughts, sleeping difficulties and dreams, and avoidance behavior (detachment from friends or past activities). Crises can turn into severe stress reactions. Refer these clients to experienced professional helpers.

CALMING AND CARING

A second major concept is providing some sense of calm and possibility in the situation, again normalizing the situation as a natural reaction. This means establishing an empathic relationship by indicating you care and will listen. Often the first priority in a crisis is that you must be calm and certain. Your personal bearing will do much to meet this first criterion of effective crisis work—calming the client.

Don't say "Calm down, it will be OK" or "You're lucky that you survived." Better calming language includes such comments as "It's safe now" (if that is true), "We will see that this situation is taken care of," "I feel bad myself, that was a terrible thing to go through," "Your reaction and what you did are common and make sense," and the critical "What would help you right now?" For those who are having flashbacks (and this occurs with all types of trauma), calming and normalizing what they are experiencing remain important.

In particular, do not minimize the crisis. Think about the survivors of Hurricane Katrina, with 20,000 people housed in the New Orleans Superdome starting August 4, 2005. It was not until September 4 that the last group was evacuated to new shelters, where their crises continued. While they were at the Superdome, portions of the roof were blown away and water poured in. Food and drinking water were in short supply, and toilet facilities failed. Sleeping was more than a challenge.

Louisiana governor Kathleen Blanco termed the use of the Superdome an "experiment." The following comments are attributed to Barbara Bush, the president's wife, after her visit to view the survivors of Katrina:

> Everyone is so overwhelmed by the hospitality. And so many of the people in the arena here, you know, were underprivileged anyway, so this is working very well for them.

Consider these two attitudes and comments as examples of the privilege of the entitled, described in the multicultural section of Chapter 2. These same attitudes toward minority and less privileged groups are often seen and heard in the media.

In some situations, you will think and even know that the client is overreacting. Be aware of your thoughts and feelings, which may be valid, but join the client where he or she is. "Enter the client's shoes," as Carl Rogers might say.

ASSURING SAFETY

Crisis and trauma survivors need to know that they are safe from the danger they have gone through. With soldiers and many others suffering from serious posttraumatic stress, developing a sense of safety and calm may not be accomplished immediately. Offer verbal reassurance that the crisis is over and they are now safe—again, if they are indeed safe. However, more than words may be needed. A woman who experiences spousal abuse or a homeless and hungry person needs to find a safe house or a place to stay and eat immediately. To some counseling and therapist supervisors, this is "violating boundaries."

This type of thinking is a relic of the field's past, but still there are some with these attitudes and beliefs. Stand up for what is right, help clients find what they need, connect them with resources. And consider the statement, with appropriate timing, "I'll be there with you to help."

TAKING ACTION

A good place to start is "What do you need now?" "What help do you want?" For yourself, what can you do that is *possible* in the here and now, and in the future? Do not overpromise. As noted above, some clients need a place to stay that night. Others need to know facts immediately. "Will I have to go through a vaginal exam after the rape?" "Are we going to be taken away by a bus?" "Where is the high ground in case the water comes again?" Answering these and other questions calmly and clearly will do much to alleviate anxiety.

The next step is to stay with clients and ensure that their needs are met. Crisis situations are often confusing. Following a major hurricane or flood, many survivors may never see you again, thus experiencing even more anxiety and tension. Prepare them for what to expect after you have finished your conversation. Volunteers in the Haiti earthquake went to help with good intentions and, indeed, provided valuable assistance. But soon they had to return home, and often people were left "up in the air" with no knowledge of what to do next.

DEBRIEFING THE STORY

Have you ever talked to a family member or friend following a difficult and traumatic hospital operation? Have you noted that they often give you detail after detail? And then, the next time you see them, they tell you the same painful story . . . and perhaps even a third or fourth time. Freud called this "wearing away the trauma." People need to tell their stories, and many must tell them again and again. Here the basic listening sequence becomes the treatment of choice. And if you paraphrase, reflect emotions, and summarize what they have said authentically and accurately, they will know that someone has finally heard them.

FOLLOW-UP

Concrete action in the immediacy of crisis is essential. Where possible, you want to arrange to meet the client again for debriefing and planning in more detail for the future. In some cases, longer-term counseling and therapy will be needed.

Watch for strengths and resilience. If given sufficient early support, most people work through their crises. They have internal strengths that will carry them through. Look for these strengths and external resources that will enable them to recover. At the same time, even the most resilient survivor needs to debrief what has happened.

IMPLICATIONS FOR YOU

There is much more to crisis counseling than what is said here, but you will find that competence and expertise in the basic listening sequence and five-stage structure will provide a map that will help carry you through some challenging situations.

All of us need to be ready to help in crisis situations. We may deal with immediate crises such as the ones we have focused on here, but we also need to understand the concepts underlying crisis counseling because so many clients will have experienced or are experiencing right now very difficult, even impossible, situations.

Also think of the need for counselors to debrief what they have seen at the scene of an accident—visualizing a dead child with a bloody mother stuck in a seat belt, the father stunned and speechless. EMTs (and police and firefighters) don't forget experiences like this; it wears on them emotionally and frequently leads to depression. There is a real need to provide counseling and support after such traumatic experiences.

There is also counselor and therapist trauma burnout. Counselors often suffer burnout when they work with a major disaster for several days, listen to endless sad stories on a crisis line, or just do daily work and intervention at a mental health center. The continuous load of people in crisis wears on helpers, who may even become traumatized themselves as they listen to horrific stories. Counselors need support when they work with these difficult situations. Counseling and therapy for the counselor needs to be considered as part of this support process.

Microskills and the five-stage interview give you a start to understanding work with crisis, but you have much more to learn to be fully helpful in such situations. At the same time, some crisis situations require many helpers and counselors. Seek some training, and offer yourself to others.

SUGGESTED SUPPLEMENTS

These textbooks are commonly used in the field, and you will find it beneficial to have at least one of them on your bookshelf:

Kanel, K. (2015). *A guide to crisis intervention* (5th ed.). Stamford, CT: Cengage Learning.

James, R. K., & Gilliland, B. E. (2013). *Crisis intervention strategies* (7th ed.). Belmont, CA: Brooks/Cole.

Miller, G. (2012). *Fundamentals of crisis counseling.* Hoboken, NJ: Wiley.

There are a number of certification programs available on the Internet. Some will be helpful, but study each with real care.

▶ Observing Crisis Counseling in Action

Each type of crisis is different; adapt your approach accordingly. Establish trust and the working relationship as quickly as possible. You will often have to act swiftly and sometimes decisively to help your clients reach the next stage beyond that first session. For an overview of what might happen in a major crisis, see Box 14.2. Community action is often the first thing before we move to individual counseling or coaching.

Below is a transcript illustrating what might happen in helping a family deal with the loss of their apartment after a fire. The fire department of a large city took Dalisay Arroyo and her children to her parents' small apartment, where they spent the night, but they obviously can't stay there more than a few days. Thus, as the counselor prepares for her session with the mother, first responders have already worked with early crisis safety and basic needs. She may need to focus more on emotional reactions and planning for the future.

Angelina meets Dalisay the morning after the fire in the office of one of the managers at the community center. Dalisay, 31, is employed as an aide in a nursing home and has two children. The father has only occasionally been involved since their birth.

This transcript is an edited and condensed version of a half-hour session. This large city has a history of preparation for crisis as it is both in a flood zone and subject to summer fires. In addition, Homeland Security has strengthened existing resources. Thus, this interview occurs in an ideal support system, something that is not available in all settings. We will review the crisis situation again after the transcript, outlining what can be done in more difficult situations with inadequate support systems.

BOX 14.2

Organizing a Crisis Team in a Major Earthquake: Plan for Long-Term Involvement

In 2011, a magnitude 6.3 earthquake struck Christchurch, a seaside city of 350,000 on the South Island of New Zealand. A total of 185 people were killed, and the central city was decimated, with about 1,000 buildings having to be demolished because of structural or land damage. Some 180,000 homes were damaged or destroyed. Serious aftershocks continued for months, adding considerably to the stress and uncertainty in people's lives.

Civil Defense met this challenge by immediately assembling 530 people into teams of four. The teams visited every home in the city over a period of two weeks, checking the safety of the damaged homes, arranging for immediate temporary housing and physical needs for residents, and checking each family for possibly needed crisis counseling.

Working with another psychologist, Robert Manthei, a professor of counseling at the University of Canterbury, helped organize, brief, and debrief the mental health members of the home visit teams. While the families welcomed the care, there appeared to be no large-scale immediate trauma among the people visited; instead, the vast majority were remarkably resilient and coping effectively.

Nevertheless, many families did need what we would term coaching support as they dealt with immediate problems of reality and key decisions that they had to make. It is possible, perhaps even likely, that this practical approach to community crisis counseling contributed to the building of wider community resiliency.

Interestingly, over the next two or three years, the major stressors that many Christchurch residents experienced centered around challenges posed by the government earthquake commission and individuals' insurance companies, which argued about how much damage existed, who should pay for it, and whether the land was safe to live on or to rebuild on. Some residents had to wait three years or longer for these decisions, leaving them in uncomfortable limbo with increasing anger and stress.

It is here that the coaching orientation could be of most use, but only if it combined empathic listening to each family's issues with effective suggestions for resolution, including referral to appropriate sources and services. Practical answers and action were needed at this stage as residents were often caught up in ongoing disputes between funding organizations.

Community resilience in crisis is clearly widespread, but over time the continuing stressors of moving, financial uncertainty, community changes, and slow-moving government/insurance processes are often the most serious. Think of the New Orleans flood, the Haitian hurricane, or the tsunami in Japan. There is the immediate crisis, and then there is the long-term aftermath.

All too often, we think a crisis is "managed" when it is immediately over and we have settled people for a day or two. However, many possible long-term issues remain, and the availability of targeted, practically oriented counseling with a coaching orientation remains necessary.

INTERVIEWER AND CLIENT CONVERSATION	PROCESS COMMENTS
Stage 1: Empathic Relationship	
1. *Angelina:* (Walks to the secretary's office, smiles warmly, and invites Ms. Arroyo in) Hello, Ms. Arroyo, I'm Angelina Knox. You and your children have had a terrible night. I'm a community counselor here in town and want to see how things are going and how I might be helpful. But before we start, are there any questions that you might want to ask me?	(Self-disclosure) Angelina is ready to spend time on developing an empathic relationship, but like many trauma survivors, the client wants to start immediately. Suggesting that the client is free to ask you questions is one way to establish transparency in the session.

(continued)

Stages 2, 3, and 4—story and strengths, goals, and the start of restorying—occur together in this case, as will often happen in crisis counseling, but all need to be addressed.

2. *Dalisay:* Angelina, I'm not sure where I should go or what I should do. I don't want to stay with my parents much longer. They are good people, but they don't have room for us and they get impatient with the kids. All my furniture is gone. I don't know what to do. (Starts crying softly)	This is common type of statement in crisis. Clients are "all over the place," topic jumping. Some clients may be unable to talk coherently, while others angrily demand that action be taken immediately. Be ready for almost any reaction, and remember they are all normal and to be expected.
3. *Angelina:* It's really hard . . . really hard. (She sits in silence for a minute until Dalisay looks up.) Your reactions make sense and are totally normal. It will take some time to sort things out, but we have some resources here that will help. But before going on, could you tell what happened?	Angelina acknowledges Dalisay's feelings and encourages her to tell her story, while seeking to normalize her thoughts and emotions. (Reflection of feeling, reframe, open question—interchangeable empathy)

Clients reacting to trauma need to tell their stories. Some will tell them at length, while others may simply describe the bare facts. Emotions will vary from a loss of control to numbness without much feeling expressed. The following is a much shortened version of what was said over 5 minutes of interaction. More tears flowed, but Dalisay also expressed relief that no one was hurt.

4. *Dalisay:* I was almost asleep and then I smelled something strange in the kitchen. I went in and there was a small fire in the wastebasket. I must not have put the cigarette out. Then, all of sudden, it went "poof," and I ran to get the children out. . . . It spread so fast, but the neighbors called the fire department right away, and only our apartment is gone. But then we got out in the cold. Firemen wrapped their blankets around us, asked if we had any help, and then they took us to my parents' house, 10 blocks away. But during all that time, the children were frantic and I couldn't quiet them. Their dolls and toys are gone. They couldn't stop crying until Grandma held them. The fire chief called this morning and said that you could likely help me figure out what to do next. My father drove me here on the way to work, the children are with their grandmother, but I guess I'll have to walk back to them, but all my warm clothes are gone. I called the nursing home today and the shift supervisor said that I could have the rest of the week off, but that likely means no pay and I can hardly pay bills now. (Serious crying) I don't have enough money to rent a new place, but I can't stay with my parents.	Throughout this longer story, previous five minutes, Angelina offered solid attention and a fair amount of natural spontaneous body mirroring. Her comments were short and usually took the form of encouragers and restatement. She did acknowledge Dalisay's emotions, but did not reflect them, believing that would be more appropriate later. Angelina's interview behavior above represented interchangeable empathy.

(continued)

5. *Angelina:* Dalisay, you and your daughters have had a terrible experience, but it is good that you got out in time and had your parents to stay with, at least for now. I can sense the horror you must have experienced and felt, even though I wasn't there. And . . . then . . . the children. I can see that you worry about them. It's great that your parents were close, even though you can't stay. As I listen to you, I get the feeling that you already have some important strengths and some clear ideas about what needs to be done. That contact with your shift supervisor was wise and will be helpful in the long run. Not everyone . . .	What we see here is a brief summary of Dalisay's situation and recognition of her emotions without pressing issues. Angelina then brings in a family focus, along with feedback supporting what Dalisay has done already to remedy her situation. Dalisay, like almost all people in crisis, has assets, strengths, and resources. (Summary, strength-oriented feedback—interchangeable empathy with some additive dimensions)
6. *Dalisay:* (Interrupts anxiously) Thanks, but I can only be gone so long. As soon as everyone was safe, I started thinking how things could work out, and I realized that I must hang onto my job. It's really scary. How am I to manage?	"What am I to do next?" While listening skills remain central, this is the time for Angelina to move to more direct influence in the session. Dalisay needs listening and emotional support, but the real issue is action.
7. *Angelina:* I hear your worry. There are some things that our office can offer. We are lucky here in that we have a trauma relief center that provide much of what you need, including some limited financial help for a few days. You won't have to stay with your parents long, as I think we can arrange for a temporary furnished apartment for you. I've already contacted the Women's Center, and they have clothes and some kitchen essentials that will help. If you are interested, I'll call and arrange for a time for you to meet with them. So you see that there are several possibilities, but we don't want to do anything until it makes sense to you. How else can we be helpful?	Angelina again acknowledges emotions and comes up with very specific directives and suggestions as to how she and her agency can help. Is Angelia offering too much material aid so soon? Certainly it would not have been too soon to offer hurricane or flood survivors with clear statements to let them know what could actually be done for them. Many felt lost in the vagueness of helping efforts, and sometimes more was promised than would ever be delivered. Here we see a large city well prepared for crisis—Homeland Security has done its job here. (Provides information and structure—potential additive empathy with action)
8. *Dalisay:* (Seeming relieved) Wow, that is more than I expected. I thought I'd be left hanging like Katrina hurricane survivors. It's terrific that I can get an apartment, and I'm amazed at the possibility of financial help. This will enable me to keep my job and take care of my children. But the next thing is, what about the children and school?	We see positive movement on the Client Change Scale. Dalisay is moving from "I can't" to "I think I can," with Angelina's help. (Dalisay was able to hear Angelina's comments, and we see that what she said was indeed additive.)
9. *Angelina:* Well, we need to talk about that. We will try to find an apartment near your old place so that they don't have to change, but that might not happen. We will do the best we can. I know that is isn't easy for you or them.	It is important not to overpromise, but to say what you and your agency really can do in a straightforward fashion. Following information giving, we see a brief acknowledgment of feeling. (Information giving—lower level interchangeable empathy. Frankness in letting clients know what one can actually do is important. Don't set up false hopes.)

(continued)

10. *Dalisay:* I feel a little better, but still a lot anxious and worried.	More movement on the CCS.
11. *Angelina:* Clearly we need to talk over the fire in more detail, the fright you experienced, what it did to the children, and how you handled it. And then there is a lot of worry over what will happen next. If we can get together tomorrow or the next day, we can do some more serious debriefing of what happened. Would you like to do that?	(As noted, this is an edited version of the longer session.) The interview has now gone on for about 20 minutes, and it is important to provide the client with security about what will happen next and how you will follow up. Debriefing of the story and the trauma needs to start as soon as possible. Note the positive feedback coupled with awareness of emotions, then an open invitation to talk, rather than telling Dalisay that she has to return. This moves toward a more egalitarian relationship.
12. *Dalisay:* Angelina, you have been so much help and so understanding. Yes, I'd like to talk about what happened. All of us, the children and me, had nightmares last night. It wasn't good. But I know that we have to get settled, so it is good that you can talk with me more later.	Dalisay is much calmer than she was at the beginning of the session. She has moved from a Level 1+ on the Client Change Scale to a beginning Level 3. Emotionally, at least in the moment, she may have reached Level 4, but don't expect this to hold unless further counseling and support are provided.

Like many crisis interviews, this one moved from topic to topic and interview stage to interview stage. But now it is important that a clear, concrete conclusion follow. Those who have gone through trauma need (1) personal supportive contact; (2) understanding and clarification of the crisis trauma; (3) awareness of their own personal strengths, as well as what external resources are available to them; (4) some short-term achievable goals; and (5) an immediate, clear, concrete action plan, with arrangements for later personal follow-up and further discussion and debriefing as soon as possible.

Stage 5. Action	
13. *Angelina:* Let's write down together where we are and what we can and need to do before we meet again. Where shall we start?	This directive brings in Dalisay as an egalitarian partner in finding solutions. Angelina could tell her client what to do, but success is much more likely if Dalisay is respected and fully involved. (Additive empathy)
14. *Dalisay:* I really appreciate that you could help us find housing. Could we begin with that? (Angelina brings out paper and pen for both of them, and they start to work.)	It takes 10 minutes for Angelina and Dalisay to write down the action plan—who will do what and when. Dalisay occasionally starts to cry, but more easily regains self-control. Out of this come workable alternatives that can be implemented in stages. Follow-up on actions and debriefing of the trauma is important, but this has to be something that Angelina wants. The children also need to tell their stories, and the school counselor needs to be consulted. At some point, it may be useful to bring the grandparents in for a family session.

► Suicide Watch: Awareness Basics[1]

> People who attempt suicide are always subject to sociological risk factors, but they need an idea or story to bring them to the edge and justify their act. If you want to prevent suicide, you want to reduce unemployment and isolation, but you also want to attack the ideas and stories that seem to justify it.
>
> —David Brooks

Suicide rates vary widely around the world, and even by state and province, with no clear patterns. Some data for the United States are shown in Table 14.1. While U.S. motor vehicle traffic deaths have declined over the past decade, mortality rates for suicide are higher (Rockett et al., 2012). Internationally, we see the same trend. Youth suicide is of particular concern, with suicide now the second major cause of death in the United States among the 15–24 age group; only a year before, it was the third leading cause. Meanwhile, the numbers in middle age are rising as well. We suggest that you search for data on your home country and region, as updated information on suicide become available as statistics are gathered, often a year or more later.

BACKGROUND THAT MIGHT LEAD TO A SUICIDE ATTEMPT

A review of research literature lists 11 key factors to consider as indicating the possibility of a suicide attempt: severe anxiety, panic attacks, depression and inability to experience pleasure, alcohol, difficulty in concentration, sleeplessness, hopelessness, employment problems, relationship loss, a history of physical/sexual abuse, and especially a history of past deliberate self-harm or prior suicide attempts (Sommers-Flannagan & Sommers-Flannagan, 2012, p. 248). To this list we would add the dangers of drug abuse, serious health issues, and serious interpersonal conflict such as bullying or harassment. Bad economic times such

TABLE 14.1   Suicide Rates in the United States

	Number	Per Day	Rate	% of Deaths
Nation	39,518	108.3	12.7	1.6
Males	31,003	84.9	20.2	2.5
Females	8,515	23.3	5.8	0.7
Whites	35,775	98.0	14.5	1.7
Nonwhites	3,743	10.3	5.8	1.0
Blacks	2,241	6.1	5.3	0.8
Elderly (65+ yrs.)	6,321	17.3	15.3	0.3
Middle-aged (45–64)	15,379	42.1	18.6	3.0
Young (15–24 yrs.)	4,882	13.2	11.0	16.5

Source: American Association of Suicidology, 2010 data, www.suicidology.org

---

[1]John Westefeld, University of Iowa, a nationally recognized expert on suicide, reviewed this section, and we thank him for his comments.

as the recent Great Recession with difficulty in finding work to match one's talents, or any work at all, can also be a factor in suicide.

The availability of guns has become an important element in suicide. About half of suicide deaths in the United States occur this way (Westefeld, Richards, & Levy, 2011). While other routes may typically be chosen in physician-assisted suicide, guns are available as a quick solution. Needless to say, discerning the availability of weapons for depressed and suicidal clients is an important issue (Westefeld et al., 2012).

RISK ASSESSMENT

Understanding and working with suicide is not the province of this book, but it seems important that a few basics and suggestions for further reading and follow-up be provided. The most central of these issues are maintaining a watchful eye for suicide potential, providing immediate crisis support, and ensuring a careful referral with follow-up to ensure that the client actually appears for sessions. An excellent next step beyond this brief section is to download the *Suicide Risk Assessment Guide* at www.mentalhealth.va.gov/docs/suicide_risk _assessment_guide.doc. Several of its useful guidelines are summarized below.

STRENGTHS AND RESOURCES

The background that may lead to suicide was outlined above, but the *Risk Assessment Guide* suggests looking for strengths and resources to build on for both the here and now of the interview and the long-term safety of the client. The *Guide* points out the following, which will be familiar to you from our emphasis on positive psychology and strength-based approaches. Use all of these as you seek to support your client while you plan for appropriate referral.

▶ Positive social support
▶ Spirituality
▶ Sense of responsibility to family
▶ Children in the home, pregnancy
▶ Life satisfaction
▶ Reality testing ability
▶ Positive coping skills
▶ Positive problem-solving skills
▶ Positive therapeutic relationship

As in decisional counseling, it can be helpful to discuss alternatives for resolution of issues, but the client needs to be able to listen to you. We recommend keeping the counseling simple rather than theoretically complex. Be with the client empathically in the here and now.

WARNING SIGNS OF IMPENDING SUICIDE

The three key warning signs are (1) actual threat to hurt or kill oneself; (2) seeking access to pills, guns, or other routes; and (3) talking or writing about death, dying, or suicide (including giving away value objects or pets to friends or family). In these cases, take immediate action. The *Risk Assessment Guide* reminds us to remove anything lethal and keep the client safe, with some caring person available. Depending on the level of risk, get immediate help if necessary and facilitate moving to a hospital.

The basic principles of crisis counseling remain. Your calmness, empathic caring, and ability to listen are central, but you are also required to make decisions. As much as possible, share the decisions with the client. Often it will be important to bring in the family or friends to help provide further support and help implement any plan.

ASK KEY QUESTIONS

The *Risk Assessment Guide* includes a pocket card that summarizes key issues and recommended questions. Being direct is something that interviewers and counselors may have trouble with, but it is essential here. Be matter of fact; show concern, but not shock or worry. One suggested way to start is:

> I appreciate how difficult this problem must be for you at this time. Some of my patients with similar problems/symptoms have told me that they have thought about ending their life. I wonder if you have had similar thoughts?

They go on to suggest:

> Are you feeling hopeless about the present or future?
> If yes, ask . . .
> Have you had thoughts about taking your life?
> If yes, ask . . .
> When did you have these thoughts, and do you have a plan to take your life?
> Have you ever made a suicide attempt?

In this process, listening, nonjudgmental warmth, respect, and caring will facilitate openness and trust. Show that you are present in the here and now and available with understanding and support.

Avoid the *why* question. This is not a time for focusing on rational explanations. The client may want to swear you to secrecy. This is not possible as it could lock out key safety procedures. Thus, be respectfully honest and open about the nature of the relationship.

RECOMMENDED RESOURCES BEYOND THE *SUICIDE RISK ASSESSMENT GUIDE*

Anika Foundation. (2013). *Explaining the rise in youth suicide.*

Rockett, I., Regier, D., Kapusta, N., Coben, J., Miller, T., Hanzlick, R., et al. (2012). Leading causes of unintentional and intentional injury mortality. *American Journal of Public Health, 102,* e84–e92.

Sommers-Flannagan, J., & Sommers-Flannagan, R. (2012). *Clinical interviewing* (4th ed., pp. 245–278). New York: Wiley.

▶ Practicing Crisis Counseling

EXERCISE 14.1 Debriefing Experience With Crisis
Crisis comes in many forms—individual, family, community. What was the crisis? It may have been an accident, the loss of a family member, a job, a fire. Your community may have experienced flood, hurricane, earthquake, or tornado. With a friend, classmate, or colleague, share your own story or that of someone you know personally. As you listen to other people's stories, use the basic listening sequence and seek to understand their thoughts, feelings, and the meanings that they take out of what happened.

EXERCISE 14.2 Prepare Mentally for Both Individual and Community Crises

Most of the people reading this paragraph will encounter crisis. Review the concepts in this chapter and develop an outline of what you think you can do to help and promote resilience. Imagine counseling a person in a specific crisis. What might be your thoughts and feelings about helping that person? Step 1 in a major community crisis is your willingness to volunteer as part of a team. What thoughts and feelings occur here? Are you willing to carry heavy sandbags for hours or follow directions to conduct routine but important tasks? More difficult to plan for—what do you imagine your counseling role would be here?

EXERCISE 14.3 Group Practice: Crisis Counseling Role-Play

With a volunteer, each of you select a possible crisis topic. It will be most helpful if observers are there to video or audio record and comment and provide feedback. Follow the five-stage model by briefly developing a relationship, drawing out story and strengths, establishing immediate goals (or longer-term goals, if the role-play occurs a few days after the crisis), and ensuring that a concrete action plan is in place. You may find it helpful to use the Microskills Interview Analysis Form (MIAF) in Chapter 7 (Box 7.1).

EXERCISE 14.4 Study Disaster Mental Health in More Depth and Become Certified

In conjunction with Red Cross, the American Counseling Association has an excellent website with 15 valuable fact sheets and instructions on how to become a certified Red Cross Disaster Mental Health Volunteer (www.counseling.org).

EXERCISE 14.5 Discussion Around Issues of Suicide

We do not believe that the brief introduction here is sufficient to make you a suicide counselor, but it still could happen. Ideally, you will want further study and training, but there are some things you can do now. Think about suicide prevention, and generate a model suicide prevention program for a school or community. It may be useful to visit relevant websites. Screening for Mental Health has a good website offering ideas and free information (www.mentalhealth-screening.org).

Case Study: Minority Experience in Counseling Training. Supervisor/Employee Conflict in Japan—Will I Be Fired? What are the issues? How would you help? How would you use the basics of crisis counseling and brief counseling to help the client?
Interactive Exercise: Microskills and Theoretical Approaches to the Interview. The two approaches in this chapter are best understood by conducting an interview.
Interactive Exercise: Basic Competence. Review and select the most appropriate response from the point of view of brief therapy.
Group Practice Exercise: Review and Discuss. Use this exercise to discuss brief counseling.

▶ Refining the Skills of Crisis Counseling and Suicide Risk

Crisis counseling demands much from you, but it also provides many rewards when you can provide concrete help and see relief start to come in for the client and the family. But imagine what it feels like in a major crisis when you don't have the resources described above and you meet with 10 or more people who have just gone through a fire, flood, or earthquake in the middle of the night in the rain. These clients have even more needs, and they could be hungry. You may only have 15 minutes and never see the person again. Thus, the calming and caring that you can provide are all the more important.

Again, burnout can be a problem for the crisis counselor. There is also your own emotional involvement. You may care for clients and their future, but follow-up to make sure that they have followed an action plan may not be possible, leaving you wondering how helpful you (and the crisis team) really were. Thus, crisis counseling can often turn into a developing crisis for the helper. This means that you need to take care of yourself throughout each interview and day. Take breaks, seek to get enough sleep, try to get a little exercise, and make sure that you debrief your experiences with understanding colleagues and/or supervisors. Many people will recover after a traumatic event, while others will develop serious symptoms and issues. Providing interventions to all may prove detrimental. It is important to conduct an initial screening of trauma victims for known symptoms or risk factors for posttraumatic responses before providing treatment. Counseling and other appropriate mental health treatments are indicated for those who are unable to overcome the trauma without assistance (Zalaquett, Carrión, & Exum, 2010).

Summarizing Key Points of Crisis Counseling and Assessing Suicide Potential

Defining Crisis Counseling

▶ Crisis counseling is an action-oriented form of helping.
▶ Everybody experience crises or traumas in their lifetime. "Normalizing" reactions to crises is key.
▶ Crises may include natural events (e.g., hurricanes, earthquakes) or human made (e.g., war, rape).
▶ Working with crises includes two phases: (1) working through the initial trauma; and (2) appropriate follow-up and further counseling. Resilient clients may not need or may not appear for further help.

Discerning Specifics of Crisis Counseling

▶ Normalizing: Survivors of a crisis need to know that however they responded to severe challenges is OK and to be expected, thus normalizing the situation as a natural reaction.
▶ Crises can turn into posttraumatic stress, which requires more long-term counseling and therapy.
▶ Calming and caring: Provide some sense of calm and possibility in the situation. Establish an empathic relationship by indicating you care and will listen. Use calming language.
▶ Safety: Offer verbal reassurance that the crisis is over and they are now safe, if they are indeed safe. Help persons find basic shelter, medical attention if injured, food, and connection to other resources.
▶ Action: Focus on clients' needs and take action to fulfill those needs. Stay with clients and ensure that their needs are met.
▶ Debrief the story: The basic listening sequence is helpful here. Clients need to know that someone is hearing them.

► Follow-up: Concrete action in the immediacy of crisis is essential. Where possible and appropriate, arrange to meet the client again for debriefing and planning in more detail for the future. Watch for strengths and resilience.

► Burnout prevention: Counselors need self-care and support when they work with these difficult situations. Teamwork is helpful.

Observing Crisis Counseling in Action

► Each type of crisis is different; adapt your approach accordingly.

► Establish trust; the working relationship with client is essential.

Suicide Watch: Awareness Basics

► Youth suicide is of real concern. Suicide was the second major cause of death among youth 15 to 24 years old, and this appears to be increasing.

► Risk assessment: Assess for suicide potential, provide immediate crisis support, and ensure a careful referral with client follow-up when needed.

► Focus on strengths and resources to build on and increase long-term safety. Connect clients with positive social support.

► Observe the presence of three key warning signs: (1) making actual threats to hurt or kill oneself; (2) seeking access to pills, guns, or other routes; (3) talking or writing about death, dying, or suicide.

► Do not hesitate to ask key questions and listen for the answers in a nonjudgmental and respectful way.

Practicing Crisis Counseling

► Use the exercises in the book and in the web resources to prepare for both individual and community disasters.

Refining the Skills of Crisis Counseling and Suicide Risk

► A calming and caring approach is helpful to the client.

► Prevent burnout by taking breaks, getting enough sleep, keeping some exercise routine, and consulting with colleagues and supervisors.

► Offer follow-up counseling and other appropriate mental health treatments to those who are unable to overcome the trauma without assistance.

Assess your current level of knowledge and competence as you complete the chapter:

1. Flashcards: Use the flashcards to check your understanding of key concepts and facilitate memorization of key information.

2. Self-Assessment Quiz: The quiz will help you assess your current knowledge and prepare for course examinations.

3. Portfolio of Competencies: Evaluate your present level of competence on the ideas and concepts presented in this chapter using the Self-Evaluation Checklist. Self-assessment of your competencies demonstrates what you can do in the real world.

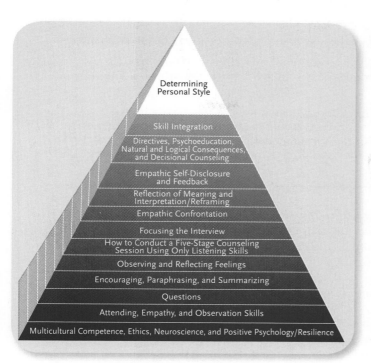

Determining
Personal Style

Skill Integration

Directives, Psychoeducation,
Natural and Logical Consequences,
and Decisional Counseling

Empathic Self-Disclosure
and Feedback

Reflection of Meaning and
Interpretation/Reframing

Empathic Confrontation

Focusing the Interview

How to Conduct a Five-Stage Counseling
Session Using Only Listening Skills

Observing and Reflecting Feelings

Encouraging, Paraphrasing, and Summarizing

Questions

Attending, Empathy, and Observation Skills

Multicultural Competence, Ethics, Neuroscience, and Positive Psychology/Resilience

Chapter 15
Determining Personal Style in a World of Multiple Theories

Intentionality is your ability to flex and cope with a multidimensional world.
—Allen and Mary Ivey

The key to change is DECIDING AND ACTING.
—Carlos Zalaquett

You can use microskills with many different styles and theories of interviewing, counseling, and psychotherapy. Intentional competence in the microskills and the five-stage interview enables you to understand and quickly master multiple approaches for facilitating your clients' growth.

But it is you who will put together and integrate the ideas of this book with your natural style of being. What skills and methods make most sense to you? Given that more than 500 theories of counseling and therapy are available for your consideration, how will you decide what comes next? What will your actions be?

Chapter Goals

Awareness, knowledge, skills, and actions developed through the concepts of this chapter and this book will enable you to:

▶ Briefly examine theories of counseling and psychotherapy as possible future systems for your own practice of helping.

▶ Visit the web resources and see more detailed information on how microskills and the five-stage interview structure are used in cognitive behavioral therapy, brief solution-based counseling, motivational interviewing, and coaching. Transcripts, web links, and practice suggestions can also be found there.

▶ Review skills, strategies, and competencies that you have developed during the term.

▶ Plan and commit to an intentional lifetime of movement, change, and constant growth as a helping professional.

Equipped with the foundational skills of listening, observing, influencing, and structuring an interview, you are well prepared to enter the complex world of theory and practice. Figure 15.1 shows how the microskills can be used in a wide variety of theoretical approaches. In effect, if you have become competent in the skills of the microskills hierarchy, you have ready access to many approaches to facilitating client growth. Note that the last column is for you to complete so that you can compare your skill competencies with 10 theoretical orientations.

Completing a full interview using only listening skills provides an introduction to Carl Rogers's (1961) person-centered theory. You likely have discovered that many clients are self-directed and, with a good listener, can do much to resolve their issues on their own. Mastery of completing a full interview using only listening skills does not make you a person-centered counselor, but it is a very important start. Study Rogers's work and complete an interview using *no questions* at all.

Decisional counseling, the least mentioned and studied approach in most texts, remains a central part of what most counselors, therapists, and coaches actually do in practice. Ultimately, whether we work in cognitive behavioral, narrative, or any other form of modern helping, understanding the decisional process remains essential. Decisional counseling (sometimes known as problem-solving counseling) may be the least recognized but most widely practiced form of helping in current practice. Samples of decisional daily practice include social work, employment counseling, placement counseling, AIDS counseling, alcohol and drug counseling, school and college counseling, and the work of community volunteers and peer helpers.

We all will deal with crisis at some point in our personal and/or professional lives. Crisis counseling, as well as issues around suicide, has been presented as we all need to think about and be prepared for the challenges we will all face at some time. As many crisis centers and hotlines are staffed with counselors who are not fully recognized as the professionals that they are, it seems vital that some introductory material on crisis be made available.

Narrative therapy, with its emphasis on storytelling and restorying, is closely allied to the emphasis in this book. A particular strength is its awareness of the importance of externalizing the story and helping clients discover that they are not the sole cause of their issues. Rather, the internal focus needs to be balanced with the cultural/environmental context.

The focusing skill is also important in multicultural counseling and therapy (MCT). With the RESPECTFUL model and the emphasis throughout on multicultural awareness, this book is sometimes used in multicultural courses.

For some of your clients, reflection of meaning will be the most useful and memorable part of their sessions with you. Finding vision, purpose, and life goals can enable the client to work and live through almost any issue. You have had an introduction to Viktor Frankl's logotherapy, discernment, and "those that have a *why* for living will find a *how*."

FIGURE 15.1 Microskills patterns of different approaches to the interview.

MICROSKILL LEAD	Decisional counseling	Person-centered	Crisis counseling	Multicultural and feminist therapy	Logotherapy	Cognitive behavioral therapy	Brief counseling	Narrative therapy	Motivational interviewing	Counseling/coaching	Psychodynamic	Eclectic	Business problem solving	Your favorite competencies
BASIC LISTENING SKILLS														
Open question	●	○	◒	◒	●	◒	●	◒	●	●●	◒	◒	◒	○
Closed question	◒	○	◒	◒	◒	●	●	◒	◒	◒	○	◒	◒	○
Encourager	●	◒	◒	◒	●	◒	◒	◒	●	●	◒	◒	◒	○
Paraphrase	●	●	◒	◒	●	◒	●	◒	●	●	◒	◒	◒	○
Reflection of feeling	●	●	◒	◒	◒	◒	◒	◒	●	◒	◒	◒	◒	○
Summarization	●	◒	◒	◒	●	◒	●	◒	●	◒	◒	◒	◒	○
INFLUENCING SKILLS														
Reflection of meaning	◒	●	○	◒	●	○	○	◒	◒	●	◒	◒	◒	○
Interpretation/reframe	◒	○	◒	◒	◒	◒	◒	●	◒	◒	●	◒	◒	○
Logical consequences	●	○	◒	●	○	◒	○	◒	●	◒	○	◒	◒	○
Self-disclosure	◒	◒	◒	◒	◒	○	○	◒	◒	◒	○	◒	◒	○
Feedback	◒	◒	○	◒	◒	◒	◒	◒	●	◒	◒	◒	◒	○
Instruction/ psychoeducation	●	○	◒	●	◒	●	○	◒	◒	●	○	◒	●	○
Directive	◒	○	◒	◒	◒	●	○	◒	◒	◒	○	◒	●	○
CONFRONTATION (Combined skill)	◒	◒	○	●	◒	◒	◒	◒	●	◒	◒	◒	◒	○
FOCUS														
Client	●	●●	◒	◒	●	●	●	●	●	◒	●	◒	◒	○
Main theme/issue	●	○	●	◒	◒	◒	●	◒	●	◒	◒	◒	●	○
Others	◒	○	◒	◒	◒	◒	◒	◒	◒	◒	◒	◒	○	○
Family	◒	○	◒	◒	◒	◒	◒	◒	◒	◒	◒	◒	○	○
Mutuality	◒	◒	○	◒	◒	◒	◒	◒	◒	◒	○	◒	○	○
Counselor/therapist	●	◒	○	◒	◒	○	○	◒	○	◒	○	◒	○	○
Cultural/environmental/ contextual	◒	○	●	●●	◒	◒	◒	●	◒	◒	○	◒	◒	○
ISSUE OF MEANING (Topics, key words likely to be attended to and reinforced)	Decision making	Self-actualization, relationship	Crisis action	Impact of Environment	Values, meaning vision for life	Thoughts, behavior	Problem solving	Values, meaning vision for life	Change	Finding and acting on life vision	Unconscious motivation	Client determined	Problem solving	
COUNSELOR ACTION AND TALK TIME	Medium	Low	High	Medium	Medium	High	Medium	Medium	Medium	Medium	Low	Medium	High	

LEGEND

● Frequent use of skill ◒ Common use of skill ○ Occasional use of skill

The web resources offer additional specifics and details of theoretical systems that your instructor may choose to add. There you will find basic information, transcripts, and exercises in the following four systems, as well as additional information on Carl Rogers's person-centered counseling, narrative therapy, and Viktor Frankl's logotherapy.

Cognitive behavioral therapy (CBT). CBT has currently become the favored mode of practice in counseling and therapy. The listening skills remain central. CBT generally gives less attention to reflecting feelings, although awareness of the centrality of emotion is increasing.

Brief solution-oriented counseling. Brief work is very closely related to decisional counseling and the five-stage interview structure. In a time of accountability and results, goal-oriented brief work will likely become very much a part of your daily practice.

Motivational interviewing (MI). MI relies on the listening skills and is easily understood with your mastery and competence in interpersonal influence.

Coaching. We have a strong belief in the coaching model. It relies heavily on the listening skills and has special strengths in its emphasis on positive psychology, therapeutic lifestyle changes, and working with clients from an egalitarian approach. With competence in the skills of this book and further study, you are well prepared for further study and later certification as a coach. The best coaches are those with a solid interviewing and counseling background.

 Group Practice Exercise: Reflect, Identify, Share, and Plan for Advancement

▶ Intentionality and Flexibility in Your Interviewing Style and in Your Future Career

DETERMINING PERSONAL STYLE AND THEORY	ANTICIPATED CLIENT RESPONSE
As you work with clients, identify your natural style, add to it, and think through your approach to interviewing and counseling. Examine your own preferred skill usage and what you do in the session. Integrate learning from theory and practice in interviewing, counseling, and psychotherapy into your own skill set.	You, as a developing interviewer or counselor, will identify and build on your natural style. You will commit to a lifelong process of constantly learning about theory and practice while evaluating and examining your behavior, thoughts, feelings, and deeply held meanings.

Video or audio recording a new interview and transcribing it gave you a chance to evaluate and classify your interviewing style. We encourage you to do this again and again over your career and obtain feedback from others, including your clients. Whether it's a mock interview or one with a real client, creating a transcript and then evaluating your microskills is one of the best ways to learn. If you haven't already done so, we urge you to go back so that you can define your skills and areas for growth more precisely.

Some interviewers and counselors have developed individual styles of helping that require their clients to join them in their view of the world. Such individuals have found the one "true and correct" formula for the session; clients who have difficulty with that formula are often termed "resistant" and "not ready" for counseling. Such counselors and therapists can and do produce effective change, but they may be unable to serve many types of clients. The complexity of humanity and the helping process is missing from their orientation. They do not recognize that some other approach may be more effective and faster than their own fixed style.

Remember, all counseling is multicultural, and your awareness, knowledge, and skills in multicultural counseling are critical to building rapport with your clients. Consequently, it is urgent that you remain aware that your client may not prefer the skills, strategies, and theories that you favor. Expand your skills and knowledge in areas where you are now less comfortable. With patience, study, and experience, you will increase your ability to work with more and more varied clients. The opportunity to learn from clients who are different from us is one of the special privileges of being an interviewer, a counselor, or a therapist.

Many of you will continue to study interviewing and counseling and will have long and useful careers in the helping fields. But it is equally likely that you may find yourself managing your own business, becoming a salesperson, or working as a physical therapist or nurse, a teacher, or in any of a myriad of other careers. Regardless of where you go, communication skills will be basic to the job and to your advancement. Microskills, the first systematic communication skills program in the world, is now used throughout that world in training settings ranging from AIDS peer counselors in Africa to executives in Sweden and Japan to Aboriginal social workers in Australia. You will find many places to use your interviewing and communication skills. And don't forget the importance of listening carefully and fully to your loved ones and friends. Research shows that relationships "wear out" over time if not renewed. Intentionality and flexibility are critical wherever we go.

Box 15.1 presents an international perspective on using microskills throughout a professional career. Note how Dr. Mary Brodhead has intentionally moved ahead through opportunities to apply her knowledge and experience in new settings. She is now in a field somewhat distant from her original work in teaching and counseling. Her experience provides an illustration that a career in counseling involves constant change and learning.

Developing your own personal approach to interviewing and counseling involves a multiplicity of factors. Be ready to learn new things and change throughout your career.

 Website: Personal Authenticity Project: Markers on the Path to Personal Authenticity

---

BOX 15.1 National and International Perspectives on Counseling Skills

Using Microskills Throughout My Professional Career
Mary Rue Brodhead, Executive, Canadian Food Inspection Agency

My first encounter with the microskills program was through my master's program in teacher education. I wanted to learn about counseling so that I could better reach teacher trainees. What I learned very rapidly was that teachers often fail to listen to their students; in fact, one research study found that out of nearly 2,000 teacher comments, there was only one reflection of feeling and, of course, most comments focused on providing information.

(continued)

| BOX 15.1 | (continued) |

Often when teachers did listen, they weren't able to recognize the verbal and nonverbal cues that were reflecting the reality of their students.

After completing my degree, I entered teacher education and found that microskills led to more student-centered teaching. I also found that teachers who matched their students' cognitive/emotional style were more effective (see Ivey et al., 2005). If a student is concrete, the teacher needs to provide specific examples and use concrete questions. If the student is more reflective, then abstract formal-operational strategies can be used. Bringing in emotional involvement via microskill strategies enriches teaching.

Microskills in Counseling: A Multicultural Experience

I next lived on an island off Vancouver where my husband Dale and I worked with the Kwakiutl Nation. I trained teachers, but I also counseled and led a group of school dropouts in a special program under the School Board. I actually taught them the same interviewing skills that you are learning in this book. Needless to say, the multicultural orientation helped sensitize me so that I adapted my counseling and teaching to better fit Kwakiutl style.

One of our first activities was a trip in the Chief's fishing boat to gather Christmas trees for the old people, a cherished tradition in the community. This was the first time that most of these young people had participated in the ritual. As they delivered the trees, the recipients, in an expression of gratitude, invited them in for something to eat and began to tell stories. These old people, thrilled to have an audience, would shower attention on the alienated youth who, in response, would listen with attention and respect. And so an upward cycle of communication began. Eventually we raised money to buy tape recorders and to capture these tales for the future, leading to further strengthening of the students' listening and questioning skills. Attending behavior increased respect and communication on both sides.

Microskills in the Workplace: Training Executives to Listen, Hear, and Understand

After three years in British Columbia, Dale and I returned to Ottawa, where I directed an employment equity program. The task centered on providing culturally sensitive counseling and opportunities to members of the Canadian four "equity groups" (visible minorities, Aboriginal peoples, persons with disabilities, and women in non-traditional occupations) to develop the skills needed for career development and success in the Federal Public Service.

As part of my work, I used microskills to train government officials to listen and really hear and understand the variety of perspectives, strengths, and styles of working found within members of our multicultural workforce. Here I learned the importance of language, and I became reasonably fluent in French, an essential skill in bilingual, multicultural Canada. I developed a multicultural counseling course used within 15 universities in Canada that included many ideas presented here.

Microskills in Management: Creating a Culture of Learning and Team Building

My current agency is responsible for animal health, plant protection, and food safety. It is one of those "science departments" involving agriculture, fisheries, scientific research, and many other issues. Given the diversity of my workforce, my first task has been to build a "culture of learning" in which vast amounts of information and knowledge can be shared effectively and efficiently. Microskills are key elements in management training, and a good communication skills workshop can be vital in team building. And, of course, all my managers need to listen to and motivate those with whom they serve.

Looking back over my career, it is amazing to find that the basic microskills have been useful in my teaching, counseling, multicultural work, and governmental leadership positions. "Training as treatment" and "teaching competence" in microskills can help us all make a difference throughout our careers.

▶ Time to Pause, Reflect, Assess, and Take Stock

Think about yourself, your values, and your personal experiences throughout this course. What have you learned about yourself? What is your current perception of interviewing? How have you changed? What else would you like to change?

> **EXERCISE 15.1 Revisiting Your First Audio or Video Recording**
> In Chapter 1 we suggested that you record an interview. This exercise documented the baseline of your natural style and skill level. Now is a good time to go back and examine your first interview. What do you notice and feel about your skills? Do you prefer some skills over others? What would you do differently now? If you were an observer, what would you suggest to the interviewer? What other skills would you use as a result of being in this course? Do the same with the interview completed in Chapter 13 (page 268). After reviewing your interviews, consider: What are your strengths and assets? How would you bring these to action?

You have been presented with more than 30 major concepts and skill categories. Within those major divisions are more than 100 specific methods, theories, and strategies. Ideally, you will commit them all to memory and be able to draw on them immediately in practice to facilitate your clients' development and progress in the interview. As you grow as an interviewer, counselor, or therapist and your competence and skill mastery increase, you will find the ideas expressed here becoming increasingly clear and a natural part of your practice.

Retaining and mastering the concepts of the microskills hierarchy may be facilitated by what is termed **chunking**. We do not learn information just in bits and pieces; we organize it into patterns. The microskills hierarchy is a pattern that can be visualized and experienced. For example, at this moment you can probably immediately recall that attending behavior has certain major concepts "chunked" under it (culturally appropriate visuals, vocal tone, verbal following, and body language—"three V's + B"). You can probably also recall the basic listening sequence, the purpose of open questions, and perhaps which questions lead to which likely outcomes (e.g., *how* questions lead to process and feelings).

Identify and classify the concept. The microskills language provides a vocabulary and communication tool with which to understand and analyze your interviewing and counseling behavior and that of others. If the concept or skill is present in an interview, can you label it? If you can identify concepts, you have most likely chunked most of the major skill points together in your mind. You may not immediately recall all the types of focus, but when you see an interview in progress you will probably recall which one is being used.

Demonstrate basic competence. Basic competence means you will be able to understand and practice the concept in an interview. Continued practice and experience become the foundation for later intentional mastery.

Demonstrate intentional competence. Skilled interviewers, counselors, and therapists can use the microskills to produce specific, concrete effects with their clients. Appendix A presents the full Ivey Taxonomy and reviews specific aspects of intentional prediction. If you reflect feelings, do clients actually talk more about their emotions? If you provide an interpretation/reframe, does your client see her or his situation from a new perspective? If you work through some variation of the positive asset search, does your client actually view the situation more hopefully? If you conduct a well-formed five-stage interview, does your client's self-concept and/or developmental level change? Does your interview produce client changes?

> **EXERCISE 15.2 Determine Your Mastery of Skills**
> Use the self-assessment in Box 15.2 to assess your competence and mastery of each concept.

 Group Practice Exercise: Share Your Reflections and Your Plans to Continue Your Journey

Take some time to review your own competence level in each of the major microskill areas and indicate your personal competencies on this chart. You may use this as a summary and as a plan for future growth. Can you identify each skill or concept and classify its place in the interview? Can you demonstrate basic competence by using the skill in the session? Most important, can you obtain specific and predictable results in client behavior as a result of your use of the skill or concept? Feel free to use the glossary to help refresh your memory.

Skill or Concept	Identification and Classification	Basic Competence	Intentional Competence	Evidence of Achieving Competence Level
1. Attending behavior				
2. Questioning				
3. Observation skills				
4. Encouraging				
5. Paraphrasing				
6. Summarizing				
7. Reflecting feelings				
8. Basic listening sequence				
9. Positive asset search				
10. Empathy				
11. Five stages of the interview				
12. Confrontation				
13. The Client Change Scale				
14. Focusing				
15. Eliciting and reflecting meaning				
16. Interpretation/reframe				
17. Logical consequences				
18. Self-disclosure				
19. Feedback				
20. Information/psychoeducation				
21. Directives				

(continued)

BOX 15.2 (continued)

Key issues and practical applications of microskills and the five-stage interview				
22. Analysis of the interview (Chapter 13)				
23. Ethics				
24. Multicultural competence				
25. Social justice				
26. Wellness				
27. Community genogram				
28. Decisional counseling				
29. Person-centered counseling				
30. Crisis counseling and suicide assessment				
31. Therapeutic lifestyle changes				
32. Neuroscience and the interview				
33. Defining personal style and self-assessment				

Other important competencies such CBT, Coaching, Motivational Interviewing, Person Centered Counseling, and Brief Counseling can be found in MindTap.

▶ Your Personal Style

There are two major factors to consider as you move toward identifying your own personal style. Unless a skill or practical application of a set of skills harmonizes with who you are, it will tend to be false and less effective; however, modifying your natural style and theoretical orientation will be necessary if you are to be helpful to many clients. Remember that you are unique, and so are those whom you would serve. We all have varying issues and problems. We all come from varying families, differing communities, and distinct views of gender, ethnic/racial, spiritual, and other multicultural issues. Please turn to Box 15.3 to review and consolidate what you consider to be your natural style. How have you developed and changed since you have been studying interviewing?

Exercise 15.3 What Is Your Story of Interviewing and Counseling?

Summarize your own story of interviewing and counseling. You may want to base it on your interview transcript completed in Chapter 13 and on your success in microskill practice exercises. What have you learned by completing this exercise? What can you say about your natural style of interviewing and counseling? Where would you like to go next?

Interactive Exercise: Self-Awareness and Emotional Understanding Self-Assessment
Interactive Exercise: Ethical Self-Assessment
Interactive Exercise: Multicultural Self-Assessment
Interactive Exercise: Comparing Your Pre and Post Self-Assessments
Interactive Exercise: Assess Your Competencies

Consider these issues as you continue the process of identifying your natural style and future theoretical/practical integration of skills and theory.

Goals	What do you want to happen for your clients as a result of their working with you? What would you *desire* for them? How do you plan to help them achieve their own goals? What else?
Skills and strategies	You have identified your competence levels. What do you see as your special strengths? What are some of your needs for further development in the future? What else?
Cultural intentionality	Consider the RESPECTFUL model. With what multicultural groups and special populations do you feel capable of working? What knowledge of diversity do you need to gain in the future? How aware are you of your own complex multicultural background? What else?
Theoretical/practical issues	What applications and what theoretical/practical story would you provide now to summarize how you view the world of interviewing and counseling? Where would you like to focus your efforts and interests next? What else?

▶ Your Journey in Interviewing and Counseling Begins Now

> Today you are You, that is truer than true. There is no one alive who is Youer than You.
> —Dr. Seuss

> Trust yourself. You know more than you think you do.
> —Benjamin Spock

You did it! You've reached the end of the book, and many of those concepts, theories, and skills that once seemed awkward and overwhelming are now more familiar and a natural part of your counseling style. We have come to the end of this phase of your interviewing and counseling journey. You have been introduced to the foundational skills and how they are applied in a variety of theoretical/practical interviews. The basic listening sequence has become an important part of your interviewing practice and way of thinking. You have a basic story of the interview to build on and work with in many types of situations.

The real journey begins now. Many of you will be moving on to individual theories of counseling, exploring issues of family counseling, becoming involved in the community, and learning the many aspects of professional practice. Others will be moving into careers in one of the many fields that use the microskills and the five stages. Among these are professional coaches, librarians, teachers, nurses, physicians, lawyers, business and salespeople, and teachers of peer helpers in a multitude of settings throughout the world. Still others of you may find this presentation sufficient for your purposes; perhaps you will use the skills in other work settings or in developing more effective partner or family communication.

We have designed this book as a clear summary of the basics of effective interviewing and counseling; a naturally skilled person can use the information here for many effective and useful helping sessions. To assist you in your continued development, we list below a few key references. Enjoy delving more deeply into our exciting field.

You've reached the end of this book, and yet your journey into the field of interviewing and counseling is just beginning. Throughout this process you have learned about your many strengths and the positive values you bring to the interview. The awareness, knowledge, and skills gained from this book will enable you to help others find their own life directions, make critical decisions, and resolve issues, all the while becoming more sure of themselves and more empowered to help others.

The National Institute of Mental Health brain-based assessment and treatment plan very likely will be another part of your continuing journey. You will be seeing major impact of neuroscience and neurobiology on counseling and therapy practice. Expect to see major changes in the Diagnostic and Treatment Manual (DSM) and the International Classification of Disease (ICD). What is wonderful about this new area is that it shows us that most of what we are already doing is already correct—and it will inform our practice so that our work will be more precise and effective. For example, we have continually emphasized the importance of listening and developing a solid relationship and working alliance. Recent fMRI research in Japan reveals that your active listening skills literally activate positive regions in the brain. Specifically, it has been found that active listening impacts the reward system—a "warm glow" for receiving positives (ventral striatum). In addition, active listening increases emotional appraisal, mentalizing, and understanding the emotional communication of others (right anterior insula and medial prefrontal cortex). Once again we see that the theologian Paul Tillich is right—"Love is listening."

As we do at the completion of our interviews, counseling, or teaching, we would like to ask you how you are going to transform your newly acquired knowledge and skills into action. Building a plan for action will help you achieve your goals. Furthermore, using it to implement your mastered skills will reduce the knowing–doing gap. Again, practice, practice, practice what you have learned by bringing your awareness, knowledge, and skills into action.

This field is full of interesting and involved people working in counseling, psychology, social work, business, medicine, and teaching, among other areas, each seeking to make a difference not only for their clients but also for their communities and the world. At the core of all of these professions is listening; it works every time and everywhere. We hope you too will use your attending and influencing skills to make a difference in our world.

We have enjoyed sharing this time with you. Many of the ideas in this book come from interaction with students. We hope you will take a moment to provide us with your feedback and suggestions for the future. Please see the evaluation forms at the end of this book or download them from the web resources. This book and the materials available through the web resources will be constantly updated with new ideas and information. You have joined a never-ending time of growth and development. Welcome to the field of interviewing and counseling!

Now get out and apply what you have learned to advance yourself, help others, continue learning, and contribute to creating a more just society. As we like to say to our students and trainees, be an agent of change! *Allen, Mary, and Carlos*

Suggested Supplementary Readings

The literature of our field is extensive, and you will want to sample it on your own. We would like to share some books that we find helpful as next steps to follow up ideas presented here. All of these build on the concepts of this book, but we have recommended several books that take different perspectives from our own.

Microskills

Evans, D., Hearn, M., Uhlemann, M., & Ivey, A. (2016). *Essential interviewing* (9th ed.). San Francisco: Brooks/Cole/Cengage. Microskills in an interactive programmed text format.

Ivey, A., Gluckstern, N., & Ivey, M. (2015). *Basic attending skills* (3rd ed.). Framingham, MA: Microtraining Associates. Brief; perhaps the most suitable book for beginners and those who would teach others microskills. Supporting videotapes are available (www.academicvideostore.com/microtraining).

Zalaquett, C., Ivey, A., Gluckstern, N., & Ivey, M. (2008). *Las habilidades atencionales básicas: Pilares fundamentales de la comunicación efectiva* [book and training videos]. Framingham, MA: Microtraining Associates and Alexander Street Press. An introduction to the microskills and the five steps of the interview in Spanish.

www.academicvideostore.com/microtraining. Visit this web resource for videos on the listening and influencing skills, as well as many theoretical orientation demonstrations. Special attention is given to multicultural counseling and therapy.

Theories of Interviewing and Counseling With an Orientation to Diversity

Ivey, A., D'Andrea, M., & Ivey, M. (2012). *Theories of counseling and psychotherapy: A multicultural perspective* (7th ed.). Thousand Oaks, CA: Sage. The major theories are reviewed, with special attention to issues. Includes many applied exercises to take theory into practice.

Sue, D., & Sue, D. M. (2007). *Foundations of counseling and psychotherapy: Evidence-based practices for a diverse society.* New York: Wiley. An excellent text, focusing on evidence-based approaches.

Suggestions for Follow-up on Specific Theories

Most of the following books are classics. In many ways they are more comprehensive, thorough, and important than more recent writings. We commend them all to you.

Decisional Counseling

D'Zurilla, T., & Nezu, A. (2012). *Problem-solving therapy* (4th ed.). New York: Springer. An entire counseling model derived from decision making.

Parsons, F. (1967). *Choosing a vocation.* New York: Agathon. (Originally published 1909). It is well worth a trip to your library to read Parsons's work. You will find that much of his thinking is still up to date and relevant.

Brief Counseling

Guterman, J. (2013). *Mastering the art of solution-focused counseling.* Washington, DC: American Counseling Association. The basics of brief counseling in brief form.

de Shazer, S., & Dolan, Y. (2007). *More than miracles: The state of the art of solution-focused brief therapy.* Binghamton, NY: Haworth.

Person-Centered Counseling and Humanistic Theory

Frankl, V. (1959). *Man's search for meaning*. New York: Pocket Books. (Originally published 1946). This is one of the most memorable books you will ever read. It describes Frankl's survival in German concentration camps through finding personal meaning. You will find that referring your future clients to this book is an excellent counseling and therapeutic tool in itself.

Rogers, C. (1961). *On becoming a person*. Boston: Houghton Mifflin. The classic book by the originator of person-centered counseling.

Cognitive Behavioral Counseling and Rational Emotive Behavioral Therapy

Alberti, R., & Emmons, M. (2008). *Your perfect right: A guide to assertiveness training* (9th ed.). San Luis Obispo, CA: Impact. The classic book by the originators of assertiveness training.

Beck, J., & Beck, A. (2011). *Cognitive behavioral therapy: Basics and beyond*. New York: Guilford Press. A fine and practical summary of Aaron Beck's lifetime contributions.

Davis, M., Eshelman, E., & McKay, M. (2008). *The relaxation and stress reduction workbook* (6th ed.). Oakland, CA: New Harbinger. One of many books on cognitive and behavioral counseling. This is clear, direct, and easily translatable into microskill approaches to the interview. If you are in practice and looking for concrete ideas, this is the book to buy.

Dobson, K. (Ed.). (2009). *Handbook of cognitive-behavioral therapies* (3rd ed.). New York: Guilford Press. A comprehensive and specific presentation of key skills, strategies, and theories.

Dobson, D., & Dobson, K. (2009). *Evidence-based practice of Cognitive-Behavioral Therapy*. New York: Guilford Press. A concise illustrated presentation of CBT practices and applications.

Ellis, A., & Ellis, D. (2011). *Rational emotive behavior therapy*. Washington, DC: American Psychological Association. Albert Ellis is the originator of cognitive behavioral approaches. His thinking is still at the center and the "emotive" is often not given enough attention in books that followed his leadership.

Multicultural Counseling and Therapy

Cheek, D. (1976). *Assertive black, puzzled white*. Atascadero, CA: Impact. This book needs to be on every interviewer's and counselor's shelf. Don't let the older date put you off. Dr. Cheek's book is as relevant today for culturally aware concrete interviewing practice as it was the day it was written. Short, but powerful and practical.

Sue, D. W., Carter, R., Casas, M., Fouad, N., Ivey, A., Jensen, M., et al. (1998). *Multicultural counseling competencies*. Beverly Hills, CA: Sage. The original and most comprehensive coverage of the necessary skills and competencies in the multicultural area.

Sue, D. W., Ivey, A., & Pedersen, P. (1999). *A theory of multicultural counseling and therapy*. Pacific Grove, CA: Brooks/Cole. A general theory of multicultural counseling and therapy with many implications for practice.

Sue, D. W., & Sue, D. (2012). *Counseling the culturally diverse* (6th ed.). New York: Wiley. The classic of the field. This book helped launch a movement.

Integrative/Eclectic Orientations

Ivey, A., Ivey, M., Myers, J., & Sweeney, T. (2005). *Developmental counseling and therapy: Promoting wellness over the lifespan*. Belmont, CA: Brooks/Cole. This book shows practice specifics for elaborating cognitive and emotional experience, how to use a positive developmental model with severely distressed clients, applications in multicultural counseling, and therapy with emphasis on cultural identity theory, spirituality, and family therapy.

Lazarus, A. (2006). *Comprehensive therapy: The multimodal way.* Baltimore: Johns Hopkins Press. The basic book for multimodal therapy. You will find the BASIC-ID model useful in conceptualizing broad treatment plans.

Books for Follow-up on Neuroscience

Grawe, K. (2007). *Neuropsychotherapy: How the neurosciences inform effective psychotherapy.* Mahway, NJ: Erlbaum. This is quite challenging reading, but it was the first to show how neuroscience relates to our practice. It remains innovative and important. Sadly, Grawe died before his book was published, but you can get a sense of his major contribution at www.psychotherapyresearch.org.

Hoffman, M. (2015). *Brain beat: An evolutionary perspective of brain health.* Seattle: Creative Space. Neuroarcheology outlines the history of the brain and body from the very beginning. You will discover how brain structures evolved over time and relationship to the natural environment—and now the changing environments of culture. This is a fascinating read that will excite and interest you in a new way to think about interviewing and counseling, as well as neuroscience and neurobiology.

Kawamichi, H., Yoshihara, K., Sasaki, A., Sugawara, S., Tanabe, H., Shinohara, R., Sugisawa, Y., Tokutake, K., Mochizuki, Y, Anme, T., & Sadato, N. (2014): Perceiving active listening activates the reward system and improves the impression of relevant experiences. (2014). Social Neuroscience, On-line publication.

Siegel, D. (2012). *The developing mind.* New York: Guilford Press. Dr. Siegel has gained the most prominence for beginning readers. Watch for his continuing outstanding writings, and visit YouTube for some excellent presentations.

Therapeutic Lifestyle Changes—The Big Seven

Don't let the unusual titles put you off. These are solid great books by the founders of the movement. As we review the best books of the past decade on therapeutic lifestyle changes, familiar names keep coming up and their classic books, some of them a few years old, remain pretty much the best sources.

General Presentation of TLCs

Ratey, J., Manning, R., & Perlmutter, D. (2014). *Go wild: Free your body and mind from the afflictions of civilization.* New York: Little, Brown. Despite the title, this is a serious book and an excellent summary of exercise, nutrition, meditation, and other key lifestyle issues. Expect a solid background in brain basics. For a fine overview of TLCs with an emphasis on exercise, visit www.johnratey.com.

Exercise

Ratey, J. (2008). *Spark: The revolutionary new science of exercise and the brain.* New York: Little, Brown. Again, despite the hoopla title, this is a solid book by the Harvard psychiatrist who has done more than anyone else to show the science of exercise. He has a valuable website with videos and extensive additional information at www.sparkinglife.org.

Nutrition

Ornish, D. (2008). *The spectrum: A scientifically proven program to feel better, live longer, lose weight, and gain health*. New York: Ballantine. No one has better research and information on diet and lifestyle. Dr. Ornish has a superb website at www .ornishspectrum.com.

Sleep

Dement, W. (2000). *The promise of sleep*. New York: Dell. Extensive work on sleep followed Dr. Dement's pioneering book. You can view him and sleep research in a brief video at the AAPMD YouTube channel.

Epstein, L., & Mardon, S. (2007). *The Harvard Medical School guide to a good night's sleep*. Cambridge, MA: Harvard University Press. Perhaps not as exciting as the ones above, but the most solid book available and one that you can share with your clients.

Meditation

Kabat-Zinn, J. (2013). *Full catastrophe living: Using the wisdom of your body and mind to face stress, pain, and illness* (Rev. ed.). New York: Bantam. The meditation master who has brought the Dalai Lama and meditation to the United States. You can learn mindfulness meditation through his excellent *Guided Mindfulness Meditation* series. Mary and Allen have used these CDs for 25 years!

Cognitive Challenge

Merzenich, M. (2013). *Soft-wired: How the new science of brain plasticity can change your life*. San Francisco: Parnassus. Don't let the title fool you. Dr. Merzenich was the first scholar to show real neuroplasticity when he demonstrated that training could change a monkey's neural structures in three days. He explores rewiring the brain in his TED Talk *Growing Evidence of Brain Plasticity* at www.ted.com.

Social Relations

See the list of supplementary readings above. This is what interviewing and counseling has historically been about.

Multicultural Pride and Cultural Identity

Cross, W. (1991). *Shades of black: Diversity in African American identity.* Philadelphia: Temple University Press. This is the foundational work on cultural self-identity and the movement toward cultural pride. Dr. Cross has inspired many authors to take his ideas into other ethic/racial groups, sexual orientation, ability/disability identity, etc. Again, don't let the date put you off. For deeper multicultural understanding, this book is essential. View his 2001 Distinguished Psychologist Address at www.youtube.com.

Sue, D. W. (2010). *Microaggressions in everyday life: Race, gender, and sexual orientation.* New York: Wiley. This book has caused a sensation in multicultural circles. It shows how everyday events shape our worldview and the harm and cumulative trauma that microagressions can cause. Out of awareness one can build a stronger cultural identity. View Dr. Sue speak in his talk *Race Talk in the Classroom: The Psychology of Racial Dialogues* on the UNM HSC Office for Diversity's YouTube channel.

APPENDIX A

The Ivey Taxonomy

Definitions of the Microskills Hierarchy and Anticipated Client Response From Using Skills

Skill, Concept, or Strategy	Anticipated Client Response When Using Skill, Concept, or Strategy
Ethics Observe and practice ethically and follow professional standards. Particularly important issues for beginning interviewers are *competence, informed consent, confidentiality, power,* and *social justice.*	Client trust and understanding of the interviewing process will increase. Clients will feel empowered in a more egalitarian session. When you work toward social justice, you contribute to problem prevention in addition to healing work in the interview.
Multicultural Issues Base interviewer behavior on an ethical approach with an awareness of the many issues of diversity. Include the multiple dimensions of the RESPECTFUL model (Chapter 2).	Anticipate that both you and your clients will appreciate, gain respect, and learn from increasing knowledge in ethics and multicultural competence. You, the interviewer, will have a solid foundation for a lifetime of personal and professional growth.
Positive Psychology and Resilience Help clients discover and rediscover their strengths through wellness assessment. Find strengths and positive assets in the clients and in their support system. Identify multiple dimensions of wellness.	Clients who are aware of their strengths and resources can face their difficulties and discuss problem resolution from a positive foundation.
Social justice Social justice is the movement toward a more equitable society. It includes interviewers' efforts to effect social change, particularly on behalf of vulnerable and oppressed individuals and groups; it also helps individual clients understand that their difficulties may not be caused by their behavior, but by an unjust or oppressive system.	Social justice information, exploration, and counseling can help make your session more effective, particularly with clients who have experienced racism, oppression, or other forms of social injustice.

(continued)

Skill, Concept, or Strategy	Anticipated Client Response When Using Skill, Concept, or Strategy
Neuroscience Quality interviewing, counseling, and psychotherapy effectively impact your brain and produce long-term outcomes. One place to begin discussion with your clients is brain plasticity. It can be useful to point out to your clients that brain plasticity means they can add new and permanent neural connections and even restory, as the two of you work together.	You will make a significant difference in your clients. Your interviewing and counseling skills will facilitate client cognitive and emotional growth, but also will impact brain's neural connections and modify memories in the hippocampus. Make behavioral health and the use of the Therapeutic Lifestyle Changes (TLCs) a central part of your work as they improve brain function and build cognitive reserve.
Therapeutic Lifestyle Changes (TLCs) Therapeutic lifestyle changes facilitate development of new connections, and brain health can be increased. The TLCs positively affect mental and physical health, self-esteem, cognitive reserve, happiness, and length of life.	Positive approaches to mental and physical health recognize the role of TLCs in interviewing and counseling. TLCs are cost effective and are supported by research, but a team approach may be required. You cannot be an expert in all therapeutic lifestyle changes; as needed, refer clients to medical personnel, nutritionists, physical therapists, personal trainers, and other behavioral health professionals.
Attending Behavior Support your client with individually and culturally appropriate visuals, vocal quality, verbal tracking, and body language.	Clients will talk more freely and respond openly, particularly about topics to which attention is given. Depending on the individual client and culture, anticipate fewer breaks in eye contact, a smoother vocal tone, a more complete story (with fewer topic jumps), and a more comfortable body language.
Client Observation Skills Observe your own and the client's verbal and nonverbal behavior. Anticipate individual and multicultural differences in nonverbal and verbal behavior. Carefully and selectively feed back observations to the client as topics for discussion.	Observations provide specific data validating or invalidating what is happening in the session. Also, they provide guidance for the use of various microskills and strategies. The smoothly flowing interview will often demonstrate movement symmetry or complementarity. Movement dissynchrony provides a clear clue that you are not "in tune" with the client.
Open and Closed Questions Open questions often begin *who, what, when, where,* or *why*. Closed questions may start with *do, is,* or *are. Could, can,* or *would* questions are considered open, but have the advantage of giving more power to the client, who can more easily say what he or she wants to respond.	Clients give more detail and talk more in response to open questions. Closed questions provide specific information, but may close off client talk. Effective questions encourage more focused client conversations with more pertinent detail and less wandering. *Could, would,* and *can* questions are often the most open of all.

Skill, Concept, or Strategy	Anticipated Client Response When Using Skill, Concept, or Strategy
Encouraging Encourage with short responses that help the client keep talking. These responses may be verbal (repeating key words and short statements) or nonverbal (head nods and smiling).	Clients elaborate on the topic, particularly when encouragers and restatements are used in a questioning tone of voice.
Paraphrasing Shorten or clarify the essence of what has just been said, but be sure to use the client's main words when you paraphrase. Paraphrases are often fed back to the client in a questioning tone of voice.	Clients will feel heard. They tend to give more detail without repeating the exact same story. If a paraphrase is inaccurate, the client has an opportunity to correct the interviewer.
Summarizing Summarize client comments and integrate thoughts, emotions, and behaviors. Summarizing is similar to paraphrasing but used over a longer time span.	Clients will feel heard and often learn how the many parts of important stories are integrated. The summary tends to facilitate a more centered and focused discussion. The summary also provides a more coherent transition from one topic to the next or a way to begin and end a full session.
Reflection of Feelings Identify the client's key emotions and feed them back to clarify affective experience. With some clients, a brief acknowledgment of feelings may be more appropriate. Reflection of feelings is often combined with paraphrasing and summarizing.	Clients experience and understand their emotional state more fully and talk in more depth about feelings. They may correct the interviewer's reflection with a more accurate descriptor.
Empathic Response Experiencing the client's world as if you were the client; understanding his or her key issues and feeding them back to clarify experience. This requires attending skills and using the important key words of the client, but distilling and shortening the main ideas. In additive empathy, the interviewer may add meaning and feelings beyond those originally expressed by the client. If done ineffectively, it may subtract from the client's experience.	Clients will feel understood and move on to explore their issues in more depth. Empathy is best assessed by the client's reaction to a statement.
Basic Listening Sequence Select and practice all elements of the basic listening sequence: using open and closed questions, encouraging, paraphrasing, reflecting feelings, and summarizing. These are supplemented by attending behavior and client observation skills.	Clients will discuss their stories, issues, or concerns, including the key facts, thoughts, feelings, and behaviors. Clients will feel that their stories have been heard.

The Five Stages/Dimensions of the Well-Formed Interview

1. *Empathic Relationship: Initiate the Session* Develop rapport and structuring. "Hello, what would you like to talk about today?"	The client feels at ease with an understanding of the key ethical issues and the purpose of the interview. The client may also know you more completely as a person and professional.
2. *Story and Strengths: Gather Data* Draw out client stories, concerns, problems, or issues. "What's your concern?" "What are your strengths and resources?"	The client shares thoughts, feelings, and behaviors; tells the story in detail; presents strengths and resources.
3. *Goals: Set Goals Mutually* "What do you want to happen?"	The client will discuss directions in which he or she might want to go, new ways of thinking, desired feeling states, and behaviors that might be changed. The client might also seek to learn how to live more effectively with situations that cannot be changed at this point (rape, death, an accident, an illness). A more ideal story ending might be defined.
4. *Restory: Explore and Create* Explore alternatives, confront client incongruities and conflict, restory. "What are we going to do about it?" "Can we generate new ways of thinking, feeling, and behaving?"	The client may reexamine individual goals in new ways, solve problems from at least those alternatives, and start the move toward new stories and actions.
5. *Action: Conclude* Plan for generalizing interview learning to "real life" and eventual termination of the interview or series of sessions. "Will you do it?"	The client demonstrates change in behavior, thoughts, and feelings in daily life outside of the interview.

Skill, Concept, or Strategy	Anticipated Client Response When Using Skill, Concept, or Strategy
Empathic Confrontation Supportively challenge the client: 1. Listen, observe, and note client conflict, mixed messages, and discrepancies in verbal and nonverbal behavior. 2. Point out internal and external discrepancies by feeding them back to the client, usually through the listening skills. 3. Evaluate how the client responds and whether it leads to client movement or change. If the client does not change, the interviewer flexes intentionally and tries another skill.	Clients will respond to the confrontation of discrepancies and conflict with new ideas, thoughts, feelings, and behaviors, and these will be measurable on the five-point Client Change Scale. Again, if no change occurs, *listen*. Then try an alternative style of confrontation.
Client Change Scale (CCS) The CCS helps you evaluate where the client is in the change process. Level 1. Denial Level 2. Partial examination Level 3. Acceptance and recognition, but no change Level 4. Generation of a new solution Level 5. Transcendence	You will be able to determine the impact of your use of skills and the creation of the New. Suggest new ways that you might try to clarify and support the change process though more confrontation or the use of another skill that might facilitate growth and development.

Skill, Concept, or Strategy	Anticipated Client Response When Using Skill, Concept, or Strategy
Focusing Use selective attention to focus the interview on the client, problem/concern, significant others (partner/spouse, family, friends), a mutual "we" focus, the interviewer, or the cultural/environmental context. You may also focus on what is going on in the here and now of the interview.	Clients will focus their conversation or story on the dimensions selected by the interviewer. As the interviewer brings in new focuses, the story is elaborated from multiple perspectives.
Reflection of Meaning Meanings are close to core experiencing. Encourage clients to explore their own meanings and values in more depth from their own perspective. Questions to elicit meaning are often a vital first step. A reflection of meaning looks very much like a paraphrase but focuses beyond what the client says. Often the words *meaning, values, vision,* and *goals* occur in the discussion.	The client discusses stories, issues, and concerns in more depth with a special emphasis on deeper meanings, values, and understandings. Clients may be enabled to discern their life goals and vision for the future.
Empathic Interpretation/Reframing Provide the client with a new perspective, frame of reference, or way of thinking about issues. Interpretations/reframes may come from your observations; they may be based on varying theoretical orientations to the helping field; or they may link critical ideas together.	The client may find another perspective or meaning of a story, issue, or problem. This new perspective may have been generated by a theory used by the interviewer, from linking ideas or information, or by simply looking at the situation afresh.
Empathic Self-Disclosure As the interviewer, share your own related past personal life experience, here-and-now observations or feelings toward the client, or opinions about the future. Self-disclosure often starts with an "I" statement. Here-and-now feelings toward the client can be powerful and should be used carefully.	The client is encouraged to self-disclose in more depth and may develop a more egalitarian interviewing relationship with the interviewer. The client may feel more comfortable in the relationship and find a new solution relating to the counselor's self-disclosure.
Empathic Feedback Present clients with clear information on how the interviewer believes they are thinking, feeling, or behaving and how significant others may view them or their performance.	Clients may improve or change their thoughts, feelings, and behaviors based on the interviewer's feedback.
Logical Consequences Explore specific alternatives with the client and the concrete positive and negative consequences that would logically follow from each one. "If you do this . . . , then. . . ."	Clients will change thoughts, feelings, and behaviors through better anticipation of the consequences of their actions. When you explore the positives and negatives of each possibility, clients will be more involved in the process of decision making.

(continued)

Skill, Concept, or Strategy	Anticipated Client Response When Using Skill, Concept, or Strategy
Information and Psychoeducation Share specific information with the client—e.g., career information, possible majors, where to go for community assistance and services. Offer advice or opinions on how to resolve issues, and provide useful suggestions for personal change. Teach clients specifics that may be useful. Help them develop a wellness plan; teach them how to use microskills in interpersonal relationships; educate them on multicultural issues and discrimination.	If information and ideas are given sparingly and effectively, the client will use them to act in new, more positive ways. Psychoeducation that is provided in a timely way and involves the client in the process can be a powerful motivator for change.
Directives Direct clients to follow specific actions. Directives are important in broader strategies such as assertiveness or social skills training or specific exercises such as imagery, thought stopping, journaling, or relaxation training. They are often important when assigning homework for the client.	Clients will make positive progress when they listen to and follow the directives and engage in new, more positive thinking, feeling, or behaving.
Skill Integration Integrate the microskills into a well-formed interview, and generalize the skills to situations beyond the training session or classroom.	Developing interviewers and counselors will integrate skills as part of their natural style. Each of us will vary in our choices, but increasingly we will know what we are doing, how to flex when what we are doing is ineffective, and what to expect in the interview as a result of our efforts.
Determining Personal Style and Theory As you work with clients, identify your natural style, add to it, and think through your approach to interviewing and counseling. Examine your own preferred skill usage and what you do in the session. Integrate learning from theory and practice in interviewing, counseling, and psychotherapy into your own skill set.	You, as a developing interviewer or counselor, will identify and build on your natural style. You will commit to a lifelong process of constantly learning about theory and practice while evaluating and examining your behavior, thoughts, feelings, and deeply held meanings.

APPENDIX B
The Family Genogram

▶ Individual Development in Family and Cultural Context

You and your clients will more easily understand the self-in-relation concept if you help them draw a family genogram. We suggest that you consider developing both family and community genograms with many of your clients (see Chapter 8 on focusing). If you keep the genograms displayed during the session, they will remind you and your clients of the cultural/environmental context in which we all live. Moreover, some clients find them comforting as the genograms bring their family history to the interview. In a sense, we are never alone; our family and community histories are always with us.

Much important information can be collected in a family genogram. Many of us have family stories that are passed down through the generations. These can be sources of strength (such as a favorite grandparent or ancestor who endured hardship successfully). These family stories are real sources of pride and can be central in the positive asset search. There is a tendency to look for problems in the family history, and of course this is appropriate. But use this important strategy positively whenever possible. Be sure to search for positive family stories as well as problems. How can family strengths help your client?

Children often enjoy the family genogram, and a simple adaptation called the "family tree" makes it work for them. The children are encouraged to draw a tree and put their family members on the branches, wherever they wish. This strategy has the advantage of allowing children to present the family as they see it, permitting easy placement of extended family and important support figures as well as immediate family members. Many adolescents and adults may also respond better to this more individualized and less formal approach to the family.

Box B-1, Drawing a Family Genogram, illustrates the major "hows" of developing a family genogram. The classic source for family genogram information is McGoldrick and Gerson (1985). Specific symbols and conventions have been developed that are widely accepted and help professionals communicate information to each other. There is a convention of placing an "X" over departed family members. Once we were demonstrating the family genogram strategy with a client and she commented, "I don't want to cross out my family members—they are still here inside me all the time." We believe that it is important to be flexible and work with the clients' view of family and their choice of symbols. The family genogram is one of the most fascinating exercises that you can undertake. You and your clients can learn much about how family history affects the way individuals behave in the here and now.

We have found family genograms helpful and use them frequently; however, there are situations in which some clients find them less satisfying than the community genogram. There is a Western, linear perspective to the family genogram that does not fit all individuals and cultural backgrounds. It is important to adapt the family genogram to meet individual and cultural differences. You will find *Ethnicity and Family Therapy* (McGoldrick, Giordano, & Garcia-Preto, 2005) a most valuable and enjoyable tool to expand your awareness of racial/ethnic issues.

This brief overview will not make you an expert in developing or working with genograms, but it will provide a useful beginning with a helpful assessment and treatment technique. First go through this exercise using your own family; then you may want to interview another individual for practice.

1. List the names of family members for at least three generations (four is preferred), with ages and dates of birth and death. List occupations, significant illnesses, and cause of death, as appropriate. Note any issues with alcoholism or drugs.
2. List important cultural/environmental contextual issues. These may include ethnic identity, religion, economic, and social class considerations. In addition, pay special attention to significant life events such as trauma or environmental issues (e.g., divorce, economic depression, major illness).
3. Basic relationship symbols for a genogram are shown in Figure B-1; an example of a genogram is shown in Figure B-2.

4. As you develop the genogram with a client, use the basic listening sequence to draw out information, thoughts, and feelings. You will find that considerable insight into one's personal life issues may be generated in this way.

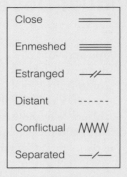

FIGURE B-1 Basic relationship symbols.

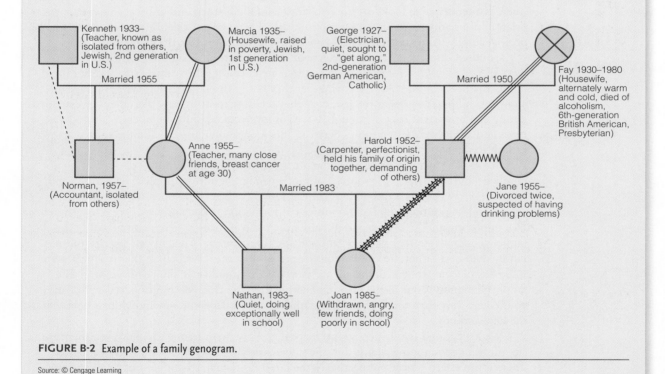

FIGURE B-2 Example of a family genogram.

Source: © Cengage Learning

The family genogram is most effective with a client who has a nuclear family and can actually trace the family over time. We developed the community genogram because some of our clients were uncomfortable with the family genogram. Clients who have been adopted sometimes find the genogram inappropriate. Single-parent families may also feel "different," particularly when important caregivers such as extended family and close community friends are not included. We have talked with gay and lesbian clients who have very differing views of the nature of their family.

Exercise B.1 Developing a Family Genogram

Develop a family genogram with a volunteer client or classmate. After the two of you have created the genogram, ask the client the following questions and note the impact of each question. Change the wording and the sequence to fit the needs and interests of the volunteer.

- ▶ What does this genogram mean to you? (individual focus)
- ▶ As you view your family genogram, what main theme, problem, or set of issues stands out? (main theme, problem focus)
- ▶ Who are some significant others, such as friends, neighbors, teachers, or even enemies, who may have affected your own development and your family's? (others focus)
- ▶ How would other members of your family interpret this genogram? (family, others focus)
- ▶ What impact do your ethnicity, race, religion, and other cultural/environmental contextual factors have on your own development and your family's? (CEC focus)
- ▶ As an interviewer working with you on this genogram, I have learned (state your own observations). How do you react to my observations? (interviewer focus)

▶ Using a Family Genogram to Understand Family Issues

Developing a genogram with your clients and learning some of the main facts of family developmental history will often help you understand the context of individual issues. For example, as you look at the family genogram in Box B-1, what might be going on at home that results in Joan's problems at school? Why is Nathan doing so well? How might intergenerational alcoholism problems play themselves out in this family tree? What other patterns do you observe? What are the implications of the ethnic background of this family? The person with a Jewish and Anglo background represents a bicultural history. Change the ethnic background and consider how this would impact counseling. Four-generation genograms can complicate and enrich your observations. (*Note:* The clients here have defined their ethnic identities as shown. Different clients will use different wording to define their ethnic identities. It is important to use the client's definitions rather than your own.)

APPENDIX C
Counseling, Neuroscience, and Microskills

The more we pulverize matter, the more it insists on its fundamental unity.
—*Teilhard de Chardin*

The brain is greater than the sum of its parts. It is a constantly interacting system within itself and in relation to the cultural/environmental context. This brief appendix discusses some central issues, many of which your clients will be aware of and understand as they come to you. We also extend this discussion with additional information in the web resources. In addition, you may want to do a web search for "Allen and Mary Bradford Ivey" and see a relatively recent presentation on basics of neuroscience and counseling.

The brain is, simultaneously, a localized and distributed system. While some of its functions are associated with specific brain structures and regions, these regions act in concert with other, sometimes distant, brain regions. What we come to experience as "mind" is the result of this intense connectivity. Each of our 100 billion neurons connects through even more synapses with an almost infinite number of receptors. The holistic connected brain is shown in Figure C-1. This illustration is from the Human Connectome Project, which is seeking to locate and specify connections among the multitudinous brain parts (Sueng, 2012). Thus, while it is valuable to speak of brain structures separately, it is the holistic connections that enable human functioning.

FIGURE C-1 The holistic connected brain.

Source: Laboratory of NeuroImaging at UCLA and Martinos Center for Biomedical Imaging at MGH, Consortium of the Human Connectome Project.

▶ Neuroplasticity

The key term for this new interviewing and counseling future is **neuroplasticity**—the brain's ability to change and reorganize itself throughout life. This means the brain can change—it is not fixed, but responds to external environmental events and actions or initiations by the individual. The old idea that the brain loses neurons and capability as we age is simply wrong. Neuroplasticity means that even in old age, new neurons, new connections, and new neural networks are born and can continue development—a brain can rewire itself. There is evidence that this occurs in adults primarily in the **hippocampus**, the main seat of memory, located in the temporal lobe's limbic system.

Therapeutic lifestyle changes (TLCs) appear to enhance neuroplasticity. For example, exercise has been shown as effective, likely through increased blood flow to the brain. Exercise is particularly relevant as a lifetime process to ensure brain and body health (Ratey, 2008; Ratey & Manning, 2014). Exercise increases blood flow to the brain and the release of positive neurotransmitters such as serotonin. Many of you reading this have experienced the serotonin "high" of a beautiful sunset, the here-and-now immediacy of a close relationship, or a physical activity such as running.

Positive stimulation underlies the importance of TLCs for brain (and body) health. Four weeks of serious meditation measurably builds brain gray matter (Tang, Fan, Yang, & Posner, 2012). At the same time, medications such as fluoxetine (Prozac) and Aricept show promise to enhance neuroplasticity (Vogel, 2012). Whether it is getting a good night's sleep, socializing with friends, learning a new language, or attending a religious service, you are improving your chances for neurogenesis and the prevention of Alzheimer's disease.

▶ Some Basic Brain Structures

Our connected brain has a vast array of structures, each with essential functions. This book and this section are oriented to the most basic ones—the ones that you are likely to encounter either now or the very near future. The terminology may be new, but over time repetition will give you familiarity and the ability to use this information to enhance your own and your client's life.

The **frontal lobe** (see Figure C-2), in association with other structures known as the TAP, is our chief executive officer (CEO); it is associated with executive functioning,

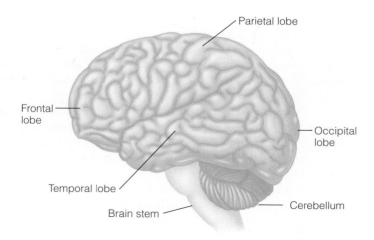

FIGURE C-2 Basic areas of the brain.

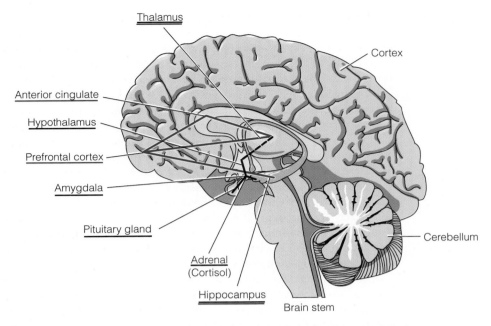

FIGURE C-3 The TAP (thalamus, anterior cingulate, and prefrontal cortex) and the limbic system. Core elements of the limbic system are the hypothalamus, pituitary, and adrenals (HPA), the amygdala, and the hippocampus. The adrenals are actually located on top of the kidneys; only the connection is shown here.

emotional regulation, abstract reasoning, and decision making. Critical for long-term memory, it is also the focus of attentional processes and much of our motor behavior, enabling us to function effectively in social systems. Emotional regulation is located here through connections with the limbic system. However, in situations of danger, emergency, or mental distress, or under the influence of drugs or alcohol, the interior limbic system may take over. Clients with frontal lobe issues may show poor emotional control, language problems, personality changes, apathy, or inability to plan.

The **TAP** (see Figure C-3) is shorthand for *thalamus*, *anterior cingulate cortex*, and *prefrontal cortex*, all associated with functions in the frontal lobe. The **thalamus** is considered the brain's communication center, sending information throughout the holistic brain. The thalamus lies just above the interior limbic system.

The thalamus is also a switching station focusing on the TAP, relaying sensorimotor signals to the cerebral cortex. It also regulates consciousness, sleep, and alertness. It is central in emotional regulation as well, because of its connections with the limbic HPA (hypothalamus, pituitary, adrenals; see Chapter 12).

The **anterior cingulate cortex** relays neural signals between the prefrontal cortex and the left and right brain. It regulates important cognitive functions, including decision making, empathy, and emotion. In addition, it plays a part in blood pressure and heart rate. "It links the body, brainstem, limbic, cortical, and social processes into one functional whole" (Ratey, 2008, p. 278).

The **parietal lobe** gives us our spatial sense, but it also serves as a critical integrating force for the senses (see/hear/feel/taste/touch) and our motor abilities. Synthesizing, putting things together, is a function of the parietal. Problems in the parietal lobe may show in personality change, lack of self-care or dressing, making things, and drawing ability. Any failure to integrate may involve the parietal lobe. Difficulties with these functions are often associated with Alzheimer's disease.

Mirror neurons, the basis of empathic understanding, are located primarily in the parietal and motor cortex (just ahead of the parietal). You can get a better sense of how mirror neurons work if you notice what occurs in your body when you see an exciting ballgame or an involving movie. Many of us find ourselves tensing up and clenching our fists in close or exciting situations. We may even sway as the pass receiver grabs the ball and heads down the field. In good movies that touch you emotionally in some way, the same thing happens. You heart rate goes up, and you may duck a swing from the villain as you sit on the edge of your seat.

The **temporal lobe** is concerned with auditory processing, language and speech production, aspects of sexuality, and memory. The limbic system containing the hippocampus, center of long-term memory, is located in the temporal lobe.

The **occipital lobe**, the smallest lobe, is the principal area for visual processing, motion recognition, and color recognition.

The **cerebellum** is approximately 10% of the brain's volume but contains more than 50% of the total number of neurons. Not so long ago, it was ignored as a vestige of the past, but recent research has revealed its centrality and significance in brain functioning— with more research to come. It is a vital part of smooth, coordinated body movement and balance. It also has a role in several cognitive functions, including attention, language processing, and the sensory modalities.

The **brain stem** connects the brain to the rest of the body via the spinal cord. It is a conduit for integrating the whole brain and is critical for central nervous system functioning such as heart rate, respiration, attention, and consciousness. It also regulates the sleep cycle.

▶ The Limbic System and the HPA Axis: Basics of Emotion

The **limbic system** deals with physical activation, emotion, and memory. Its key structures related to interviewing and counseling are the amygdala, the hippocampus, and the HPA stress system—hypothalamus, pituitary, adrenals (see Figure 12.3). The adrenal glands are located just above the kidneys.

The **amygdala** is the energizer of emotive strength. It is the power of emotions that place information in memory in the hippocampus; thus, the interrelationship of the amygdala, hippocampus, and frontal TAP are central. Drawing sensory information (what is seen, heard, felt, tasted, smelled) from other parts of the brain, the amygdala signals intensity. Nothing happens in other parts of the brain or body (or memory) unless there is sufficient external or internal stimulation.

The hippocampus, as noted earlier, is our memory "organ" and works closely with the amygdala and the cerebral cortex. The hippocampus is key in short-term and working memory, and it distributes information throughout the brain for long-term memory. Stimulation from the amygdala tells the hippocampus which information might be remembered, and this information is distributed throughout the interconnected brain. A highly stressful event (such as war or rape) can overwhelm the whole system like a lightning bolt and result in destruction of neural nets, resulting in memory loss or serious distress.

HPA hormones are activated by the amygdala; in times of emergency, fear, or anger, they may bypass executive functioning. As described in Chapter 12, the hypothalamus is the master gland controlling hormones that affect heart rate, biological body responses to emotions such as hunger, sleep, aggression, and other biological factors. The pituitary is another control gland that receives messages from the hypothalamus and influences growth, blood pressure, sexual functioning, the thyroid, and metabolism. The adrenal glands produce corticosteroids, including cortisol (potentially damaging due to stress, but also necessary for stimulating learning and memory). They also produce the neurotransmitters epinephrine (also known as adrenaline) and norepinephrine (also known as noradrenaline) that regulate body response, heart rate, and the fight-or flight-response.

The **nucleus accumbens** is our pleasure center; it plays a part in reward, laughter, addiction, aggression, fear, and the placebo effect. The nucleus accumbens is significant in sexual functioning and the "high" from certain recreational drugs, which increase the supply of pleasurable dopamine. Key in understanding addiction, the nucleus accumbens is particularly responsive to marijuana, alcohol, and related chemicals. One of our great challenges is helping these clients examine and rewrite their stories and find new, healthy alternative highs to replace the immediate rewards of addiction.

▶ The Holistic Interacting System of the Executive TAP and the Emotional Limbic System

We need to return to the concept of executive control and emotional regulation within the TAP (thalamus, anterior cingulate cortex, and prefrontal cortex). Under most conditions, we expect the TAP to be in charge, but obviously it is often overruled by the limbic and HPA systems. It is equally important to note that nothing enters either learning or memory unless there is sufficient stimulus to energize the amygdala.

It has been argued that the six basic emotions described in Chapter 6 can be reduced to only two fundamental behavioral reactions—approach and avoidance. At all levels of development, from small organisms through snakes and mammals to humans, survival depended on knowing when and how to obtain food and reproductive opportunities while avoiding danger of all types. From this point of view, it is helpful to think again of the limbic HPA and executive TAP. The limbic area is the location of the amygdala and the seat of protective emotions (e.g., fear, anger), which are also seen as negative emotions. The evolutionarily more recent TAP is deeply involved in basic positive emotions, as well as defining the social/cultural emotions.

Accepting this explanation, you can see the critical importance of taking a positive approach, becoming skilled in stress management, and encouraging therapeutic lifestyle changes (TLC). Building on this foundation, neuroscience research has offered exciting findings. For example, we can now identify specific neurons in the amygdala that affect anxiety, depression, and posttraumatic stress. These harmful networks remain in place unless treated effectively. Using classic behavioral methods derived from Pavlov's work, we can change the power of these neurons through a process of presenting the feared object in the absence of danger. Medications can do the same thing as counseling and therapy if targeted to specific "intercalated neurons" (Ekaterina et al., 2008). These are very clear examples of the approach/avoidance hypothesis. The implication for counseling and therapy is one that we have frequently made: A positive wellness approach is the most effective way to help clients deal with their difficult issues of mental and physical health.

The Nobel Prize–winning psychologist Daniel Kahneman (2011) carries this discussion a bit further. He comments that our emotional likes and dislikes determine what we believe about the world—about politics, irradiated food, global warming, motorcycles, or tattoos. Once we are settled emotionally, it is difficult for us or our clients to change. Many times— perhaps most of the time—our feelings and emotions guide our cognitions. Again, this reinforces the importance of exploring and reflecting feelings and emotions. When we see a picture of a scary spider or view blood and gore, the amygdala and negative feelings are activated, both verbally and nonverbally. With things that we like or have a deep interest in, our pupils dilate with positive feelings.

▶ Social Justice, Stress Management, and Therapeutic Lifestyle Changes

> Poverty in early childhood poisons the brain. . . . neuroscientists have found that many children growing up in very poor families with low social status experience unhealthy levels of stress hormones, which impair their neural development. The effect is to impair language development and memory—and hence the ability to escape poverty—for the rest of the child's life.
>
> —Paul Krugman

Stress is a factor in virtually all client issues and problems. It may not be the presenting issue, but be prepared to assess stress levels and provide education and treatment as needed, using some of the TLC strategies of Chapter 2 and stress management concepts of Chapter 12. Stress will show in body tension and nonverbal behavior. Cognitive/emotional stress is demonstrated in vocal hesitations, emotional difficulties, and the conflicts/discrepancies clients face in their lives.

A moderate amount of stress, if not prolonged, is required for development and for physical health. For example, repeated stressing of a muscle through weight lifting breaks down muscle fibers, but after rest the rebuilding muscle gains extra strength. A similar pattern occurs for running and other physical exercise. If there is no stress, neither physical development nor learning will occur. Stress should also be seen as a motivator for focus and as a condition for change. The amygdala needs to be energized. For example, the skill of confrontation can result in stress for the client. But used as a supportive challenge, this stressor becomes the basis for cognitive and emotional change. In all of the above, note the word *moderate* and the need for rest between stressors.

Psychiatrist John Ratey of Harvard Medical School points out both the benefits and the toxic dangers of damaging stress:

> *Cortisol* is the long-acting stress hormone that helps to mobilize fuel, cue attention and memory, and prepare the body and brain to battle challenges to equilibrium. Cortisol oversees the stockpiling of fuel, in the form of fat, for future stresses. Its action is critical for our survival. At high or unrelenting concentrations such as post-traumatic stress, cortisol has a toxic effect on neurons, eroding the connections between them and breaking down muscles and nerve cells to provide an immediate fuel source. (Ratey, 2008, p. 277)

An article titled "Excessive Stress Disrupts the Architecture of the Developing Brain" (National Scientific Council on the Developing Child, 2005) includes the following useful points:

1. In the uterus, the unborn child responds to stress in the mother, while alcohol, drugs, and other stimulants can be extremely damaging.
2. For the developing child neural circuits are especially plastic and amenable to growth and change, but again excessive stress results in lesser brain development and in adulthood that child is more likely to have depression, an anxiety disorder, alcoholism, cardiovascular problems, and diabetes.
3. Positive experiences in pregnancy seem to facilitate child development.
4. Caregivers are critical to the development of the healthy child.
5. Children of poverty or who have been neglected tend to have elevated cortisol levels.

If you review the sections above, particularly those describing the frontal cortex and limbic system, you can obtain some sense of what poverty and challenges such as racism and oppression do to the brain. Incidents of racism place the brain on hypervigilance, thus producing significant stress, with accompanying hyperfunctioning of the amygdala and interference with memory and other areas of the brain. We need to be aware that many environmental stressors, ranging from poverty to toxic environments to a dangerous community, all work against neurogenesis and the development of full potential. And let us expand this list to include trauma.

Recall that the infant and young children can only pay attention to what is happening in the immediate environment. Adolescent brains have not fully **myelinated** and are also more vulnerable. Again, think of the varying positive and negative environments that your clients come from. One of the purposes of the community genogram is to help you and the client understand how we as individuals relate and are related to other individuals, family, groups, and institutions around us. The church that welcomes you helps produce positive development; the bank that refuses your parents a loan or peers that tease and harass you harm development.

Clients need to be informed about how social systems affect personal growth and individual development. Our work here is to help clients understand that the problem does not lie in them but in a social system or life experience that treated them unfairly and did not allow an opportunity for growth.

Finally, there is social action. What are you doing in your community and society to work against social forces that bring about poverty, war, and other types of oppression? Are you teaching your clients how they can work toward social justice themselves? A social justice approach includes helping clients find outlets to prevent oppression and work with schools, community action groups, and others for change.

▶ Looking to the Future

Neuroscience research provides a biological foundation for understanding the impact of our work. The very act of counseling and therapy produces changes in client memory (and your own). Always be aware that new ideas and learning are being constructed in the session. We suggest that you continue to study and learn about brain structures and functions, as new findings may provide further support for your work and suggest specific guidelines for practice.

Brain research is not in opposition to the cognitive, emotional, behavioral, and meaning emphasis of counseling and therapy. Rather, it will help us pinpoint types of interventions that are most helpful to the client. In fact, one of the clearest findings is that the brain needs environmental stimulation to grow and develop. You can offer a healthy atmosphere for client growth and development. We advocate the integration of counseling, psychotherapy, neuroscience, molecular biology, and neuroimaging and the infusion of knowledge from such integrated fields of study to practice, training, and research.

▶ To Be Continued in the Web Resources

 The above, we believe, contains the most basic information important for interviewing and counseling practice. However, should you wish more information, please visit a continuation of this discussion in the online web resources. There you will find information on neurotransmitters and how various treatment strategies might affect their impact on the brain and behavior. You will also find our thoughts on how the microskills affect neurotransmitter transmission.

GLOSSARY

The number in parentheses at the end of each entry indicates the chapter in which the term is introduced.

Acknowledging feelings is a brief acknowledgment of feeling that is just as helpful as a full reflection of feeling. In acknowledging feelings, you state the feeling briefly ("You seem to be sad about that," "It makes you happy") and move on with the rest of the conversation. (6)

Active listening is a communication process that requires active participation, decision making, and responding on our part. (5)

Additive empathy means that the interviewer response adds a link to something the client has said earlier or a new idea or frame of reference that helps the client see a new perspective. Wellness and the positive asset search can be vital parts of additive empathy. (3)

Adrenal glands produce corticosteroids, including cortisol. They also produce the neurotransmitter epinephrine (also known as adrenaline or norepinephrine) that regulates heart rate and the fight-or-flight response. (12)

Advocacy is speaking out for your clients; working in the school, community, or larger setting to help clients; and also working for social change. (8)

Amygdala is the energizer of emotive strength. Drawing sensory information (what is seen, heard, felt, tasted, smelled) from other parts of the brain, the amygdala signals intensity. Nothing happens in other parts of the brain or body (or memory) unless there is sufficient external or internal stimulation. (C)

Anterior cingulate cortex relays neural signals between the prefrontal cortex and the left and right brain. It regulates important cognitive functions, including decision making, empathy, and emotion. In addition, it plays a part in blood pressure and heart rate. (C)

Attending behavior is defined as supporting your client with individually and culturally appropriate visuals, vocal quality, and body language. (3)

Basic empathy means that the interviewer response is roughly interchangeable with that of the client. The interviewer is able to say back accurately what the client has said. (3)

Basic listening sequence (BLS) is used to draw out clients' stories by using open and closed questions, encouraging, paraphrasing, reflecting feelings, and summarizing. Your goal is to draw out the key facts, thoughts, feelings, and behaviors of the client's story. (4)

Being-in-relation is similar to self-in-relation, but points out that all of us are beings who grow in the here-and-now moment as we interact with others. (8)

Body language is nonverbal communication via the movements, gestures, or postures of the body. It is essential to build trust, show interest, express warmth, and demonstrate attentiveness and authenticity. (3)

Bombardment/grilling means using too many questions. This may give too much control to the interviewer and tend to put many clients on the defensive. (4)

Boundaries of competence are limits within which counselors practice, based on their education, training, supervised experience, state and national professional credentials, and appropriate professional experience. Counselors are expected to gain knowledge, personal awareness, sensitivity, and skills pertinent to working with a diverse client population. (2)

Brain stem connects the brain to the rest of the body via the spinal cord. It is a conduit for integrating the whole brain and is critical for central nervous system functioning such as heart rate, respiration, attention, and consciousness. It also regulates the sleep cycle. (C)

Brainstorming is a technique to gather multiple ideas to find a possible solution. It encourages clients to "loosen up" and let any thought come to their mind. You can help through your own creativity, but focus on helping clients generate their own solutions. (12)

Cerebellum is approximately 10% of the brain's volume but contains more than 50% of the total number of neurons. It is a vital part of smooth, coordinated body movement and balance. It also has a role in several cognitive functions, including attention, language processing, and the sensory modalities. (C)

Checklist is a list of specific items that need to be considered in the interview, particularly the first interview. Even the most experienced interviewer may forget to cover important basics during a session. (13)

Checkout, or **perception check**, is a brief question at the end of the communication, asking the client for feedback on whether what you said was relatively correct and useful. Periodically checking with your client (a) helps to communicate that you are listening and encourages her or him to continue and (b) allows the client to correct any wrong assumptions you may have. You can also paraphrase or reflect feelings and use a checkout by ending your comment/sentence with a raised tone, indicating a questioning voice to check for accuracy. (4)

Chunking refers to learning information not just in bits and pieces, but by organizing it into patterns. (15)

Client Change Scale (CCS) is a five-level model that provides you with a systematic way to evaluate the effectiveness of interventions and to track client change in the here and now of the interview. At Level 1, clients may deny issues, at Level 2 deal with them partially, and at Level 3 acknowledge fairly accurately what is going on. Levels 4 and 5 indicate significant changes. (9)

Closed questions enable you to obtain specific information and tend to be answered in very few words. Closed questions often begin with *is*, *are*, or *do*. (4)

Cognitive and emotional balance sheet is an extension of logical consequences, but with each alternative written down in a list of gains and losses. (12)

Community genogram is a visual map of the client in relation to the environment showing both stressors and assets within the person's life. It is a useful way to understand your client's history and identify strengths and resources. It can be used by itself or to supplement the family genogram. (8)

Confidentiality is respecting your client's right to privacy. Trust is built on your ability to keep client information confidential. (2)

Corrective feedback is a delicate balance between negative feedback and positive suggestions for the future. When clients need to seriously examine themselves, corrective feedback may need to focus on things that clients are doing wrong or behavior that may hurt them in the future. (11)

Counseling is a more intensive and personal process than interviewing. It is less about information gathering and more about listening to and understanding clients' personal issues, challenges, and opportunities and helping them develop strategies for further growth and change. (1)

Crisis counseling is the most pragmatic and action oriented form of helping. Even more than decisional counseling, crisis counseling is concerned with action and useful, practical results for the client. Both are provided with a caring attitude and with respect for the client's response. (14)

Cultural intentionality is acting with a sense of capability and deciding from among a range of alternative actions. The intentional individual can generate alternatives in a given situation and approach a problem from different vantage points, using a variety of skills and personal qualities and adapting styles to suit different individuals and cultures. (1)

Cultural trauma is possible in each of the RESPECTFUL dimensions. Large historical events and daily microaggressions in the form of insults to one's color, ethnicity, gender, sexual orientation, religious beliefs, or disability result in personal and group trauma, frustration, anger, hopelessness, and depression. (2)

Decisional counseling may be described as a practical model that recognizes decision making and the microskills as a foundation for most—perhaps all— systems of counseling. Another term for decisional counseling is problem-solving counseling. (12)

Determining personal style and theory means that as you work with clients, identify your natural style, add to it, and think through your approach to interviewing and counseling. Examine your own preferred skill usage and what you do in the session. Integrate into your own skill set your learning from theory and practice in interviewing, counseling, and psychotherapy. (1)

Directives direct clients to follow specific actions. Directives are important in broader strategies such as assertiveness or social skills training or specific exercises such as imagery, thought-stopping, journaling, or relaxation training. They are often important when assigning homework for the client. (12)

Dual relationships occur when a helper has more than one relationship with a client. Another way to think of this is as a conflict of interest. (2)

Eliciting meaning often precedes reflection. Clients do not often volunteer meaning issues, even if they are central to their concerns; through careful attending and questions, you can bring out client stories and meaning. (10)

Emotional regulation involves using our cognitive capacity to regulate impulsive emotions and act appropriately in complex situations. It relates to reflection of feelings and the brain's limbic system, but it is also a key part of executive functioning. (5)

Empathic confrontation is not a direct, harsh challenge but is a more gentle skill that involves listening to clients carefully and respectfully and helping them examine themselves or their situations more fully. Empathic confrontation is not "going against" the client; it is "going with" the client, seeking clarification and the possibility of a new resolution of difficulties. Think of empathic confrontation as a supportive challenge. (9)

Empathic relationship—story and strengths—goals—restory—action is a five-stage model of the interview; it includes a basic decision-making model and adds relationship building, issues of confidentiality, and ethical concerns. (7)

Empathy is defined as experiencing the world as if you were the client, but with awareness that the client remains separate from you. Part of empathy is communicating that you understand.

Encouragers are verbal and nonverbal expressions the counselor or therapist can use to prompt clients to continue talking. Encouragers are minimal verbal utterances ("ummm" and "uh-huh"), head nods, open-handed gestures, and positive facial expressions that encourage the client to keep talking. Silence, accompanied by appropriate nonverbal communication, can be another type of encourager. (5)

Encouraging is comprised of short responses that help clients keep talking. They may be verbal (repeating key words and short statements) or nonverbal (head nods and smiling). (5)

Ethics means observing, practicing, and following professional standards. Particularly important issues for beginning interviewers are competence, informed consent, confidentiality, power, and social justice. (2)

Executive (brain) functioning refers to the ability of the prefrontal cortex to organize cognition and make decisions. (5)

External conflict refers to conflict with others (friends, family, coworkers, or employers) or with a difficult situation, such as coping with failure, living effectively with success, having difficulty achieving an important goal financially or socially, or struggling to navigate one's cultural or ethnic identity. (9)

Family genogram is a visual map of the client in relation to his or her family and brings out additional details of family history. It can be used by itself or to supplement the community genogram. (8)

Feedback presents clients with clear information on how the interviewer believes they are thinking, feeling, or behaving and how significant others may view them or their performance. (11)

Focusing is a skill that enables multiple perspectives on the client's story and helps you and your client understand contributing factors and think of new possibilities for change. Usually, clients come with a one-dimensional perception of their issues. Focusing help them clarify their issues, engage in more holistic restorying, and find more satisfying decisions and actions. Focusing also encourages you, the interviewer or counselor, to think about alternative frames of reference, broader systems, and cultural issues that might affect the client. (8)

Frontal lobe is associated with executive functioning, emotional regulation, abstract reasoning, and decision making. Critical for long-term memory, it is also the focus of attentional processes and much of our motor behavior, enabling us to function effectively in social systems. Clients with frontal lobe issues may show poor emotional control, language problems, personality changes, apathy, or inability to plan. (C)

HIPAA privacy refers to the Health Insurance Portability and Accountability Act (HIPAA) requirements regarding the protection and confidential handling of protected health information. (2)

Hippocampus, located in the temporal lobe's limbic system, is the main seat of memory. (C)

Hypothalamus is the master gland controlling hormones that affect heart rate, biological body responses to emotions such as hunger, sleep, aggression, and other biological factors. (12)

Incongruity occurs when clients have difficulty making important decisions, they feel confusion or sadness, or they have mixed feelings and thoughts about themselves or about their personal or cultural background. (9)

Influencing is part of all interviewing and counseling. Through selective attention and the topics you choose consciously or unconsciously to emphasize (or ignore), you influence what the client says. It is impossible not to influence what happens in the interview. Confrontation, focusing, reflection of meaning, and interpretation/ reframing have been identified as strategies of interpersonal influence, but listening skills can be as or more influential. Help clients set their own direction as much as possible. *Influencing* can be defined as "a power affecting a person, thing, or course of events, especially one that operates without any direct or apparent effort" (*Free Dictionary*, 2010). (8)

Information includes advice, opinions, and suggestions; it can be an important part of interviewing and counseling. (12)

Informed consent is the consent of a client to participate in the interview or therapy after being informed of its benefits, limitations, potential risks, and other pertinent information. (2)

Intentionality is acting with a sense of capability and deciding from among a range of alternative actions. (1)

Interchangeable empathy. See **basic empathy.** (3)

Internal conflict is a synonym for *incongruity*, with a focus on conflict inside the person. It may or may not involve external conflict as well. (9)

Interpretation is an explanation of the meaning or significance of something. Closely allied to reframing, it is an important influencing skill to help clients discover new ways of thinking. It often comes from a specific theoretical framework. (10)

Interviewing is the basic process used for gathering data, helping clients resolve their issues, and providing information and advice to clients. (1)

Ivey Taxonomy provides definitions of the microskills and predicted results from using each skill. (A)

Limbic system deals with physical activation, emotion, and memory. Its key structures related to interviewing and counseling are the amygdala, the hippocampus, and the HPA stress system—hypothalamus, pituitary, adrenals. (C)

Linking is the bringing together of two or more ideas to provide the client with a new insight. (10)

Listening is the core of developing a relationship and making real contact with clients. (3)

Logical consequences is a strategy in which you explore with the client specific alternatives and the concrete positive and negative consequences that would logically follow from each one. "If you do . . . , then . . . will possibly result." (12)

Meaning refers to purpose and significance. (10)

Microaggressions are small insults and slights that accumulate and magnify over time. They are usually associated with discriminatory remarks and over time can become traumatic and injurious to physical and mental health. (2)

Microskills are communication skill units that help you to interact more intentionally with a client. (1)

Microskills hierarchy consists of specific skill steps that will enable you to work with a multitude of clients and anticipate which skill will likely be most effective with what anticipated result. In addition, competence in the microskills hierarchy will give you a clear sense of how to organize and structure a session. (1)

Mirror neurons, the basis of empathic understanding, are located primarily in the parietal and motor cortex (just ahead of the parietal); they enable us to understand the actions of others and thus learn from them. (C)

Movement complementarity is paired movements that may not be identical but are still harmonious. (3)

Movement dissynchrony is lack of harmony in movement, common between people who disagree markedly or who may not be aware they have subtle conflicts. (3)

Movement synchrony refers to the situation in which people who are communicating well "mirror" each other's body language. (3)

Multicultural competence is your awareness of yourself as a cultural being and your knowledge of, and skills to work effectively with, people different from you. (2)

Multiculturalism includes race/ethnicity, gender, sexual orientation, language, spiritual orientation, age, physical ability/disability, socioeconomic status, geographical location, and other factors. (2)

Multiple questions are another form of bombardment. Throwing out too many questions at once may confuse clients; however, it may enable clients to select which question they prefer to answer. (4)

Myelin is a fatty sheath around nerve fibers that speeds information with less loss between brain areas. Myelination is a slow process; typically, we are not fully myelinated until 20 to 30 years of age. Lack of full myelination tends to mean less effective judgment. Interestingly, women tend to myelinate earlier than men. (C)

Negative consequences are undesirable, painful, or aversive results of our decisions or actions. Potential negative consequences could include the harm caused by smoking while pregnant, the disruption of long-term friendships, or the problem of moving children to a new school. (12)

Negative feedback may be necessary when the client has not been willing to hear corrective feedback. However, it is to be used rarely and with care and empathy. Otherwise, the client is almost certain to reject what you say. (11)

Neural networks are large numbers of interconnected neurons working together to achieve a specific outcome. (5)

Neuroplasticity is the brain's ability to change and reorganize itself throughout life. (C)

Occipital lobe, the smallest lobe of the brain, is the principal area for visual processing, motion recognition, and color recognition. (C)

Open questions are those that can't be answered in a few words. They encourage others to talk and provide you with maximum information. Typically, open questions begin with *what*, *how*, *why*, or *could*. (4)

Oppression is the continued mistreatment and exploitation of a group of people by another person, organization, or culture. (2)

Paraphrasing helps to shorten and clarify the essence of what has just been said, but be sure to use the client's main words when you paraphrase. Paraphrases are often fed back to the client in a questioning tone of voice. (5)

Parietal lobe gives us our spatial sense, but it also serves as a critical integrating force for the senses (see/hear/feel/taste/touch) and our motor abilities. Synthesizing, putting things together, is a function of the parietal. Problems in the parietal lobe may show in personality change, lack of self-care or dressing, making things, and drawing ability. (C)

Perception check. See **checkout.** (4)

Person-as-community points out that our family and community history live within each of us. (8)

Person-centered counseling was founded by Carl Rogers, who revolutionized the helping field by clearly identifying the importance of listening and empathy as the foundation for human change. A major assumption of person-centered theory is that the skills are used to help the client uncover internal strength and resilience. (7)

Pituitary is a control gland that receives messages from the hypothalamus and influences growth, blood pressure, sexual functioning, the thyroid, and metabolism. (12)

Political correctness (PC) is a term used to describe language that is calculated to provide a minimum of offense, particularly to the racial, cultural, or other identity groups being described. It has developed negative interpretations in parts of society. The issue is providing RESPECT for the client and using the language the client prefers. Avoid the politics and pay attention to the client before you. (2)

Portfolio of competencies is a practical system developed to document your learning, practice, and outcomes. The portfolio will help you track your personal progress throughout your journey toward mastering intentional interviewing and counseling. (1)

Positive consequences are desirable and enjoyable results of our decisions or actions. Potential positive consequences could include a pay raise, a better school system, or a healthier environment. It is part of the logical consequences skill. (12)

Positive feedback is concrete, strengths-based feedback that helps clients restory their problems and concerns. Whenever possible, find things right about your client. (1)

Positive psychology is a strength-based approach in which the interviewer seeks out and listens for times when clients have succeeded in overcoming obstacles. Find something that the client is doing right, and discover the client's resources and supports. Use these strengths and positive assets as a foundation for personal growth and resolving issues. (2)

Power is part of the act of helping. Simply by being a client, one begins counseling with perceived lesser power than the interviewer. This power relationship may remain, or you can work toward a more equal relationship with the client. (11)

Primary emotions are those that have been validated throughout the world in all cultures: *sad, mad, glad, scared, disgust,* and *surprise.* (6)

Privilege is power given to people through cultural assumptions and stereotypes. There are many types of privilege including, among others, White privilege, social class privilege, economic privilege, religious privilege, and male privilege. (2)

Problem solving consists of defining the problem, generating or brainstorming alternatives, and then deciding among the alternatives. Benjamin Franklin developed the first working model. (7)

Psychoeducation is similar to information giving, but the counselor takes on more of a teaching role and provides systematic ways for clients to increase their life skills. Psychoeducation has become prominent in cognitive behavioral counseling in recent years, as a way to prepare the client to meet challenging situations. (12)

Psychotherapy focuses on deep-seated personality or behavioral difficulties. (1)

Questions are an essential component in many theories and styles of helping and are used for gathering needed information. Open questions tend to draw out the most information, while closed questions can be useful for specifics. (4)

Questions as statements are a way in which some interviewers use questions to sell their own points of view. If you are going to make a statement, do not frame it as a question. (4)

Referral includes identifying the needs of the client that cannot be met by the counselor or agency and helping the client find appropriate help for those needs. When referral is necessary, provide sufficient support during the transfer process. (2)

Reflection of feeling identifies a client's key emotions and feeds them back to clarify affective experience. With some clients, the brief acknowledgment of feeling may be more appropriate. Often combined with paraphrasing and summarizing. (6)

Reflection of meaning focuses beyond what the client says. Often the words *meaning, value*s, and *goals* will appear in the discussion. Clients are encouraged to explore their own meanings in more depth from their own perspective. It provides more clarity on values and deeper life meanings. (10)

Reframing provides the client with a new perspective, frame of reference, or way of thinking about issues. This is closely related to interpretation, but reframing more often comes from the interviewer's frame of reference. Careful work with clients often enables them to do the reframing themselves. (10)

Resilience is the ability to learn, grow, recover from, or adapt to stress, trauma, or challenges. (1)

RESPECTFUL model (D'Andrea & Daniels, 2001) points out that all of us are multicultural beings and identifies ten dimensions of multiculturalism to consider. (2)

Restatement is type of extended encourager in which the counselor or interviewer repeats short statements of two or more words exactly as used by the client. (5)

Restorying is the generation of new meanings and new ways of thinking, which leads to more effective executive functioning in the brain and more effective action, as seen in overt behavior. (5)

Selective attention is what we choose to listen to, consciously or unconsciously. Watch your patterns. Therapists and clients can miss issues because of selective attention. (3)

Self-disclosure means that you as interviewer share your own related past life experience, here-and-now observations or feelings toward the client, or opinions about the future. Self-disclosure often starts with an "I" statement. Here-and-now feelings toward the client can be powerful and should be used carefully. (11)

Self-in-relation defines the "individual self" as existing in relationship to others. It is an expansion of the former view of an "independent self-constructing self." (8)

Social justice is the movement toward a more equitable society. It includes interviewers' efforts to effect social change, particularly on behalf of vulnerable and oppressed individuals and groups, as well as to help individual clients understand that their difficulties may not be caused by their behavior, but by an intolerant or oppressive system. Social justice information and exploration can also be part of an effective interview, particularly with clients who have experienced oppression. (8)

Subtractive empathy refers to an interviewer response to the client that is less than what the client has said and perhaps even distorts or misunderstands the client. In this case, the listening skills are used inappropriately and take the client off track. (3)

Summarizing uses client comments and integrates thoughts, emotions, and behaviors. Summarizing is similar to paraphrasing but used over a longer time span. It typically, but not always, includes a summary of emotions as well. (5)

TAP is shorthand for *thalamus, anterior cingulate cortex,* and *prefrontal cortex,* all associated with functions in the frontal lobe. (C)

Temporal lobe is concerned with auditory processing, language and speech production, aspects of sexuality, and memory. The limbic system containing the hippocampus, center of long-term memory, is located in the temporal lobe. (C)

Thalamus is considered the brain's communication center, sending information throughout the holistic brain. The thalamus lies just above the interior limbic system. (C)

Therapeutic lifestyle changes (TLCs) are a key route to identifying and encouraging a healthier and more engaged lifestyle. Examples of lifestyle changes include increased exercise, better nutrition, meditation, and helping others, which in turn helps us feel better about ourselves. (2)

Treatment planning refers to specific written goals and objectives, which are increasingly standard and often required by agencies and insurance companies, that represent the important third stage of the interview. Treatment planning also stresses the critical importance of action to ensure that these goals are actually implemented. (13)

Verbal tracking is effectively following the client's story. Don't change the subject; stay with the client's topic. (3)

Verbal underlining is giving louder volume and increased vocal emphasis to certain words and phrases. (3)

Visual/eye contact refers to looking at people when you speak to them. Cultural and individual variations are important to note. (3)

Vocabulary of emotions is the words that people use to name and understand emotions. (6)

Vocal qualities refer to your vocal tone and speech rate, which should be appropriate to the client and the situation.

***Why* questions** can put interviewees on the defensive and cause discomfort. As children, most of us experienced some form of "Why did you do that?" Any question that evokes a sense of being attacked can create client discomfort and defensiveness. On the other hand, "why" is often basic to helping a client explore issues of meaning and life vision. The person who has a why will find a how. (4)

REFERENCES

Alberti, R., & Emmons, M. (2008). *Your perfect right: A guide to assertiveness training* (9th ed.). San Luis Obispo, CA: Impact. (Original work published 1970)

Alim, T., Feder, A., Graves, R., Wang, Y., Weaver, J., Westphal, M., et al. (2008). Trauma, resilience, and recovery in a high-risk African-American population. *American Journal of Psychiatry, 165,* 1566–1575.

American Counseling Association. (2005). *ACA code of ethics.* Alexandria, VA: Author.

American Psychiatric Association. (2013). *Diagnostic and statistical manual of mental disorders* (5th ed.). Arlington, VA: American Psychiatric Publishing.

American Psychological Association. (2002). *Ethical principles of psychologists and code of conduct.* Washington, DC: Author.

American Psychological Association. (2009). Teaching resilience, sense of purpose in schools can prevent depression, anxiety and improve grades, according to research. http://www.apa.org/news/press/releases/2009/08/positive-educate.aspx

Anika Foundation. (2013). Explaining the rise in youth suicide. http://www.anikafoundation.com/rise_in_suicide.shtml

Asbell, B., & Wynn, K. (1991). *Touching.* New York: Random House.

Banks, J. A. (Ed.). (2012). *Encyclopedia of diversity in education.* Thousand Oaks, CA: Sage.

Banjo, O., Hu, Y., & Sundar, S. (2008). Cell phone usage and social interaction with proximate others: Ringing in a theoretical model. *Open Communication Journal, 2,* 127–135.

Barnett, J. E. (2011). Psychotherapist self-disclosure: Ethical and clinical considerations. *Psychotherapy, 48,* 315–321.

Barrett, M., & Berman, J. (2001). Is psychotherapy more effective when therapists disclose information about themselves? *Journal of Consulting and Clinical Psychology, 69,* 597–603.

Beck, J., & Beck, A. (2011). *Cognitive behavioral therapy: Basics and beyond.* New York: Guilford Press.

Benson, H., & Proctor, W. (2010). *Relaxation revolution: The science and genetics of mind body healing.* New York: Simon & Schuster.

Bergland, C. (2013). 25 Studies confirm: Exercise prevents depression. *Psychology Today.* http://www.psychologytoday.com/blog/the-athletes-way/201310/25-studies-confirm-exercise-prevents-depression

Blonna, R., Loschiavo, J., & Watter, D. (2011). *Health counseling: A microskills approach for counselors, educators, and school nurses* (2nd ed.). Sudbury, MA: Jones & Bartlett.

Bonanno, G. (2004). Loss, trauma, and human resilience: Have we underestimated the human capacity to thrive after extremely aversive events? *American Psychologist, 59,* 20–28.

Bourdieu, P., & Passeron, J. (1990). *Reproduction in education, society and culture.* London: Sage.

Boyle, P., Buchman, A., Barnes, L., & Bennett, D. (2010). Effect of a purpose in life on risk of incident Alzheimer disease and mild cognitive impairment in community-dwelling older persons. *Archives of General Psychiatry, 67,* 304–310.

Brooks, A. W. (2013). Get excited: Reappraising pre-performance anxiety as excitement. *Journal of Experimental Psychology: General, 143,* 1144–11558.

Buckholtz, J. W., & Marois, R. E. (2012). The roots of modern justice: Cognitive and neural foundations of social norms and their enforcement. *Nature Neuroscience, 15,* 655–661.

Campbell, S. W., & Kwak, N. (2014). Mobile communication and civic life: Implications of private and public uses of the technology. In G. Goggin & L. Hjorth (Eds.), *The Routledge companion to mobile media* (pp. 409–418). New York: Routledge.

Camus, A. (1955). *The myth of Sisyphus.* New York: Vintage Books.

Canadian Counselling Association. (2007). *Code of ethics.* Ottawa, Ontario: Author.

Carkhuff, R. (2000). *The art of helping in the 21st century.* Amherst, MA: HRD Press.

Carl, J. R., Soskin, D. P., Kerns, C., & Barlow, D. H. (2013). Positive emotion regulation in emotional disorders: A theoretical review. *Clinical Psychology Review, 33,* 343–360.

Center for Credentialing and Education. (n.d.). *Distance Credentialed Counselor* (DCC). http://www.cce-global.org/home

Chapin, T., & Russell-Chapin, L. (2014). *Neurotherapy and neurofeedback: Brain-based treatment for psychological and behavioral problems.* New York and London: Routledge.

Cheek, D. (1976). *Assertive black, puzzled white.* Atascadero, CA: Impact.

Collura, T. F. (2014). *Technical foundations of neurofeedback.* New York: Routledge.

Collura, T. F., Zalaquett, C., Bonnstetter, R. & Chatters, S. (in press). Towards an operational model of decision making, emotional regulation, and mental health impact. *Advances in Mind/Body Medicine.*

Conoley, C. W. (2014, May). *A positive psychology training manual for counseling psychology.* Paper presented at the Counseling Psychology Conference on Counseling Psychology in Action: Future Opportunities and Challenges, Atlanta, GA.

Cross, W. (1991). *Shades of black: Diversity in African American identity.* Philadelphia: Temple University Press.

D'Andrea, M., & Daniels, J. (2001). RESPECTFUL counseling: An integrative model for counselors. In D. Pope-Davis & H. Coleman (Eds.), *The interface of class, culture and gender in counseling* (pp. 417–466). Thousand Oaks, CA: Sage.

Daniels, T. (2010). A review of research on microcounseling: 1967–present. In A. Ivey, M. Ivey, & C. Zalaquett, *Intentional interviewing and counseling: Your interactive resource* (CD-ROM) (7th ed.). Belmont, CA: Brooks/Cole.

Daniels, T. (2014). A review of research on microcounseling: 1967–present. In A. Ivey, M. Ivey, & C. Zalaquett, *Intentional interviewing and counseling: Your interactive resource CourseMate.* Belmont, CA: Brooks/Cole.

Daniels, T., & Ivey, A. E. (2006). *Microcounseling: Making skills work in a multicultural world.* Springfield, IL: Thomas.

Danner, D. D., Snowdon, D. A., & Friesen, W. V. (2001). Positive emotions in early life and longevity: Findings from the nun study. *Journal of Personality and Social Psychology, 80,* 804–813.

Davidson, R. J. (2004). Well-being and affective style: Neural substrates and biobehavioural correlates. *Philosophical Transactions of the Royal Society B: Biological Sciences, 359,* 1395–1411.

Davidson, R., Pizzagalli, D., Nitschke, J., & Putnam, K. (2002). Depression: Perspectives from affective neuroscience. *Annual Review of Psychology, 53,* 545–574.

Davis, M., Eshelman, E., & McKay, M. (2008). *The relaxation and stress reduction workbook* (6th ed.). Oakland, CA: New Harbinger.

Decety, J., & Jackson, P. (2004). The functional architecture of human empathy. *Behavioral and Cognitive Neuroscience Reviews, 3,* 71–100.

Dement, W. (2000). *The promise of sleep.* New York: Dell.

de Shazer, S., & Dolan, Y. (2007). *More than miracles: The state of the art of solution-focused brief therapy.* Binghamton, NY: Haworth.

Dobson, K. (Ed.). (2009). *Handbook of cognitive-behavioral therapies* (3rd ed.). New York: Guilford Press.

Duncan, B. L., Miller, S. D., & Sparks, J. A. (2004). *The heroic client: A revolutionary way to improve effectiveness through client-directed outcome-informed therapy.* New York: Jossey-Bass/Wiley.

D'Zurilla, T., & Nezu, A. (2012). *Problem-solving therapy* (4th ed.). New York: Springer.

Egan, G. (2014). *The skilled helper: A problem-management and opportunity-development approach to helping* (10th ed.). Belmont, CA: Brooks/Cole.

Ekman, P. (2007). *Emotions revealed.* New York: Holt Paperbacks.

Ekman, P. (2013). Facial expressions of emotion are not universal? http://www.ekmaninternational.com/paul -ekman-international-plc-home/news/facial-expressions -of-emotion-are-not-universal.aspx

Ekman, P., & Cordaro, D. (2011). What is meant by calling emotions basic. *Emotion Review, 3,* 364–370.

Ellis, A., & Ellis, D. (2011). *Rational emotive behavior therapy.* Washington, DC: American Psychological Association.

Epstein, L., & Mardon, S. (2007). The Harvard Medical School guide to a good night's sleep. Cambridge, MA: Harvard University Press.

Ericsson, A. K., Charness, N., Feltovich, P., & Hoffman, R. R. (2006). *Cambridge handbook on expertise and expert performance.* Cambridge, UK: Cambridge University Press.

Evans, D., Hearn, M., Uhlemann, M., & Ivey, A. (2011). *Essential interviewing* (8th ed.). Belmont, CA: Brooks/Cole.

Fall, K., Fang, F., Mucci, L. A., Ye, W., Andrén, O., Johansson, J. E., et al. (2009). Immediate risk for cardiovascular events and suicide following a prostate cancer diagnosis: Prospective cohort study. *PLoS Medicine, 6*(12), e1000197.

Fang, F., Keating, N. L., Mucci, L. A., Adami, H. O., Stampfer, M. J., Valdimarsdóttir, U., et al. (2010). Immediate risk of suicide and cardiovascular death after a prostate cancer diagnosis: Cohort study in the United States. *Journal of the National Cancer Institute, 102*(5), 307–314.

Farnham, S., Gill, J., McLean, R., & Ward, S. (1991). *Listening hearts.* Harrisburg, PA: Morehouse.

Frankl, V. (1959). *Man's search for meaning.* New York: Simon & Schuster.

Frankl, V. (1978). *The unheard cry for meaning.* New York: Touchstone.

Fredrickson, B. L., Tugade, M. M., Waugh, C. E., & Larkin, G. R. (2003). What good are positive emotions in crisis? A prospective study of resilience and emotions following the terrorist attacks on the United States on September 11th, 2001. *Journal of Personality and Social Psychology, 84,* 365–376.

Fukuyama, M. (1990, March). *Multicultural and spiritual issues in counseling.* Workshop presentation at the American Counseling Association Convention, Cincinnati.

Gawande, A. (2009). *The checklist manifesto.* New York: Holt.

Gergen, K., & Gergen, M. (2005, February). The power of positive emotions. *The Positive Aging Newsletter.* http://www .healthandage.com

Goldberg, J. (2012) The effects of stress on your body. *WebMD.* http://www.webmd.com/mental-health /effects-of-stress-on-your-body

Goldin, P. R., Ziv M., Jazaieri, H., Hahn, K., Heimberg, R., Gross, J. J. (2013). Impact of cognitive behavioral therapy for social anxiety disorder on the neural dynamics of cognitive reappraisal of negative self-beliefs: Randomized clinical trial. *JAMA Psychiatry, 10,* 1048-56.

Goleman, D. (2013, October 5). Rich people care less. *New York Times.* opinionator.blogs.nytimes.com/2013/10/05 /rich-people-just-care-less/?_r=0

Graham, A. M., Fisher, P. A., & Pfeifer, J. H. (2013). What sleeping babies hear: A functional MRI study of interparental conflict and infants' emotion processing. *Psychological Science, 24,* 782–789.

Grawe, K. (2007). *Neuropsychotherapy: How the neurosciences inform effective psychotherapy.* Mahway, NJ: Erlbaum.

Grazianot, M. (2013). *Consciousness and the social brain.* Oxford, UK: Oxford University Press.

Greene, J. D (2009). The cognitive neuroscience of moral judgment. In M. S. Gazzaniga (Ed.), *The cognitive neurosciences* (4th ed., pp. 987–1001). Cambridge, MA: MIT Press.

Greene, J. D., Nystrom, L. E., Engell, A. D., Darley, J. M., & Cohen, J. D. (2004). The neural bases of cognitive conflict and control in moral judgment. *Neuron, 44,* 389–400.

Guterman, J. (2013). *Mastering the art of solution-focused counseling.* Washington, DC: American Counseling Association.

Haley, J. (1997). *Problem-solving therapy* (2nd ed.). San Francisco: Jossey-Bass.

Hall, E. (1959). *The silent language.* New York: Doubleday.

Hall, R. (2007). Racism as a health risk for African Americans. *Journal of African American Studies, 11,* 204–213.

Hargie, O., Dickson, D., & Tourish, D. (2004). *Communication skills for effective management.* Basingstoke, UK: Palgrave.

Hermans, E., van Marle, H., Ossewaarde, L., Henckens, A., Qin, S., Kesteren, M., et al. (2012). Stress-related noradrengic activity prompts large-scale neural network configuration. *Science, 334,* 1151–1153.

Herring, M. P., Jacob, M. L., Suveg, C., Dishman, R. K., & O'Connor, P. J. (2012). Feasibility of exercise training for the short-term treatment of generalized anxiety disorder: A randomized controlled trial. *Psychotherapy and Psychosomatics, 81,* 21–28.

Hölzel, B., Carmody, J., Vangel, M., Congleton, C., Yerramsetti, S. M., Gard, T., et al. (2011). Mindfulness practice leads to increases in regional brain gray matter density. *Psychiatry Research: Neuroimaging, 191,* 36–43.

Ishiyama, I. (2006). *Anti-discrimination Response Training (A.R.T.) program.* Framingham, MA: Microtraining Associates.

Ivey, A. (1973). Media therapy: Educational change planning for psychiatric patients. *Journal of Counseling Psychology, 20,* 338–343.

Ivey, A., D'Andrea, M., & Ivey, M. (2012). *Theories of counseling and psychotherapy: A multicultural perspective* (7th ed.). Thousand Oaks, CA: Sage.

Ivey, A. E., Ivey, M. B., Myers, J., & Sweeney, T. (2005). *Developmental counseling and therapy: Promoting wellness over the lifespan.* Boston: Lahaska/Houghton Mifflin.

Ivey, A., Ivey, M., & Zalaquett, C. (2012, March). *Therapeutic lifestyle changes for mental and physical health.* Presentation at the American Counseling Association Conference, Cincinnati.

Jack, R., Garrod, O., & Schyns, P. (2014). Dynamic facial expressions of emotion transmit an evolving hierarchy of signals over time. *Current Biology, 24,* 187–192.

James, R. (2007). *Crisis intervention strategies.* Belmont, CA: Brooks/Cole.

Jordan, J., Hartling, L., & Walker, M. (Eds.). (2004). *The complexity of connection: Writings from the Stone Center's Jean Baker Miller Training Institute.* New York: Guilford Press.

Kabat-Zinn, J. (2013). *Full catastrophe living: Using the wisdom of your body and mind to face stress, pain, and illness* (Rev. ed.). New York: Bantam Books.

Kabat-Zinn, J., & Davidson, R. (2012). *The mind's own physician: A scientific dialogue with the Dalai Lama on the healing power of meditation.* New York: New Harbinger.

Kahneman, D. (2011). *Thinking fast and slow.* New York: Farrar, Strauss, & Giroux.

Kanel, K. (2011). *A guide to crisis intervention* (4th ed.). Belmont, CA: Brooks/Cole.

Karlsson, H. (2011, August 11). How psychotherapy changes the brain. *Psychiatric Times.* http://www.psychiatrictimes .com/psychotherapy/how-psychotherapy-changes-brain

Kim, E. S., Park, N., & Peterson, C. (2011). Dispositional optimism protects older adults from stroke: The health and retirement study. *Stroke, 42,* 2855–2859.

King Center. (2014). Six steps of nonviolent social change. *The King philosophy.* http://www.thekingcenter.org /king-philosophy#sub3

Krumboltz, J. (2014, March). *Practical career counseling: Applications of happenstance learning theory.* Presentation to the Society of Counseling Psychology, Atlanta.

Lazarus, A. (2006). *Comprehensive therapy: The multimodal way.* Baltimore: Johns Hopkins University Press.

Lee, J. R., & Hopkins, V. (2009). Cortisol and the stress connection. *Virginia Hopkins Test Kits.* http://www .virginiahopkinstestkits.com/cortisolstress.html

Likhtik, E., Popa, D., Apergis-Schoute, J., Fidacaro, G. A., & Paré, D. (2008). Amygdala intercalated neurons are required for expression of fear extinction. *Nature, 454,* 642–645.

Liu, Y., Zhang, D., & Luo, Y. (2014). How disgust facilitates avoidance: An ERP study on attention modulation by threats. *Social Cognitive and Affective Neuroscience.* doi: 10.1093/scan/nsu094

Logothetis, N. K. (2008). What we can do and what we cannot do with fMRI. *Nature, 453,* 869–878.

Mann, L. (2001). Naturalistic decision making. *Journal of Behavioural Decision Making, 14,* 375–377.

Mann, L., Beswick, G., Allouche, P., & Ivey, M. (1989). Decision workshops for the improvement of decision making skills. *Journal of Counseling and Development, 67,* 237–243.

Marshall, J. (2010, February 22). Poverty during early childhood may last a lifetime. *Discovery News.* Retrieved May 10, 2010, from http://news.discovery.com/human /poverty-children-income-adults.html

McGoldrick, M., & Gerson, R. (1985). *Genograms in family assessment.* New York: Norton.

McGoldrick, M., Giordano, J., & Garcia-Preto, N. (2005). *Ethnicity and family therapy* (3rd ed.). New York: Norton.

McIntosh, P. (1988). *White privilege and male privilege: A personal account of coming to see correspondences through work in women's studies.* Wellesley, MA: Wellesley College Center for Women.

Meara, N., Pepinsky, H., Shannon, J., & Murray, W. (1981). Semantic communication and expectation for counseling across three theoretical orientations. *Journal of Counseling Psychology, 28,* 110–118.

Meara, N., Shannon, J., & Pepinsky, H. (1979). Comparisons of stylistic complexity of the language of counselor and client across three theoretical orientations. *Journal of Counseling Psychology, 26,* 181–189.

Merzenich, M. (2013). *Soft-wired: How the new science of brain plasticity can change your life.* San Francisco: Parnassus.

Miller, G. (2012). *Fundamentals of crisis counseling.* Hoboken, NJ: Wiley.

Miller, S. D., Duncan, B. L., & Hubble, M. A. (2005). Outcome-informed clinical work. In J. C. Norcross & M. R. Goldfried (Eds.), *Handbook of psychotherapy integration* (2nd ed., pp. 84–102). New York: Oxford University Press.

Miller, W. R., & Rollnick, S. (2013). *Motivational interviewing: Helping people change.* (3rd ed.). New York: Guilford Press.

Montes, S. (2013). The birth of the neurocounselor? *Counseling Today,* pp. 33–40.

Myers, J. E., & Sweeney, T. J. (2004). The indivisible self: An evidence-based model of wellness. *Journal of Individual Psychology, 60,* 234–244.

Myers, J. E., & Sweeney, T. J. (Eds.). (2005). *Counseling for wellness: Theory, research, and practice.* Alexandria, VA: American Counseling Association.

National Heart, Lung, and Blood Institute. (2013). Reduce screen time. http://www.nhlbi.nih.gov/health/educational /wecan/reduce-screen-time

National Organization for Human Services. (1996). *Ethical standards for human service professionals.* http://www.nationalhumanservices.org/ethical -standards-for-hs-professionals

National Scientific Council on the Developing Child. (2005/2014). *Excessive stress disrupts the architecture of the developing brain: Working paper 3* (Updated ed.). http:// developingchild.harvard.edu/resources/reports_and _working_papers/working_papers/wp3

Nes, L. S., & Segerstrom, S. C. (2006). Dispositional optimism and coping: A meta-analytic review. *Personality and Social Psychology Review, 10,* 235–251.

Neukrug, E. S. (2012). *Theory, practice, and trends in human services* (5th ed.). Belmont, CA: Brooks/Cole.

Nezu, A., Nezu, C., & D'Zurilla, T. (2012). *Problem-solving therapy: A treatment manual.* New York: Springer.

Norcross, J. (2012). *Psychotherapy relationships that work: Evidence-based responsiveness* (2nd ed.). New York: Oxford University Press.

Obonnaya, O. (1994). Person as community: An African understanding of the person as intrapsychic community. *Journal of Black Psychology, 20,* 75–87.

Ornish, D. (2008). *The spectrum: A scientifically proven program to feel better, live longer, lose weight, and gain health.* New York: Ballantine Books.

Parsons, F. (1967). *Choosing a vocation.* New York: Agathon. (Originally published 1909)

Peterson, S., & Posner, M. (2012). The attention system of the human brain: 20 years after. *Annual Review of Neuroscience, 35,* 73–89.

Power, S., & Lopez, R. (1985). Perceptual, motor, and verbal skills of monolingual and bilingual Hispanic children: A discrimination analysis. *Perceptual and Motor Skills, 60,* 1001–1109.

Ratey, J. (2008). *Spark: The revolutionary science of exercise.* New York: Hatchette/Little, Brown.

Ratey, J., & Hagerman, E. (2013). *Spark: The revolutionary new science of exercise and the brain.* New York: Little, Brown.

Ratey, J., & Manning, R. (2014). *Go wild: Free your body and mind from the afflictions of civilization.* New York: Little, Brown.

Ratts, M., Toporek, R., & Lewis, J. (2010). *ACA advocacy competencies: A social justice framework for counselors.* Washington, DC: American Counseling Association.

Rigazio-DiGilio, S. A., Ivey, A. E., Grady, L. T., & Kunkler-Peck, K. P. (2005). *Community genograms: Using individual, family, and cultural narratives with clients.* New York: Teachers College Press.

Rockett, I., Regier, D., Kapusta, N., Coben, J., Miller, T., Hanzlick, R., et al. (2012). Leading causes of unintentional and intentional injury mortality. *American Journal of Public Health, 102,* e84–e92.

Rogers, C. (1957). The necessary and sufficient conditions of therapeutic personality change. *Journal of Consulting Psychology, 21,* 95–103.

Rogers, C. (1961). *On becoming a person.* Boston: Houghton Mifflin.

Rosenthal, R., & Jacobson, L. (1968). *Pygmalion in the classroom.* New York: Holt, Rinehart & Winston.

Roskies, A. L. (2012). How does the neuroscience of decision making bear on our understanding of moral responsibility and free will? *Current Opinion in Neurobiology, 22*, 1022–1026.

Schlosser, L. Z. (2003). Christian privilege: Breaking a sacred taboo. *Journal of Multicultural Counseling and Development, 31*, 44–51.

Schwartz, M. S., & Andrasik, F. (2005). *Biofeedback: A practitioner's guide.* New York: Guilford Press.

Schwartz Gottman, J., & Gottman, J. M. (2014). *10 Principles for doing effective couples therapy.* New York: Norton.

Seligman, M. (2009). *Authentic happiness.* New York: Free Press.

Sherin J. E., & Nemeroff, C. B. (2011). Post-traumatic stress disorder: The neurobiological impact of psychological trauma. *Dialogues in Clinical Neuroscience, 13*(3), 263–278.

Shostrum, E. (1966). *Three approaches to psychotherapy* [Film]. Santa Ana, CA: Psychological Films.

Siegel, D. (2012). *The developing mind.* New York: Guilford Press.

Singer, T., Seymour, B., O'Dougherty, J., Kaube, H., Dolan, R., & Frith, C. (2004). Empathy for pain involves the affective but not sensory components of pain. *Science, 303*, 1157–1161.

Sommers-Flannagan, J., & Sommers-Flannagan, R. (2012). *Clinical interviewing* (4th ed.). New York: Wiley.

Steele, C. M. (1997). A threat in the air: How stereotypes shape the intellectual identities and performance of women and African-Americans. *American Psychologist, 52*, 613–629.

Stephens, G., Silbert, L., & Hasson, U. (2010). Speaker–listener neural coupling underlies successful communication. *Proceedings of the National Academy of Sciences of the United States of America, 107*(32), 14425–14430.

Stoet, G., & Geary, D. C. (2012). Can stereotype threat explain the gender gap in mathematics performance and achievement? *Review of General Psychology, 16*, 93–102.

Sue, D. W. (2010). *Microaggressions and marginality: Manifestation, dynamics, and impact.* New York: Wiley.

Sue, D. W., Carter, R., Casas, M., Fouad, N., Ivey, A., Jensen, M., et al. (1998). *Multicultural counseling competencies.* Beverly Hills, CA: Sage.

Sue, D. W., Ivey, A., & Pedersen, P. (1999). *A theory of multicultural counseling and therapy.* Pacific Grove, CA: Brooks/Cole.

Sue, D., & Sue, D. W. (2007). *Foundations of counseling and psychotherapy: Evidence-based practices for a diverse society.* New York: Wiley.

Sue, D. W., & Sue, D. (2013). *Counseling the culturally diverse* (6th ed.). Hoboken, NJ: Wiley.

Sueng, S. (2012) *Connectome: How the brain's wiring makes us who we are.* New York: Houghton Mifflin Harcourt.

Tang, Y., Lu, Q., Fan, M., Yang, Y., & Posner, M. (2012). Mechanisms of white matter changes induced by meditation. *Proceedings of the National Academy of Science, 109*(26), 15649–15652.

Taub, E. (1977). Self-regulation of human tissue temperature. In G. Schwartz & J. Beatty (Eds.), *Biofeedback: Theory and practice* (pp. 265–300). New York: Academic Press.

Uchino, B. N., Bowen, K., Carlisle, M., & Birmingham, W. (2012). What are the psychological pathways linking social support to health outcomes? A visit with the "ghosts" of research past, present, and future. *Social Science and Medicine, 74*, 949–957.

Vogel, G. (2012). Can we make our brains more plastic? *Science, 338*, 336–338.

Wang, S. (2012, August 30). Coffee break? Walk in the park? Why unwinding is hard. *Wall Street Journal.* online.wsj .com/news/articles/SB100014240531119041994045765382603269657724

Westefeld, J. S., Casper, D., Lewis, A., Manlick, C., Rasmussen, W., Richards, A., et al. (2012). Physician assisted death and its relationship to the human services professions. *Journal of Loss and Trauma, 18*, 539–555.

Westefeld, J. S., Richards, A., & Levy, L. (2011). Protective factors. In D. Lamis & D. Lester (Eds.), *Understanding and preventing college student suicide* (pp. 170–182). Springfield, IL: Thomas.

White, W., & Miller, W. (2007). The use of confrontation in addiction treatment: History, science and time for change. *Counselor, 8*(4), 12–30.

Xu, P., Gu, R., Broster, L. S., Wu, R., Van Dam, N. T., Jiang, Y., et al. (2013). Neural basis of emotional decision making in trait anxiety. *Journal of Neuroscience, 33*(47), 18641–18653.

Zalaquett, C., Carrión, I., & Exum, H. (2010). Counseling survivors of national disasters: Issues and strategies for a diverse society. In J. Webber & B. Mascari (Eds.), *Terrorism, trauma, and tragedies: A counselor's guide to preparing and responding* (3rd ed., pp. 19–24). Alexandria, VA: American Counseling Association.

Zalaquett, C. P., Ivey, A. E., Gluckstern-Packard, N., & Ivey, M. B. (2008). *Las habilidades atencionales básicas: Pilares fundamentales de la comunicación efectiva.* [Manuals and videos]. Framingham, MA: Microtraining Associates.

Ziv-Beiman, S. (2013). Therapist self-disclosure as an integrative intervention. *Journal of Psychotherapy Integration, 23*, 59–74.

Zur, O., Williams, M. H., Lehavot, K., & Knapp, S. (2009). Psychotherapist self-disclosure and transparency in the Internet age. *Professional Psychology: Research and Practice, 40*, 22–30.

Zur, O. (2011). *The HIPAA compliance kit.* http://www .zurinstitute.com/hipaakit.html

NAME INDEX

▲ SUBJECT INDEX

contracts
   in five-stage interview process, 139
   sample practice, 33
corrective feedback, 224, 225
corticosteroids, 245
cortisol, 245, 319
could questions, 80
counseling
   client vs. counselor experience of, 264–265
   cultural intentionality in, 9–11
   definition of, 4–5
   demystifying process of, 151–153
   goals of, 11–12
   vs. interviewing, 4, 5
   microskills approach to, 6–9
   natural style of, 15–17
   neuroscience of, 12–13
   physical setting for, 14–15
   supplementary readings on, 300–303
   theories of, 212–214
   See also specific approaches and aspects
counseling theories
   microskills used in, 290–292
   supplementary readings on, 300–301
   See also specific theories
creativity, in counseling, 157
crisis, definition of, 273–274
crisis counseling, 272–288
   challenges of, 286–287
   counselor burnout in, 278, 287
   definition of, 273–274
   essential components of, 276–278
   in example interview, 278–282
   microskills used in, 290, 291
   practice exercises on, 285–286
   suicide prevention in, 283–285
   team approach to, 274, 279
crying, 219
cultural differences. *See* multicultural issues
cultural/environmental context (CEC), in focusing, 161, 163–164
cultural identity, therapeutic lifestyle changes in, 45
cultural intentionality, 9–11
cultural trauma, 22

D
deaf clients, 53–54, 57
debriefing, in crisis counseling, 277
decisional counseling, 237–240
   in example interviews, 141–146, 247–250
   five-stage interview process in, 141–147, 239–240
   interpretation/reframing in, 212
   methods of, 238–240
   microskills used in, 290, 291
   practice and self-assessment of, 147–151
   supplementary readings on, 300
decision making
   as essential to counseling process, 141
   ethical, 214
de-identified health information, 34
demystifying, of helping process, 151–153
denial stage, 184–186
depression, brain activity in, 124–125
Diagnostic and Statistical Manual of Mental Disorders
   (DSM-5), 10, 299

direct action, in conflict mediation, 194–195
directives, 233–235
   anticipated client response to, 233, 310
   definition of, 234, 310
   practice exercises on, 250–251
   strategies for, 245–247
disabilities, clients with, attending behavior for,
   53–54, 57
discerning, definition of, 52
discernment
   definition of, 210
   of meaning of life, 210–212
discrepancies, confronting, 181–183
discrimination
   cumulative stress of, 104–105
   internalization of, 182
disgust, as primary emotion, 113–115
disorders, use of term, 10, 275
Distance Credentialed Counselors (DCCs), 15
diversity
   and ethics, 31
   in multicultural competence, 25, 28–29
   supplementary readings on, 300
   See also multicultural issues
drug abuse, 46
dual relationships with clients, 31

E
earthquakes, 279
eclectic orientations, supplementary readings on, 301–302
education, continuing, 47
education, patient. *See* psychoeducation
electromyography (EMG), 263
emotion(s)
   of bilingual clients, 106
   cognitions in relation to, 110
   nonverbal communication of, 112, 113–116
   primary, 113–115
   techniques and strategies of observing, 113–116
   vocabulary of, 113–114, 123, 128–129
   See also reflection of feelings
emotional balance sheet, 239, 240
emotional regulation
   definition of, 11, 75, 112
   in executive brain functioning, 75, 93, 104
   and reflection of feelings, 112–113
empathic confrontation. *See* confrontation
empathic feedback. *See* feedback
empathic relationships
   in conflict mediation, 194
   in decisional counseling, 239
   in five-stage interview process, 133–135, 142
   in treatment planning, 260
empathic self-disclosure. *See* self-disclosure
empathy, 59–61
   anticipated client response to, 59, 307
   definition of, 59, 307
   levels of, 59–60
   in microskills hierarchy, 7–8
   multicultural issues in, 67–70
   neuroscience of, 61
   and positive expectations, 60–61
   practice exercises on, 66–67, 70
encouragers, 96–97

Health Insurance Portability and Accountability Act (HIPAA), 34, 141
heart rate feedback, 263
helping
    demystifying process of, 151–153
    major forms of, 5–6
hippocampus, 315, 317
Hispanic clients. *See* Latino/Latina clients
historical moment, 161
homework assignments
    in five-stage interview process, 139
    positive emotions in, 125
hormones, in stress response system, 244–245
hostile statements, 191
how questions, 79
HPA, 244–245, 317–318
humanistic theory, supplementary readings on, 301
humor, sense of, 47
Hurricane Katrina, 276
hypothalamus, 244–245, 317–318

I

imagery, in psychoeducation, 246–247
immediacy, in self-disclosure, 221
incongruity
    confronting, 181–183
    definition of, 180
indecision, 115–116
individualism, vs. collectivism, 208
Indivisible Self wellness model, 40
influencing skills
    definition of, 157
    in five-stage interview process, 133
    in microskills hierarchy, 7, 8
    1-2-3 pattern of, 220
    overview of, 157–158
    See also specific skills
information gathering, in conflict mediation, 194–195
information sharing, with clients, 233–234, 250, 310
informed consent, 32–34
integrative orientations, supplementary readings on, 301–302
intentional competence, 295
intentionality
    cultural, 9–11
    in personal style, 292–294
interchangeable empathy, 60
internal conflict, 180, 181–182, 189–190
internalization of problems, 182
International Classification of Disease (ICD), 299
Internet counseling, 5, 14–15
Internet resources, 3–4
interpersonal distance, 57–58
interpretation/reframing, 198–216
    anticipated client response to, 200, 309
    definition of, 199, 200, 309
    in example interview, 204–207
    practice exercises on, 207–208
    in psychoeducation, 246
    vs. reflection of meaning, 201–203
    theories of counseling and, 212–214
interviewing, 2–19
    vs. counseling, 4, 5
    cultural intentionality in, 9–11

definition of, 4–5
demystifying process of, 151–153
goals of, 11–12
microskills approach to, 6–9
natural style of, 15–17
neuroscience of, 12–13
physical setting for, 14–15
supplementary readings on, 300–303
See also specific approaches and aspects
Inuits, 55, 153, 192
Islam, 208
issues, use of term, 10
"I" statements, 218–219
Ivey Taxonomy, 305–310

J

journaling
    in five-stage interview process, 139
    in psychoeducation, 247
Judaism, 208–209
judgmental feedback, 224
judgmental vocal tones, 191

K

Katrina, Hurricane, 276
key questions, in suicide prevention, 285
key words
    in active listening, 96–98
    of bilingual clients, 106
    in paraphrasing, 97
    verbal underlining of, 56
King Center, 194
knowledge, in multicultural competence, 25–26

L

language, of bilingual clients, 105–106.
    See also vocabulary
Latino/Latina clients
    bilingual, 106
    empathic confrontation with, 192–193
    and focusing, 168
    nonverbal behavior of, 54, 57–58
leading questions, 89, 99
life, discernment of mission and goals in, 210–212
life skills training, 235–236
lifestyle changes. *See* therapeutic lifestyle changes
limbic system, 317–318
linking, 203
listening skills
    in attending behavior, 51–52
    with children, 99–102
    identifying and defining, 51–52
    1-2-3 pattern of, 220
    See also active listening; attending behavior; basic listening sequence
logical consequences, 237–238
    anticipated client response to, 237, 309
    practice exercise on, 250
love, as positive emotion, 115

M

mad, as primary emotion, 113–114
Maintaining Change Worksheet, 266–267

management training, microskills in, 294
Man's Search for Meaning (Frankl), 198n
meaning
    as core of human experience, 199
    discernment of, 210–212
    eliciting, 200–204
    in multicultural context, 208–209
    in resilience, 209–210
    value of, 198n
    See also reflection of meaning
mediation, 193–194
meditation, 44–45
microaggressions, 26, 104–105
microcounseling, 9
microskills, 6–9
    definition of, 6
    demystifying, 152–153
    hierarchy of, 6–9, 295, 305–310
    research on, 9
    self-assessment of, 296–297
    supplementary readings on, 300
    use in different theoretical approaches, 290–292
    use in different workplaces, 293–294
    See also specific skills
microskills integration. *See* skill integration
Microskills Interview Analysis Form (MIAF), 147–151
minority clients
    self-disclosure with, 222
    See also specific groups
mirroring, of body language, 62–63, 121
mirror neurons, 61, 317
mixed feelings, 115–116
mixed messages, confronting, 181–183
morals, 29–37
    definition of, 29
    See also ethics
motivational interviewing (MI), microskills used in, 291, 292
movement complementarity, 62
movement dissynchrony, 62–63
movement synchrony, 62–63
multicultural competence, 21–29
    anticipated client response to, 21, 305
    definition of, 21, 305
    guidelines for, 25–29
    in microskills hierarchy, 6–7
    practice exercises on, 25, 26–27, 28, 29
    privilege in, 23–25
    RESPECTFUL model in, 22–23, 26–27
multicultural counseling and therapy (MCT)
    interpretation/reframing in, 213
    microskills used in, 290, 291, 294
    supplementary readings on, 301
multiculturalism
    in counseling, 27
    definition of, 25
    and ethics, 31
    rise of, 21
    terminology of, 28–29
multicultural issues
    in active listening, 103–107
    in attending behavior, 54–59, 67–70
    in conflict mediation, 194–195
    in empathic confrontation, 192–193, 194–195
    in empathy, 67–70

    in focusing, 168
    in nonverbal communication, 68–69
    in observation, 67–70
    in questions, 82–83
    in reflection of feelings, 122–123
    in reflection of meaning, 208–209
    in self-disclosure, 218, 222–223
    supplementary readings on, 300
    in use of microskills, 293–294
multicultural pride, 45
multiple questions, 82
Muslims, 208
mutuality
    in focusing, 164
    in goal setting, 136
myelin, 320

N

narrative therapy
    internalization of problems in, 182
    microskills used in, 290, 291
National Association of Social Workers, 35
National Institute of Mental Health, 153, 299
nationality, use of terms for, 29
National Organization for Human Services, 31
National Scientific Council on the Developing Child, 319
Native American clients
    active listening with, 107
    empathic confrontation with, 192–193
    microskills with, 294
    nonverbal behavior of, 54–55
natural consequences, 237–238, 250
natural talent, 71
nature breaks, 46
negative consequences, 237–238
negative emotions, 113–115
negative expectations, 61
negative feedback, 225
negotiation, in conflict mediation, 194–195
neural networks, 104
neurocounselors, 13
neurofeedback (EEG), 263–265
neurons, mirror, 61, 317
neurogenesis, 13, 315, 320
neuroplasticity, 13, 315
neuroscience, 314–320
    on active listening, 153, 299
    anticipated client response to, 306
    on conflict, 193
    definition of, 306
    in demystifying counseling process, 153
    on effects of counseling, 12–13
    on emotions, 123–125
    on empathy, 61
    on ethical decision making, 214
    future of, 299, 320
    on neurofeedback, 264
    on stress, 13, 241–245, 319–320
    supplementary readings on, 302
New, creation of the, 157
new solutions stage, 184–186
nonempathic confrontation, 190–192
nonverbal communication
    of emotions, 112, 113–116

in encouraging, 96
multicultural issues in, 68–69
observation of, 62–63, 121
See also body language
normalizing, in crisis counseling, 275–276
note taking
in five-stage interview process, 140–141, 260–261
in treatment planning, 260–261
nucleus accumbens, 318
nutrition, 42–43

O

observation skills, 62–71
anticipated client response to, 62, 306
on body language, 62–63, 121
definition of, 62, 306
on emotions, 113–116
in microskills hierarchy, 7–8
multicultural issues in, 67–70
on nonverbal communication, 62–63, 121
1-2-3 pattern of, 220
practice exercises on, 66–67, 70
occipital lobe, 317
Occupational Safety and Health Administration (OSHA), 242
office setting, 14
1-2-3 influencing pattern, 220
online resources, 3–4
opening conversations, 67–68
open questions, 78–81
anticipated client response to, 78, 306
definition of, 78, 306
example of, 85–87
how to use, 79–81
with less verbal clients, 89
in person-centered counseling, 147
oppression, 22, 105
optimism, 37–38, 46

P

paraphrasing, 93–109
in active listening, 95, 97–98
anticipated client response to, 95, 307
in basic listening sequence, 131–132
with children, 99–102
definition of, 97, 307
in microskills hierarchy, 8
multicultural issues in, 103–107
practice exercises on, 102–103
in reflection of feelings, 116
vs. reflection of feelings, 111–113
vs. reflection of meaning, 204
techniques and strategies of, 97–98
parents, informed consent by, 33
parietal lobe, 316
partial acceptance stage, 184–186
pejorative statements, 191
perception checks. *See* checkouts
personal commitment, in conflict mediation, 194–195
personal style, 289–303
anticipated client response to, 292, 310
discovering, 15–17, 292–293, 297–298, 310

intentionality and flexibility in, 292–294
in microskills hierarchy, 8
reviewing and analyzing, 268–269, 294–297
self-assessment of, 296–297
person-as-community, 163
person-centered counseling, 147
interpretation/reframing in, 212
microskills used in, 290, 291
supplementary readings on, 301
person factor, 264
phone(s)
cell, 69
counseling by, 14–15
physical disabilities, clients with, 54
physical exercise, 41–42, 246
pituitary glands, 244–245, 317–318
planning
for first interviews, checklist for, 257–260
for relapse prevention, 266–268
for treatment, 260–262
Platinum Rule, 195
political correctness (PC), 28–29
poor listening, 51
portfolio of competencies, 15
positive asset search, 38
positive consequences, 237–238
positive emotions, 115, 123–125
positive empathy, 60
positive expectations, 60–61
positive feedback, 225
positive psychology, 37–39
anticipated client response to, 38, 305
definition of, 38, 305
meaning and purpose in, 209–210
in microskills hierarchy, 6–7
resilience through, 6–7, 37–39
positive reflection, strategies for, 125
positive stress, 242
positive thinking, 46
posttraumatic stress (PTS), 22
posttraumatic stress disorder (PTSD), 22, 275
poverty
social justice action on, 35
and stress, 319–320
power balance, 217, 218
practice, importance of, 70–71
pragmatism, in crisis counseling, 273
prejudice, cumulative stress of, 104–105
preliminary goals, 135
present tense, 116
pride, multicultural, 45
primary emotions, 113–115
privacy, under HIPAA, 34
privilege, 23–25, 47
problems
alternatives to term, 10
internalization vs. externalization of, 182
use of term, 10
problem solving
three phases of, 141
See also decisional counseling
problem-solving therapy. *See* decisional counseling
protected health information (PHI), 34
psychodynamic/interpersonal theory, 213